Environmental and Human Security in the Arctic

This is the first comprehensive exploration of why human security is relevant to the Arctic and what achieving it can mean.

Change in the Arctic is occurring at an unprecedented rate, placing an immense strain on the many factors that contribute to human well-being and security, such as the health of the environment, identity of peoples, supply of traditional foods, community health, economic opportunities and political stability.

The traditional definition of security has already been actively employed in the Arctic region for decades, particularly in relation to natural resource sovereignty issues, but how and why should the human aspect be introduced? What can this region teach us about human security in the wider world?

The book reviews the potential threats to security, putting them in an analytical framework.

Contributions come from natural, social and humanities scientists, hailing from Canada, Russia, USA, Finland and Norway.

Environmental and Human Security in the Arctic is an essential resource for policy-makers, community groups, researchers and students working in the field of human security, particularly for those in the Arctic regions.

Gunhild Hoogensen Gjørv is Associate Professor in International Relations at the University of Tromsø, Norway.

Dawn R. Bazely is Professor in Biology and Director of the Institute for Research and Innovation in Sustainability at York University, Canada.

Marina Goloviznina is a post-doctoral fellow at the University of Tromsø in the Department of Sociology, Political Science and Community Planning.

Andrew J. Tanentzap is a Banting Fellow at York University in the Department of Biology.

Environmental and Human Security in the Arctic

Edited by Gunhild Hoogensen Gjørv, Dawn R. Bazely, Marina Goloviznina and Andrew J. Tanentzap

Routledge
Taylor & Francis Group
LONDON AND NEW YORK

earthscan
from Routledge

First published in paperback 2024

First edition published 2014
by Routledge
4 Park Square, Milton Park, Abingdon, Oxon OX14 4RN

and by Routledge
605 Third Avenue, New York, NY 10158

Routledge is an imprint of the Taylor & Francis Group, an informa business

Publisher's Note
The publisher has gone to great lengths to ensure the quality of this reprint but points out that some imperfections in the original copies may be apparent.

British Library Cataloguing in Publication Data
A catalogue record for this book is available from the British Library

Library of Congress Cataloging-in-Publication Data
Environmental and human security in the Arctic / edited by Gunhild Hoogensen Gjørv, Dawn Bazely, Marina Goloviznina, Andrew Tanentzap.
pages cm
Includes bibliographical references and index.
1. Environmental policy--Arctic regions. 2. Human security--Arctic regions. 3. Arctic regions--Environmental conditions. 4. Global environmental change--Arctic regions. 5. Environmental degradation--Arctic regions. 6. Arctic peoples. 7. Human ecology--Arctic regions. I. Hoogensen Gjørv, Gunhild, 1966- author, editor of compilation.
GE190.A75E68 2013
304.2091 1'3--dc23
2013018315

ISBN: 978-1-84407-549-2 (hbk)
ISBN: 978-1-03-292345-1 (pbk)
ISBN: 978-1-315-88538-4 (ebk)

DOI: 10.4324/9781315885384

Typeset in Garamond
by GreenGate Publishing Services, Tonbridge, Kent

Contents

Illustrations

Figures

Table

Contributors

Gunhild Hoogensen Gjørv is Associate Professor in political science at the University of Tromsø, specializing in security theory and politics, civil–military interaction and international relations theory. She has been the International Principal Investigator and lead in the International Polar Year Project 'GAPS' (The Impacts of Oil and Gas Activities on Peoples of the Arctic Using a Multiple Securities Perspective), which examines multiple understandings of security in the Arctic region, particularly in Norway, Canada and Russia, cooperating with such institutions as the Norwegian Polar Institute, the Russian Academy of Sciences (Ural Division), York University (Canada) and the University of Cambridge (UK). She has been increasingly focusing upon the relationships between health, economic and environmental security.

Dawn R. Bazely is Professor of Biology at York University, Toronto, and has been Director of IRIS, the university's Institute for Research and Innovation in Sustainability, since 2006. Trained in the ecological field of plant–animal interactions at the universities of Toronto and Oxford, she has carried out extensive field research in grasslands and forests, from temperate to Arctic regions. She led the Canadian section of the International Polar Year project, 'GAPS', and recently spent her sabbatical as a Charles Bullard Fellow at Harvard Forest, Harvard University.

Marina Goloviznina received her PhD in Sociology from the European University at Saint Petersburg (Russia) in 2005. She has conducted research and written on governance of security in crime control. She also worked as a research consultant for Penal Reform International, DanChurchAid and with various non-government and human rights organizations in the North-West of Russia. In 2011, she joined the IPY GAPS project team and has focused her studies on the linkages of oil, gas and health impacts in the Komi Republic. She works as a post-doctoral fellow at the University of Tromsø in the Department of Sociology, Political Science and Community Planning.

Andrew J. Tanentzap is an ecologist with broad interests in biology, conservation science and statistical modelling. He is currently an inaugural Banting Postdoctoral Fellow at York University in Canada, and will take up his new position as University Lecturer in the Department of Plant Sciences at the University of Cambrige in the autumn of 2013. He previously held a Gates Cambridge Scholarship, established by the Bill and Melinda Gates Foundation, to study for a PhD in the Department of Plant Sciences at the University of Cambridge, UK. Andrew's current research focuses on predicting how a changing world will affect the world's biodiversity and the many important services that society derives from natural ecosystems.

Kristian Åtland works as a Senior Research Fellow at the Norwegian Defence Research Establishment, located at Kjeller, Norway. He holds a PhD in political science from the University of Tromsø and an MA degree in Russian cultural studies from the University of Oslo. In the period from 1994 to 2002, he was employed in the Norwegian Foreign Service.

Chad Michael Briggs is Strategy Director at GlobalInt and a consultant on Arctic security issues. He was formerly a Minerva Defense Fellow and Chair of Energy and Environmental Security at the US Air Force, Senior Advisor at the US Department of Energy and a Fulbright professor to Berlin, Germany, and Budapest, Hungary. He holds a PhD from Carleton University in Canada.

Julia Christensen is currently a SSHRC Postdoctoral Research Fellow in Geography at the University of British Columbia and a Research Fellow at the Institute for Circumpolar Health Research. She will be assuming a new position as Assistant Professor in the Department of Environmental, Social and Spatial Change, Roskilde University in the autumn of 2013. She is a former Trudeau Scholar and the author of several articles on northern housing security and homelessness, community-based research and research storytelling. She was born and raised in Yellowknife, Northwest Territories, Canada.

Lassi Heininen is a Professor (of Arctic Politics) and Docent (of International Relations) at the Faculty of Social Sciences, University of Lapland, and Docent (of Northern Geopolitics) at the University of Oulu, Finland. His research fields include IR, geopolitics, security studies, environmental politics, Russian studies, northern and Arctic studies and political history. He is the author of more than 200 scientific publications and the editor of the *Arctic Yearbook*.

Geir Hønneland is Research Director at the Fridtjof Nansen Institute and Adjunct Professor in political science at the University of Tromsø, Norway. He has published widely on Russian politics and East–West relations in the European North. His latest books are *Making Fishery Agreements Work: Post-Agreement Bargaining in the Barents Sea* (Edward Elgar, 2012) and *Borderland Russian: Identity, Narrative and International Relations* (Palgrave Macmillan, 2013).

Stephanie Irlbacher-Fox is Adjunct Professor at the Canadian Circumpolar Institute at the University of Alberta, Canada, and is the author of *Finding Dahshaa: Self Government, Social Suffering and Aboriginal Policy in Canada* (UBC Press, 2009). She lives in Yellowknife, Northwest Territories, Canada.

Line Linstad graduated as Cand.polit from the University of Oslo in 1994. She worked as a regional health planner in northern Norway before she started working in the field of telemedicine. She is currently the Department Manager of the Prevention and Self-management Department at the Norwegian Center for Integrated Care and Telemedicine (NST).

Maria Lvova is an Associate Professor in the Institute of Humanities, Social and Political Science, at Northern (Arctic) Federal University (Arkhangelsk, Russia). She teaches courses in modern philosophy, Russian politics and security. Her research interests are in the field of human and energy security, identity issues and postmodern philosophy. She additionally manages the Research Office of the University of the Arctic.

David Malcolm is Senior Partner at the consulting firm Malcolm and Associates in Yellowknife, Northwest Territories, Canada. He holds a PhD from the University of Warwick in the UK. He specializes in studies of Entrepreneurial Sustainable Living in remote communities, with special emphasis on engineering solutions in climate change adaptation and reducing community dependency in all areas.

Torbjørn Pedersen is an international relations reporter with *Aftenposten* newspaper. He has previously been a senior advisor to the Norwegian Defence Staff, a Senior Research Fellow with the University of Tromsø, where he defended his doctoral dissertation *Conflict and Order in Svalbard Waters* in 2008, and a Fulbright fellow at the Paul H. Nitze School of Advanced International Studies, Johns Hopkins University, Washington, DC, where he graduated as Master of Arts in international relations and international economics in 1997. Torbjørn has also worked as a reporter and correspondent for Reuters, a financial reporter at the Norwegian Broadcasting Corporation and the editor of *Svalbardposten*.

Jackie Price is an Inuk from Nunavut. She considers Rankin Inlet and Iqaluit her home community. Jackie is a doctoral student with the Scott Polar Research Institute, University of Cambridge. Her research explores Inuit governance within the context of settlement life, set within increasingly present international debates on Arctic governance. She enjoys teaching at the early college level.

Lars Rowe is a Senior Research Fellow at the Fridtjof Nansen Institute in Norway. He holds a PhD in history, and his main research interests include Norwegian–Soviet/Russian bilateral relations, Soviet/Russian industrial and environmental history and Soviet/Russian foreign policy. He has published widely on these subjects in Norwegian and English.

Gabrielle Slowey is an Associate Professor in the Department of Political Science at York University (Toronto) where she teaches courses in Canadian, Aboriginal and Arctic politics. Her research investigates the intersection between resource extraction, indigenous development and the state in multiple regions (Northern Alberta, Northern Quebec, Yukon, NWT, Ontario and New Zealand). She is the author of *Navigating Neoliberalism: Self-Determination and the Mikisew Cree First Nation* (UBC Press, 2008) and numerous book chapters.

Kirsti Stuvøy is Associate Professor in international studies at Lillehammer University College, Norway. As a member of the Human Security Research Group at the University of Tromsø, Norway, she conducted her PhD research on security theory and women's groups in Russia addressing violence against women. In 2009 she defended her PhD thesis *Security Under Construction: A Bourdieusian Approach to Non-State Crisis Centres in Northwest Russia*. Since 2010 she has been Associate Professor at Lillehammer University College, and has continued her work on civil society groups in collaboration with the Norwegian Institute of Foreign Affairs (NUPI) in the project 'Modernizing the Russian North', funded by the Norwegian Research Council.

Elana Wilson Rowe is a Senior Research Fellow at the Norwegian Institute for International Affairs (NUPI) and an adjunct professor at the University of Nordland. Her research interests include the Russian North, climate policy, international relations in the Arctic and the politics of expert knowledge. She is the author of *Russian Climate Politics: When Science Meets Policy* (Palgrave, 2013) and editor/co-editor of *Russia's Encounter with Globalization* (with Julie Wilhelmsen, Palgrave, 2011), *Russia and the North* (University of Ottawa Press, 2009) and *The Multilateral Dimension of Russian Foreign Policy* (with Stina Torjesen, Routledge, 2009).

Oran R. Young is Research Professor at the Bren School of Environmental Science and Management, University of California (Santa Barbara). His research focuses on theories of environmental governance with applications to climate change, the oceans and the polar regions. His most recent book is *On Environmental Governance: Sustainability, Efficiency, and Equity* (Paradigm, 2013). Oran has played a leading role for many years in the global environmental change research community.

Foreword

Remarks made by former Canadian Ambassador to Norway, Shirley Wolff Serafini, at the Human Security in the Arctic Seminar (University of Tromsø, 3 May 2004, Tromsø, Norway)

Ladies and gentlemen,

I am very pleased to address this ground-breaking seminar on Human Security in the Arctic. Let me begin by bringing you warm greetings from Canada's Minister of Foreign Affairs, the Honourable Bill Graham.

As you well know, both Canada and Norway are northern nations. Our history and development have and continue to be forged by the strong Arctic winds that blow across our countries and our shared commitment to ongoing dialogue on matters of importance to Northerners. I want to congratulate the Political Science Department of the University of Tromsø for organizing this seminar, and to express our appreciation to Gunhild Hoogensen, who has been the initiator and driving force in bringing you all together here. The Canadian Embassy in Norway is glad to be a participating sponsor for this seminar along with the Research Council of Norway and the University of Tromsø. Three years ago, in April 2001, we partnered with the Research Council of Norway, the Norwegian Ministry of Foreign Affairs and the Fridtjof Nansen Institute to organize a Canadian–Norwegian Arctic Co-operation Seminar. I believe there is much to be gained through dialogue between Canadian and Norwegian human security researchers and Arctic researchers. In this regard, I believe that you will find common themes in your respective work that will lead to new insights and new avenues for co-operative research.

What unites your common efforts is the concept of human security in the broadest sense. During the UN Millennium Summit in 2000, UN Secretary General Kofi Annan called upon the international community to advance the twin goals of 'freedom from want' and 'freedom from fear'. The Commission on Human Security, in its report *Human Security Now*, elaborated further on the interrelated dimensions of human security. Let me remind you of how they described what human security means. 'Human security means protecting vital freedoms. It means protecting people from critical and pervasive threats and situations, building on their strengths and aspirations. It also means creating systems that give people the building blocks of survival, dignity and livelihood. Human security connects different types of freedoms – freedom from want, freedom from fear and freedom to take action on one's own behalf.'

In my remarks this morning, I would like to provide some context for your deliberations by outlining Canadian and Norwegian co-operative efforts to address Arctic and human security issues, and by sharing with you the Government of Canada's perspectives on key challenges in these two spheres.

Canada's northernness

Canada's sense of northernness has long been central to our identity. This extends even to Canadians who have never been to the North. Canadians increasingly understand the North's unique rhythms and its environmental fragility. They also understand its growing importance to our security and our prosperity. Internationally, the Arctic is emerging as an organized and recognized region in world affairs. Global awareness about the importance of the Arctic to the health of the planet and its vulnerability to new threats is growing. Issues such as the long-range transport of persistent organic pollutants, climate change and increased navigation in ice-covered waters make it abundantly clear that circumpolar co-operation is not only desirable, but necessary.

Governments are in turn responding to demands from an increasingly confident northern population that is determined to address the serious environmental, economic, social and cultural threats facing their communities.

The Arctic Council

To address these threats, northerners are reaching out to communities around them. Canadians and Norwegians are engaging their circumpolar neighbours to find effective ways of dealing with common issues. One demonstration of this is the unprecedented co-operation and institution building in the circumpolar North over the last decade, designed to meet the region's challenges and aspirations. The wide variety of organizations such as the Arctic Council, Barents Euro-Arctic Council, Nordic Council and the Northern Forum are all contributing to the strong circumpolar community that continues to emerge.

The Arctic Council for example was created in 1996 to advance circumpolar co-operation. The Council's mandate is to protect the Arctic environment and promote the economic, social and cultural well-being of northern peoples. It is made up of the eight Arctic states and international indigenous peoples' organizations as Permanent Participants. Several non-Arctic states, international organizations and non-governmental organizations participate as observers. Working groups of the Arctic Council have focused on sustainable development,

Arctic monitoring and assessment, conservation of Arctic flora and fauna, emergency prevention, preparedness and response, and the protection of the Arctic marine environment. Landmark work of the Arctic Council includes the Arctic Monitoring Assessment Programme (AMAP) report which was the backbone of the 'Stockholm Convention on Persistent Organic Pollutants'.

The major priority of the Arctic Council this year is on climate change. Because discussion of climate change usually tends to focus on political, economic and technical issues, we frequently lose sight of the serious human impacts and consequences for northerners, especially the indigenous peoples of the Arctic who are facing social and cultural dislocation in the coming years. Human-induced climate change is undermining the ecosystem upon which the Inuit of Canada, Alaska, Greenland and Russia depend for their physical and cultural survival. Climate change in the Arctic is not just an environmental issue with unwelcome economic consequences. It is also a matter of livelihood, food and individual and cultural survival. In short it is a 'human security' issue. That is why climate change is such an important priority in the work of the Arctic Council. This autumn the results of the Arctic Council initiative, the Arctic Climate Impact Assessment (ACIA), will be released. It will evaluate and synthesize knowledge on climate variability, climate change and increased UV radiation and their consequences. The assessment will include environmental, human health, and social and economic impacts of climate change. This assessment will be conducted in the context of other developments and pressures on the Arctic environment, its economy, regional resources and peoples. It will provide powerful and reliable information to governments, organizations and peoples of the Arctic region to support policy development. The ACIA promises to be a landmark body of work just as the AMAP report was a few years ago.

But multilateral bodies such as the Arctic Council are not the only avenues for co-operation. Strengthened bilateral relations between Canada and our northern neighbours – Norway in particular – have also been important in forwarding our common agendas, both on human security and on the Arctic.

The Lysøen accord

In the same period when the Arctic Council was being established, the 'Ottawa Convention on Anti-personnel Landmines' was being negotiated. In this process a close partnership developed between Canada and Norway. Following the signing of the Ottawa Treaty in December 1997, then-Canadian Foreign Minister Lloyd Axworthy and then-Norwegian Foreign Minister Knut Vollebaek found they shared a similar ambition: to see if the winning formula that produced that landmark treaty could be replicated again with other issues.

Minister Vollebaek invited Minister Axworthy to a retreat in Bergen in May 1998, and what emerged was 'A Canada–Norway Partnership for Action:

The Lysøen Declaration'. The Lysøen Declaration was the first time that the concept of 'human security' appears in an official agreement, negotiated between Canada and another government. More importantly, it outlined a 'partnership agenda' of nine human security issues for further bilateral collaboration. These nine issues were: landmines; the International Criminal Court; human rights; international humanitarian law; gender dimensions in peacebuilding; small arms proliferation; children in armed conflict, including child soldiers; child labour; and Arctic and northern co-operation.

Both countries were keen to use the Lysøen Declaration to anchor a strong bilateral partnership. Since then we have worked together on each of the agenda items, identifying common goals in specific multilateral fora and undertaking a number of joint initiatives.

The Human Security Network

The Lysøen Declaration led to the creation of an important multilateral process as well: the Human Security Network. The Human Security Network was an attempt to institutionalize a 'coalition of the willing' to motivate international action on a broad range of human security initiatives. It was formally launched at a Ministerial meeting in Bergen, Norway, in 1999.

The commitment made by Human Security Network countries in Bergen appeared exceedingly bold at the time. Collectively, it called for a 'humane world where people can live in security and dignity, free from poverty and despair', and committed member states to work towards 'freedom from pervasive threats to people's rights, their safety or even their lives'. Ministers said that this approach should 'become a new measure of global security and a new agenda for global action'. While the world is certainly not yet one in which all people can live in security and dignity, remarkable progress has been made in a few short years.

As I mentioned earlier, the promotion of 'freedom from fear' was a core element of the 'UN Secretary General's Millennium Report'. The UN Security Council has now formally accepted its responsibility to address the protection of civilians in situations of armed conflict. The G8 is promoting core elements of the human security agenda through the 'Miyazaki Initiative for Conflict Prevention', and in its support of the 'New Partnership for African Development' (NePAD), where strengthening peace and security is seen as a precursor to new investment and economic development. Freedom from fear is also the underlying logic to the new 'OECD Development Assistance Committee (OECD/DAC) Guidelines on Helping Prevent Violent Conflict'.

Remarkable progress has also been made on advancing work on priority issues identified by the Network – obvious examples include addressing the needs of war-affected children, enhancing the role of women on issues of peace and security, reducing the widespread availability of small arms and strengthening the capacity for peace support operations.

In some ways, perhaps the greatest achievement of them all is the entry-into-force of the 'Rome Statute' creating the International Criminal Court (ICC). The creation of the ICC will ensure that those responsible for the most serious crimes known to humankind – genocide, war crimes and crimes against humanity – will be held responsible for their actions. An effective ICC will represent a historic triumph over tyranny and terror.

In the wake of the events of September 11th, some would suggest that in the face of pressing terrorist threats what is most urgently required is an emphasis on national security instead of human security. Canada would disagree.

In combination with military, intelligence and law enforcement efforts, the human security agenda is a crucial dimension of the broader response to terrorism. Resolving festering conflicts and rebuilding failed-states, ending impunity and promoting security sector reform, combating transnational organized crime and stemming the proliferation of small arms and materials of weapons of mass destruction are all essential elements of an effective post-September 11th strategy. The overarching objectives of the Human Security Network are now more important than ever.

Progress on core human security issues such as landmines, or the ICC, succeeded in large part due to the effectiveness of the 'like-minded' groups that shepherded the negotiations. In the face of multilateral efforts held hostage by rigid regional negotiating blocs, the Human Security Network provides a model for a transparent cross-regional way of working. The Network can be a unique tool in helping to restore trust in, and to strengthen, multilateralism. Bringing together countries from different regional groupings, it has the potential to bridge the gap between regions and players on the international scene.

As Canada takes on the Chairmanship of the Human Security Network at the close of the upcoming Ministerial in Mali in May 2004, we look forward to continued close collaboration with Norway and other Human Security Network partners. A key priority of our Chairmanship will be to advance work on the Responsibility to Protect.

The focus on the safety of individuals raises the question about the relationship between human security and national security. When states act in the security interests of their people, state security and human security are mutually supportive.

States, however, are not necessarily guarantors of human security. When states are externally aggressive, internally repressive or too weak to govern effectively, they threaten the security of people. In the face of massive state-sponsored murders, appalling violations of human rights and the calculated brutalization of people, the humanitarian imperative to act cannot be ignored and can outweigh concerns about state sovereignty. Ultimately, state sovereignty is not an end in itself – it exists to serve citizens and to protect their security.

The Charter of the United Nations embodies the view that security cannot be achieved by a single state in isolation. The phrase 'international peace and security' implies that the security of one state depends on the security of other states. A human security perspective builds on this logic by noting that the security of people in one part of the world depends on the security of people elsewhere. A secure and stable world order is built both from the top down and from the bottom up. The security of states, and the maintenance of international peace and security, are ultimately constructed on the foundation of people who are secure.

It is sometimes argued that attending to the violent threats that people face diverts funds from the more basic priority of development. But human security and human development are opposite sides of the same coin. Together they address the twin goals of *freedom from fear* and *freedom from want*. Human security provides an enabling environment for human development. Where violence or the threat of violence makes meaningful progress on the developmental agenda impossible, enhancing the safety of people is a prerequisite. Promoting human development is also an important strategy for furthering human security. By addressing the root causes of violent conflict, by strengthening governance structures and by providing humanitarian relief, development assistance complements political, legal and military initiatives in enhancing human security.

In many ways, the concept of human security attempts to do for the theory and practice of security what human development did for approaches to development. Through the 1950s and 1960s development was defined in narrow economic terms and measured through abstract national figures such as growth in GDP. The notion of human development challenged both the scope and the level of analysis – a broader range of indicators was employed including health and education, and an attempt was made to measure the effects on people rather than states. Similarly, human security attempts to reconceive security by examining a broader range of threats and by focusing directly on the effects on people.

Towards a new diplomacy

An emphasis on human security determines not only the objectives of our foreign policy, but also the manner in which we pursue those objectives. While this so-called 'new diplomacy' is not exclusively linked to our human security agenda, the mixture of powerful ideas, persistent persuasion, public advocacy and partnership with civil society has proven remarkably effective.

Developing innovative global partnerships helps link countries, institutions and non-governmental organizations with like-minded objectives. Such coalitions between governments and civil society helped make the campaign to ban landmines a success and were instrumental to progress in adopting the statute of the International Criminal Court and the establishment of the Arctic Council. They are harbingers of the future, demonstrating the power of good ideas and pooled resources.

At its core, the human security agenda is an effort to construct a global society in which the safety and well-being of the individual is an international priority and a motivating force for international action; a society in which international humanitarian standards and the rule of law are advanced, woven into a coherent web protecting the individual, where those who violate these standards are held fully accountable; and finally, a society in which our global, regional and bilateral institutions – present and future – are built and equipped to promote and enforce these standards.

The Northern Dimension of Canada's Foreign Policy (NDFP)

In this vein and in response to the heightened engagement of circumpolar states and northern residents, which I talked about earlier, Canada developed and released the Northern Dimension of Canada's Foreign Policy, or NDFP, in 2000. By defining broad goals and focused objectives, this policy creates a firm foundation that assists Canada in determining priorities for action. The broad goals are:

- first, to enhance the security and prosperity of Canadians, especially northerners and Indigenous peoples;
- second, to assert and ensure the preservation of Canada's sovereignty in the North;
- third, to establish the circumpolar region as a vibrant geopolitical entity integrated into a rules-based international system; and
- finally, to promote the human security of northerners and the sustainable development of the Arctic.

To help realize these goals, we work with northern residents, academic institutions, northern stakeholder organizations and, of course, our circumpolar neighbours in the four following priority areas:

- first, strengthening the Arctic Council and promoting a central place in circumpolar relations for it;
- second, helping to establish a University of the Arctic designed to foster academic excellence and sustainability – including traditional knowledge;
- third, developing and expanding opportunities to work with Russia in addressing its northern challenges; and
- finally, promoting sustainable development through the pursuit of economic development and trade opportunities across the circumpolar region.

A key objective of this policy is to build international co-operation that promotes sustainable development in the circumpolar region. This includes working with our traditional circumpolar partners, but also reaches beyond, for example to the European Union, which exerts important influence on the domestic policies of its northern members, and to multilateral organizations that have largely ignored the Arctic in the past. Indeed, Canada, like Norway, is committed to taking the Arctic voice to the wider world, and we continue to promote co-operation that focuses on integral elements of human security, which are sustainable economic development, environmental protection, social and cultural well-being and knowledge-based decision making.

To achieve this, we focus much of our effort in the Arctic Council because it is the body best placed to identify and promote circumpolar co-operation and priorities. Information technology, innovative approaches to sharing information and new research partnerships are important features in all of the Council's work. We place special value in the Arctic Council not only because we were the first country to assume its chairmanship, from 1996 to 1998, or that we were one of the main proponents of its creation, but because nation states *and* indigenous peoples' organizations participate fully in Council decisions around the same table. That is why we see the Council as a primary focus of our northern foreign policy.

Domestic relevance

Our domestic work in the Arctic also complements and contributes to our international efforts. In the 2004 federal budget, Canada's Minister of Finance announced that a 'whole of government' northern strategy will be developed. Its aim is to ensure that economic development opportunities are developed in partnership with northern Canadians. Furthermore, six government-funded northern research chairs in universities across Canada are focusing on issues such as the study of permafrost in the Yukon and Northwest Territories (NWT), the study of environmental contaminants in the food supply, environmental changes, northern ecosystems and the marine environment. It is clear that these commitments are important not only for our domestic requirements but form an integral part of our contribution to a more thorough understanding of the circumpolar world by the international community.

Sustainable development and health

We are convinced of the value of circumpolar co-operation for achieving common goals. We can point to concrete examples of how beneficial and productive collective action can be, such as the success of bringing Arctic issues to the World Summit on Sustainable Development and the newly formed 'Northern Dimension Partnership in Public Health and Social Wellbeing'. The launching of the Northern Dimension Health Partnership in Oslo last October saw Canada and Norway, as well as 15 other countries and several regional organizations, become founding members. The aim of the Partnership is to deal more effectively with serious health issues such as HIV and tuberculosis in northern Europe, with a particular focus on the Baltics and Russia. We, in Canada, saw an opportunity to tie in the work we already undertake in the area of northern health and share our expertise and experiences with others facing similar Arctic conditions.

With respect to the World Summit for Sustainable Development, we took a two-track approach. The first was through our national position where Canadian negotiators pushed to include an Arctic clause in the final declaration of the Summit. The second was through our membership in the Arctic Council where, together with our Council partners, we worked throughout the process in the lead-up to the Summit in Johannesburg to sensitize the larger world community to important Arctic issues.

Continued Arctic co-operation

We can say with confidence that we have seen measurable progress in the circumpolar region over the past decade on these issues, not the least of which is the solid partnerships among nations, implementing agencies, non-governmental organizations and – most importantly – with northerners themselves.

Canada has taken the lead on many issues and projects that contribute to sustainability. I would draw your attention to our work in promoting capacity building as a process where people, ideas, information and resources intersect to produce concrete actions to increase human well-being and to enhance a sustainable circumpolar environment. One such area that is progressing is an innovative project on the Future of Children and Youth of the Arctic. It focuses on health, internships and youth networking. We are gathering baseline data on the physical and socio-psychological health of young people to develop a better understanding of what promotes and deters good health in the North, including environmental factors. The internship and networking component allows young graduates to gain work experience in areas associated with sustainable development, biodiversity conservation and promotion of traditional knowledge. As a matter of fact, over the last few years, a number of northern Canadian interns have had assignments in Norway where they worked at organizations such as GRID-Arendal, World Wildlife Fund for Nature and the Peace Research Institute. The Future of Children and Youth programme with its different streams builds capacity and confidence in our northern youth, and stimulates healthy communities built on environmental security, respect for cultural diversity and equity.

Conclusion

The Lysøen Declaration established a framework for consultation and concerted action to enhance human security, promote human rights, strengthen humanitarian law, prevent conflict, foster democracy and work on Arctic co-operation. Much has been accomplished toward these goals. Our challenge now is to move forward on our shared values and interests and to capitalize on the possibilities of intensified co-operation to make better use of the communities of existing organizations and Arctic and human security research networks, and to draw on our collective resources to address the human security challenges for the benefit of our shared North.

I wish you well in your deliberations. Thank you.

Acknowledgements

The process of creating this book, which has been a long time in coming, and its allied projects could not have been done without the support and help of many people. In particular, we thank David Dewitt and Alf Håkon Hoel for their crucial leadership and support at York University and the University of Tromsø respectively, and Annette Dubreuil and Bahareh Toghiani Rizi for administrative and editing support, good advice and maintaining enthusiasm for the project. A special acknowledgement and thanks goes to Brigt Dale and Berit Kristoffersen whose PhD research and resulting work, stemming from project proposals emanating from the original 2004–2005 discussions, provided crucial insights into this burgeoning field. As well, we thank Kari Anne Bråthens, Alana Kronstal, Zoe Wilson, Erin Baines, Taylor Owen, Geir Gabrielsen, Evgeny Bojko, Olga Ketkina, Rajiv Rawat and Jon Øyvind Odland for stimulating discussions that aided in the development of ideas and concepts. Research funds from the University of Tromsø, the Leiv Eiriksson Mobility Programme of the Norwegian Research Council, and the Norwegian and Canadian International Polar Year programmes allowed many of these discussions to take place.

Chapter 1

Introduction

Can we broaden our understanding of security in the Arctic?

Gunhild Hoogensen Gjørv and Marina Goloviznina

Introduction

Security is a concept about power, as well as a powerful concept. Invoking the concept is a political act. It makes a claim to power. The debate around what the concept means is itself a practice in power. It is a concept that has been, and continues to be, invoked to draw attention to 'something' that is or should be valued above all other things. Often this something is 'existence' and a continued existence (it is about the future), and a core feature of the 'security debate' is a debate as to the existence of what. So-called 'narrower' or 'traditional' approaches to security demand restrictions upon how the concept is used, who can use it and how. These narrower approaches demand that the existence we should be most preoccupied with is the existence of the state, assuming that without the state, nothing else really matters (Buzan *et al.* 1998). These approaches assume that security is provided through the use or threat of use of military force (Walt 1991). They assume that the state establishes and maintains security for itself as well as for its 'contents' (people and territory), partly reflecting, for example, social contract theory and related traditions in western political thought that assume a relationship of security between the individual and the state. What is lost in state-focused approaches to security, however, is the important role of the individual, which has also had a significant role in western philosophical thinking about security (Rothschild 1995; Hoogensen 2005a).

The concept of security can invoke a sense of urgency and sometimes fear. At the same time, it can also invoke a sense of stability, comfort and hope. The concept is used to draw attention to that which we consider most important to my, your or 'our' existence, depending on how the security referent is defined. The identification of a security referent (the 'object' or 'subject' that should be made secure) depends further on the values held by the security or 'securitizing' actor (the one who wishes to alert a public or an audience to a security problem). What that person or organization/institution values most will play a role in how that same actor interprets and identifies security and insecurity. Narrow conceptions of security are the purview of elite state interests, and thus the values and priorities of the state as an actor are those that give content to the concept of security. Opening the space for competing (and possibly complementary) security perspectives means recognizing and allowing that different actors other than the state are given the power to have their primary concerns and values heard.

Studying security and identifying security actors and practices is, therefore, both a scholarly/analytical activity as well as a political activity. The very act of defining security and making a claim for that definition is an act of power, supporting the politics that depend on that definition, or making a normative claim for why security ought to be defined in a particular way. When security analysts 'observe' acts of security or security moves, the analyst has immediately contributed to the politics of the process by recognizing (or not recognizing) an actor as a security actor and a securitizing move as being successful or not.

These dynamics of security are well illustrated in the Arctic. The notion of security has had a constant presence in the Arctic, but its meaning has changed over time, based on alterations in values and priorities for the region. The Cold War has played a particularly significant role in the ways in which security has been understood in the Arctic, whereby individual and community perceptions of security have, to be fair, never really played a role. This was largely a result of policy makers and security analysts focusing almost exclusively upon the tensions between two superpowers, reducing Arctic security to the balance of power played out with the threat of nuclear force (Heininen 2004). Since the Cold War, however, understanding security in the Arctic has broadened its scope, including environmental, energy, economic and, as we wish to argue here, human security (Hoogensen *et al.* 2009; Broadhead 2010; Offerdal 2010; Stokke 2011).

The focus of this book is therefore not re-articulation of 'traditional' Arctic security focusing solely upon protection of borders and the power dynamics between nation states, linked to a classic geopolitical framework. These security manoeuvres matter and are always a relevant factor in the security dynamic, but are limited in their capabilities to provide a more holistic and comprehensive security picture of the region. Although recognizing the importance and role of more limited perceptions of security, the work in this book draws on a wide variety of intellectual traditions, of which many, if not most, break beyond the narrow concept. Some of these traditions remain close to the narrow limitations of traditional security, like the Copenhagen school on securitization, while others explore a deeper connection between security actors and security practices as informed by feminist security studies, critical security studies and human security studies.

As many of the authors in this volume acknowledge, human security was initially conceived as a concept for the 'global South', with the implication that the concept has no relevance for global North states. Indeed, it appeared to be a concept designed for global North states to act for or in the global South, identifying human security needs, if not potentially rectifying them.

This approach seemed to take a form of 'virtuous imperialism' where the North intervened in the South with the best of intentions (see Hoogensen 2006 and Hoogensen Gjørv, this book). The activities of northern-based actors (such as NGOs or other non-state actors, or state-based actors) work to provide human security in other states, such as humanitarian or development assistance, as well as military interventions that claim a humanitarian cause (see Anderson 1999 and Polman 2010). However, this is equally the case in the North, such as Norwegian actors attempting to affect health security in Russia (see Stuvøy, and Rowe, Rowe and Hønneland, this book), which is not always received as a purely noble activity in all instances. The legitimacy and authority of security actors are thus equally relevant in North–North interaction as they would/should be in North–South interaction. This point is emphasized with examples where local security actors, with or without additional support mechanisms, attempt to address their own local health/food/environmental/personal security needs on the basis of their own capacities and definitions about what the potential threat is and how it can be addressed (such as women in Northwest Russia, Inuit in Canada). In all cases a multi-actor security process is operating, where in each context the power dynamics between actors (state and non-state) will differ. The question is to what extent the different 'voices' of these actors are heard (the power between them). In a narrow security approach, the position of different actors does not matter, as there is only one 'legitimate' voice that comes from the state (and, if necessary, acted out through the tool of the military). The work in this book is meant to make visible the voices and values that are relevant to understanding security in a more comprehensive sense, including a human security approach that centres on the experiences and practices of individuals and communities.

The contributors to this volume come from a variety of backgrounds within the social and natural sciences. The contributions thus vary in the ways in which security is conceptualized and the extent to which, and the complexity with which, each author engages the debates. Not every author is convinced of the efficacy of using the term 'human security', or security at all, particularly in the Arctic setting. This scepticism was reflected when this project had its modest beginnings in 2004 when researchers of different disciplines, ranging from international relations/political science, to anthropology, ecology and ecotoxicology, met in Tromsø, Norway, to discuss if the concept of 'human security' was at all relevant to the Arctic context. There was by no means agreement at that time, and these same, and other, researchers will undoubtedly continue to debate this question. However, the question has not died out. In fact, it is arguable that an increasing acknowledgement of the complexity of security in the Arctic region, not least including human security perspectives, has become both relevant and necessary to better understand the complex concept of security.

The 2004 debate in Tromsø led to the development of the International Polar Year (IPY) project, *Impacts of Oil and Gas Activity in the Arctic Using a Multiple Securities Perspective* (GAPS) (2006–2012), which examined the relationship between hydrocarbon extractive activities and the people living in the regions where these activities are taking place. This relationship has been examined through a multiple security approach (Hoogensen Gjørv 2012), whereby security is understood comprehensively (see also Heininen, this book), including but reaching beyond solely state interests, making visible processes by which individuals, communities, states and the international community ensure that that which they value in life, that which defines life for them, is maintained in the future. The GAPS project team examined the ways in which hydrocarbon extractive industries have affected/influenced the future expectations/security of communities in the Arctic, focusing in particular in northern Norway (Lofoten), northern Canada (Northwest Territories) and in Russia (Komi Republic).

Perhaps one of the most significant results of the project was the extent to which the project extended out to communities in an attempt to bridge a gap between communities and academia. The extensive collaboration that took place between scientists and community members ranged from participation by Grade 7 school children researching impacts of oil and gas development for their community (at the annual Forskningdagene in 2009), to talking with people in oil and gas impacted regions about their health (Komi Republic) and listening to people about the potential effects of oil and gas development in communities with long traditions connected to nature and the seas (Lofoten and fisheries; Dale 2012). Security in the Arctic has been demonstrated to be much more than 'states and borders'.

It is our general hope that this book will act to strengthen further development of multi-actor-based debates towards complex challenges facing Arctic peoples, communities and the environment today, with a special space for the voices of ordinary people living in the Arctic.

Overview of the book and its contents

As it was envisioned by the editors' team from the beginning, this book is a forum, in which all have come to share their thoughts, experiences and research results that have a bearing upon understanding security in the Arctic. The contributors to this book come from different disciplines, ranging from political science and philosophy to engineering and ecology, and from different nationalities, including Russian, Canadian, Norwegian, American and Finnish, as well as indigenous and non-indigenous. This results in a combination of works that understand security from different vantage points. Some authors concentrate more on the concept of security itself, while for others it lies in the background of the topic they are exploring. Taken together, the book provides one glimpse into an emerging security dynamic that is developing in the Arctic and how this is discursively conceptualized and practically developed, organized and delivered by multiple security actors, both state and non-state.

An examination of the security dynamics between Russia and Norway opens the discussion, where Kristian Åtland and Torbjørn Pedersen use securitization theory to analyse the ways in which Russian security politics has developed in relation to the Svalbard archipelago. The authors note that despite the thawing of relations due to the end of the Cold War, Norway's continued membership in the North Atlantic Treaty Organization (NATO) has been a source of suspicion for Russian leaders, influencing their perceptions of Norway's Svalbard policies in the region. The authors argue that a historical analysis demonstrates that Russia's policies towards Svalbard post-Cold War have much in common with the Soviet policies during the Cold War, demonstrating that Russian attitudes have not changed much despite the fall of the Soviet Union. Thus today's fears that Svalbard will be used for military purposes by Norway (military security), that Russian mining will be hindered by Norwegian environmental policy (societal security) and that Russian fishing will likewise be negatively affected by Norwegian fisheries policy (economic security) are rooted in Russian perceptions developed during the Cold War. The authors show however that many of the security moves (securitizing an issue) that took place after the fall of the Iron Curtain were done by upper-level government officials or high-ranking military personnel who were addressing a likeminded audience, taking their 'historical baggage' with them into the post-Cold War period. Åtland and Pedersen provide a strong historical analysis using securitization theory to show how perceptions of security develop, how such perceptions can have staying power despite changes in global politics and how such perceptions influence state policy.

The chapter by Lassi Heininen makes a case for comprehensive security, where state-based security policy and practices are recognized as playing a central role in our understanding of security, but that to obtain a holistic picture of security in the Arctic, this perspective must be accompanied by human and environmental security perspectives. Heininen does so by looking at the Arctic as a region and developing a regional security approach that integrates different perspectives. He takes as his departure point the military and military security, but demonstrates that even this actor has had to expand its understanding of security due to the influences of priorities and values generated by environmental issues, such as nuclear waste, environmental degradation and pollution. Further, other issues in addition to environmental degradation are generating opportunities for cooperation between traditional and non-traditional security actors, such as in the monitoring and use of transportation in the Arctic, the development of hydrocarbon industries, the impacts of climate change, processes of transnationalism and globalization, and the development of governance.

Chapter 4 moves deeper into the concept of human security itself and its possible relevance to the global North. Gunhild Hoogensen Gjørv examines the use of the concept when it first became popular in the 1990s, where it was perceived as having a relevance to the global South, and assumed a role for the global North as a 'human security provider'. The assumptions embedded within this concept led, in Hoogensen Gjørv's view, to a type of 'virtuous imperialism' where northern states were perceived as having the capacity to both identify as well as rectify human security concerns in the global South, endowing the South with the benefit of northern skills and knowledge. This approach or understanding of human security has a one-way character that, Hoogensen Gjørv argues, disguises both human security issues relevant in the North, but also any shared human security concerns that transcend the constructed boundary between the North and South. She argues that the human security issues in northern states are not equated to or necessarily of the same character as some human security issues in the South (not least, large-scale conflict and its impacts on civilians), but that being open and listening to the priorities and security concerns of northern local actors is an imperative part of any security debate. Hoogensen Gjørv looks specifically to what we can learn from indigenous scholars and feminist debates in a pursuit of understanding human security in the Arctic.

The chapter by Oran R. Young argues that the intricate linkages between social systems and ecosystems demand a re-thinking of the way that we understand and guide or govern these systems if our end goal is sustainability. Young argues that the dominant political approaches to governing the human relationship with the environment are based upon scientific approaches that perpetuate the assumption that the human impact on the environment, or ecosystems, is marginal (and vice versa), and that social systems can be governed largely independently of an understanding of, or connection to, ecosystems. Young argues that such an approach cannot and will not lead to a sustainable future, and that we need new models of analysis that are better at combining the intertwined relationship between systems, acknowledging the significant impacts that social systems have on ecosystems, as well as how those impacts loop back and further impact social systems. Young focuses on the political institutions that need to reflect sustainability-oriented policy making. He recommends policy makers to draw upon diverse sources of knowledge, develop an adequate range of possible responses to uncertainty, think more seriously about the long-term consequences, improve monitoring, improve learning about the social consequences of ecosystem change and use crisis as learning opportunities.

Starting with a critical overview of security and climate change discourses, the chapter by Chad Michael Briggs examines the vulnerability and adaptability of social systems to abrupt changes in climate. Briggs argues that we need to understand the ways in which social and political systems are able to adjust under variable changes to the climate, occurring either rapidly or more slowly. He claims that the understanding of adaptability is not dependent upon a binary 'failed' or 'successful' state, but upon understanding how new forms of stability develop in response to change, and what consequences such stability have on the social world. Briggs focuses on the different elements of vulnerability – risk, sensitivity, resilience and fragility – to provide a tool set for understanding social adaptability to abrupt changes in climate. Briggs uses these tools to examine the impacts of climate change in the Arctic, drawing on evidence of change occurring with the melting Greenland ice sheet, and the increasing release of methane from Arctic oceans contributing to GHG emissions. Briggs goes on to demonstrate the ways in which human social and political systems are vulnerable to changes in the environment, and makes clear the dependencies human social systems have upon ecosystems, which in turn has an impact on human, state and international security.

David Malcolm brings an engineering and technology background to his chapter on technological adaptation to climate change. Malcolm links changes to the environment in the Arctic to adaptations and improvements in engineering and technology, arguing that this is an important part of the adaptation process, increasing human security under conditions of instability in the environment. Malcolm draws on long-term data regarding climate change in the Arctic, mapping some of the key impacts that have taken place so far. As Malcolm states, despite what appears to be a lack of interest from southern-based communities (in this case, southern Canada), climate change in the Arctic cannot be understood as a phenomenon that affects this region alone. What happens in the North has, or will have, an impact in the South. Malcolm describes an integrated approach to planning and engineering in the North, where technological expertise must meet not only technological demands, but also comply with sustainable development goals and community values.

The next chapter addresses the interdisciplinary nature of human and environmental security concepts, where the authors Dawn R. Bazely, Julia Christensen, Andrew J. Tanentzap and Gunhild Hoogensen Gjørv discuss the ways in which the human security concept can be useful to natural scientists to make their work relevant to the social and political world. In other words, the authors go after what can be considered a core challenge of communication and dissemination: the science–policy gap. As the authors note, there is no question that there remains significant challenges in the ability for social scientists and natural scientists to collaborate effectively, but suggest that a human security framework that recognizes the interconnections between social systems and ecosystems provides a solid foundation for interdisciplinary cooperation and research. Ecology has a number of 'uncomfortable' messages for human societies based on decades of research. This includes the message that there are indeed limits to growth; that due to the complexity of the systems we are speaking of there is always uncertainty regarding the outcomes and impacts of human–environmental interactions, but human systems are not well equipped to deal with uncertainties; and that ecosystems are subject to curvilinear relationships, which means that change can occur very abruptly and suddenly. The authors discuss how to make ecology relevant to a broader audience, and that as a social science concept, human security is appealing due to its longer-term, broader dimensions.

The chapter by Line Linstad looks at health security as an aspect of human security, arguing in favour of technological innovations such as telemedicine to improve access for people living in remote communities. Linstad uses the concept of regional security as a way of understanding the interconnectedness between remote communities and their mutual health security needs. Linstad highlights a number of issues where states and communities need to work together to strengthen health security. The first regards epidemics, both regional and international. The second issue she addresses is the lack of access to health care services, and lastly a deficiency in numbers of health care personnel. As Linstad states, health care became primary when the health of populations is concerned, and here international efforts towards cooperation are of paramount concern. The threat of disease does not stop at the border. She shows the ways in which this is relevant in the Arctic, particularly with regard to tuberculosis. Further, part of disease management and health security is supporting better controls, which means access to health care. This includes access to well-trained health care personnel, who are not always present in remote locations. Taking into account special Arctic/sub-Arctic factors and conditions including low density of population, difficulties of communication and lack of transportation logistics, the chapter argues that telemedicine should be considered as a crucial tool that could increase the personal well-being of Arctic peoples, especially in isolated communities, and enhance the health security in the region, in general.

Although telemedicine has recommended itself as a proven tool to provide primary medical care, it is still a quite exceptional and expensive practice which cannot largely respond to everyday health problems from the Arctic communities, in forms of mental health, violence and accidental death. Lars Rowe, Elana Wilson Rowe and Geir Hønneland analyse the Arctic cross-border cooperation between health providers regarding the threats posed by chronic and communicable diseases, predominantly HIV/AIDS and tuberculosis. Based on interviews with people working in the health care sector in the Barents region (Russia and Baltic countries), the authors present conflicting evidence on how health care cooperation, which could be perceived as working towards health/human security, actually operates. Although many of the respondents to the interviews seemed overall to be positive towards the efforts of the Task Force on Communicable Disease Control, a foreign-led, World Health Organization (WHO) programme originally designed for developing countries and focused upon pre-determined vulnerable groups (prisoners, drug-users, prostitutes), there were also significant expressions of resistance to the programme. The authors demonstrate the different cross-cultural perceptions of the notions of 'health risks', 'health care' and 'treatment' from the position of post-Soviet medical staff, who took issue with the idea that foreigners could identify and tackle health security issues better than they could. The mixed 'success' of the programme demonstrated a problem with the human security concept if the concept refers to a so-called competent 'security provider' (developed country) providing services and skills to an assumed 'security recipient' (developing country). The chapter raises an important question about the legitimacy of how human security is provided and by whom.

The chapter by Gabrielle Slowey moves to indigenous issues, where she focuses readers' attention on the development of self-government, and analyses practices of local governance and co-management of resources in terms of human security in two Arctic communities in Canada (Inuvialuit and Vuntut Gwitchin). Slowey's empirically based analysis links the notion of self-government to human security, arguing that more control over decision-making power has significant effects upon the human security of a community. As Slowey argues, self-government, like human security, is not an end goal but is a work in progress through which the Aboriginal peoples are working in order to increase their overall quality of life, raise their standard of living and increase an Aboriginal voice in the local and regional governance structures, as well as in relation to the national state. Looking comparatively at experiences from two different communities, the results of empirical research demonstrate the ways in which self-government along with land claims, especially in areas of increased oil and gas activity, dramatically improve the ability of indigenous communities to address their own human security issues.

Continuing on the 'work in progress' of indigenous self-government and governance, the authors Stephanie Irlbacher-Fox, Jackie Price and Elana Wilson Rowe explore further links of human security to political representation, communication vs consultation, and women's voices. In this chapter we see the linkages between representation and participation and human security, where inadequate attention (if not state-based negligence) to the needs of the community have resulted in serious human insecurities amongst indigenous populations. The development of politics and political institutions in the Canadian North has, according to these authors, generated and imposed unequal power relationships between men and women that were not a part of the northern indigenous cultures prior to these moves. The authors demonstrate how the security of indigenous women is deeply connected with the security of the community, both men and women. The authors introduce a consultation model in the interest of creating a more representative, reflective and responsive political system, including effective gender-balanced representation of indigenous peoples, moving indigenous governance beyond the confines of colonial political systems. This model allows for different levels of community and regional engagement, reflecting the multi-actor security perspective introduced in Part I of this book.

Violence in a variety of forms is a certain threat that jeopardizes many women in the Arctic in their daily lives. Domestic violence, sex abuse, trafficking and prostitution form the spectrum of violence the women in Arctic communities, including in Russia, Finland, Sweden and Norway, were experienced and listed among risks to a personal (in)security. Two chapters examine gender in Russia – the use and abuse of sex, and the ability for women to combat violence in their own life situations or life worlds.

In order to examine both violence against women and democracy development on the local level, Kirsti Stuvøy, based on the results of an empirical case study of Northwest Russia (Archangelsk, Murmansk and Saint Petersburg regions, 2004–2009), offers a detailed analysis of non-state crisis centres for victims of domestic violence at both conceptual and empirical levels. The chapter asks it a central question as to whether or not the gendered/domestic violence in the Northwest of Russia is being securitized, and, if so, by whom, why and how. The research results demonstrated the important shifts occurred within the field of security for women in Russia during the last decades, including the diversification of security providers, and expansion in this domain of non-state actors. The crisis centres are considered to be one of these new agents, an agent of change, which challenged the dominant state position and power to define legitimate social problems and decisions relevant to address women's security situations.

Maria Lvova examines the relationship between democratization processes and the position of women in post-Soviet Russia, using the development of attitudes towards sex (particularly through prostitution and trafficking) as an illustration of how women's security has not been a core concern. Lvova discusses the development of prostitution, and the concurrent surge in trafficking, after the fall of the Soviet Union. She argues that women in Russia have been subjected to contradictory expectations, particularly with the advent of a more 'democratized' Russia, whereby on the one hand women were expected to return to the home and tend to families (in opposition to the socialist demand for women to be equal contributors to the public domain), but at the same time were 'free' to make all manner of career choices. When women were not welcome into the public domain (they should be at home), their choices have often been reduced to jobs that commodified the body, not least through prostitution. Trafficking also became a severe problem when women looked to overseas solutions for finding employment. Lvova argues that democratization in Russia has relied more so on market economy ideals without including a necessary focus upon the protection of individual/human rights.

The last chapter by Marina Goloviznina and Gunhild Hoogensen Gjørv provides some concluding comments on human security in the Arctic. The authors review the rationale behind using the concept in the Arctic setting, and focus on issues of climate change and the impacts of extractive industries, and how the impacts of both can be understood using a human security lens. The authors conclude by discussing how the use of the human security concept in the Arctic is both a political choice as well as an active decision, meaning that those who employ the concept do so consciously, recognizing it as a move to power for individuals and communities. The concept claims a voice for those who have usually been left out of the security debate, as it was the arena for 'high politics' that did not involve actors from 'below'. The research processes needed to learn more about human security from the 'bottom up' demand communication and engagement with communities. Indeed, as the authors argue, security is, in the end, about community.

Bibliography

Acemoglu, D., Johnson, S. and Robinson, J. (2001) 'The colonial origins of comparative development: An empirical investigation', *The American Economic Review*, 91(5):1369–401.

ACIA – Arctic Climate Impact Assessment Report (2005) ACIA Secretariat and Cooperative Institute for Arctic Research, University of Alaska Fairbanks.

AGPR – Arctic Governance Project Report (2010) Cambridge: Cambridge University Press.

AHDR – Arctic Human Development Report (2004) Akureyri: Stefansson Arctic Institute.

AMAP – AMAP Assessment (1998) *Arctic Pollution Issues*, Oslo: Arctic Monitoring and Assessment Programme.

—— (2004) *Persistent Toxic Substances, Food Security and Indigenous Peoples of the Russian North: Final Report*, Oslo: Arctic Monitoring and Assessment Programme.

Anderson, M. (1999) *Do No Harm: How Aid can Support Peace – or War*, Boulder, CO: Lynne Rienner.

Åtland, K. (2008) 'Mikhail Gorbachev, the Murmansk initiative, and the desecuritization of interstate relations in the Arctic', *Cooperation and Conflict*, September, 43(3):289–311.

Ayoob, M. (1995) *The Third World Security Predicament: State Making, Regional Conflict, and the International System*, Boulder, CO/London: Lynne Rienner.

Bardhan, P. and Mookherjee, D. (2000) 'Capture and governance at local and national levels', *American Economic Review*, 90(2):135–9.

Berkes, F. (1994) 'Co-management: Bridging the two solitudes', *Northern Perspectives*, 22(2–3):18–20.

Boemcken, M. (1994) '10 years of convention: The main achievements', in M. Brozska and P. Croll (eds) *Promotion Security: But How and For Whom?*, BICC 10-Year Anniversary Conference. Online. Available at: http://www.bicc.de/publications/publicationpage/publication/promoting-security-but-how-and-for-whomcontributions-to-biccs-ten-year-anniversary-conference (accessed 7 August 2011).

Broadhead, L.-A. (2010) 'Canadian sovereignty versus northern security: The case for updating our mental map of the Arctic', *International Journal*, 65(4):913–30.

Burgess, P., Amicelle, A., Bartels, E., Bellanova, R., Cerami, A., Eggum, E., Hoogensen, G., Kittelsen, S., Knibbe, K., de Koning, M., Koser, K., Krause, K. and Salemink, O. (2007) *Promoting Human Security: Ethical, Normative and Educational Frameworks in Western Europe*, Paris: Section of Human Security, Democracy and Philosophy, Social and Human Sciences Sector, UNESCO.

Buzan, B., Waever, O. and de Wilde, J. (1998) *Security: A New Framework for Analysis*, Boulder, CO: Lynne Rienner.

Campante, F. and Ferreira, F. (2007) 'Inefficient lobbying, populism and oligarchy', *Journal of Public Economics*, 91(5–6):993–1021.

Campbell, T. (1996) 'Co-management of aboriginal resources', *Information North*, 22(1):14–20.

Commission on Global Governance (1995) *Our Global Neighbourhood*, Oxford: Oxford University Press.

Dale, B. (2012) *Securing the Contingent Future*. Doctoral thesis, University of Tromsø.

Fjellheim, R. (2006) 'Arctic oil and gas: Corporate social responsibility', in R. S. Fjellheim and J. B. Henriksen (eds) *Oil and Gas Exploitation on Arctic Indigenous Peoples' Territories: Human Rights, International Law and Corporate Social Responsibility*, Kautokeino: Gáldu Resource Centre for the Rights of Indigenous Peoples' Guovdageaidnu.

Galasso, E. and Ravallion, M. (2005) 'Decentralized targeting of an antipoverty program', *Journal of Public Economics*, 89(4):705–27.

Gautier, D., Bird, K., Charpentier, R., Grantz, A., Houseknecht, D., Klett, T., Moore, T., Pitman, J., Schenk, C., Schuenemeyer, J., Sørensen, K., Tennyson, M., Valin, Z. and Wandrey, C. (2009) 'Assessment of undiscovered oil and gas in the Arctic', *Science*, 324(5931):1175–9.

Gorenburg, D. (2011) 'Russia's Arctic security strategy', *Russian Analytical Digest (RAD)*, 96:7–10.

Haufler, V. (2001) *A Public Role for the Private Sector: Industry Self-Regulation in the Global Economy*, Washington, DC: Carnegie Endowment for International Peace.

Heininen, L. (2004) 'Circumpolar international relations and geopolitics', in N. Einarsson, J. Nymand Larsen, A. Nilsson and O. Young (eds) *AHDR (Arctic Human Development Report)*, Reykjavik: Akureyri Stefansson Arctic Institute.

Hoogensen, G. (2005a) *International Relations, Security and Jeremy Bentham*, London/New York: Routledge.

—— (2005b) 'Bottoms up! A toast to regional security?' *International Studies Review*, 7(2):269–74.

—— (2006) *Virtuous Imperialism: Human Security and the Exporting of the 'Wisdom' of the North to the South*, San Diego, CA: International Studies Association. Online. Available at: http://citation.allacademic.com/meta/p_mla_apa_research_citation/0/9/9/3/7/pages99374/p99374-1.php (accessed 20 March 2013).

Hoogensen, G., Bazely, D., Christensen, J. and Bojko, E. (2009) 'Human security in the Arctic: Yes, it is relevant!', *Journal of Human Security*, 5(2):1–10.

Hoogensen Gjørv, G. (2012) 'Security by any other name: Negative security, positive security and a multi-actor security approach', *Review of International Studies*. Online. Available at: http://journals.cambridge.org/abstract_S0260210511000751 (accessed 22 February 2012).

Huebert, R., Exner-Pirot, H., Lajeunesse, A. and Gulledge, J. (2012) *Climate Change and International Security: The Arctic as a Bellwether*, Arlington, Virginia: Center for Climate and Energy Solutions. Online. Available at: http://www.c2es.org/publications/climate-change-international-arctic-security (accessed 2 November 2012).

Jones, T. (2007) 'The governance of security: Pluralization, privatization and polarization in crime control', in M. Maguire, R. Morgan and R. Reiner (eds) *The Oxford Handbook of Criminology*, 4th edition, New York: Oxford University Press.

Juda, L. (1999) 'Consideration in developing a functional approach to the governance of large marine ecosystems', *Ocean Development and International Law*, 30(2):89–125.

Kaufmann, D., Kraay, A. and Mastruzzi, M. (2008) *Governance Matters VII*. Policy Research Working Paper 4654. World Bank Development Research Group Macroeconomics and Growth Team & World Bank Institute Global Governance Program. Online. Available at: http://papers.ssrn.com/sol3/papers.cfm?abstract_id=1148386 (accessed 3 November 2012).

Keck, M. and Sikkink, K. (1998) *Activists Beyond Borders: Advocacy Networks in International Relations*, Ithaca, NY: Cornell University Press.

Krause, K. (2004) 'Is human security "more than just a good idea"?', in M. Brozska and P. Croll (eds) *Promotion Security: But How and for Whom?*, BICC 10-Year Anniversary Conference.

Online. Available at: http://www.bicc.de/publications/publicationpage/publication/promoting-security-but-how-and-for-whomcontributions-to-biccs-ten-year-anniversary-conference (accessed 11 August 2011).
Lippman, W. (1943) *US Foreign Policy: Shield of the Republic*, Boston, MA: Little, Brown.
Loader, I. and Walker, N. (2007) *Civilizing Security*, Cambridge: Cambridge University Press.
Nye, J. (1974) 'Nuclear learning and the U.S.–Soviet security regimes', *International Organization*, 41:371–402.
Offerdal, K. (2010) *The Politics of Energy in the European High North: Norway and the 'Petroleum Dialogue' with the US and the EU*, PhD, Oslo: University of Oslo.
Parson, E. (2003) *Protecting the Ozone Layer: Science and Strategy*, Oxford: Oxford University Press.
Picciotto, R. and Weaving, R. (eds) (2006) *Security and Development: Investment in Peace and Prosperity*, New York: Routledge.
Polman, L. (2010) *What's Wrong with Humanitarian Aid? The Crisis Caravan*, New York: Picador.
Rhodes, R. (1997) *Understanding Governance: Policy Networks, Governance, Reflexivity and Accountability*, Buckingham: Open University Press.
Rothschild, E. (1995) 'What is security?', *Daedalus*, 124(3):53–98.
Ruggie, J. (2012) *Large-Scale Industrial Methods of Extraction, Production and Development and their Impacts on the Rights of Indigenous People and Local Communities*, UN Working Group on Issues of Human Rights and Transnational Corporations and Other Business Enterprises, New York, 16–20 January.
Serreze, M., Barrett, A., Stroeve, J., Kindig, D. and Holland, M. (2009) 'The emergence of surface-based Arctic amplification', *The Cryosphere*, 3:11–19.
Simpson, A. (2007) 'The environmental–energy security nexus: Critical analysis of an energy "love triangle" in Southeast Asia', *Third World Quarterly*, 28(3):539–54.
Smith, M. (2010) *NATO in the First Decade after the Cold War*, Dordrecht: Kluwer Academic.
Stokke, O. S. (2011) 'Environmental security in the Arctic: The case for multi-level governance', *International Journal*, 66(4):835–48.
Stroeve, J., Serreze, S., Drobot, S., Gearhead, M., Holland, J., Maslanik, W., Meier, W. and Scambo, T. (2008) 'Arctic sea ice extent plummets in 2007', *Eos, Trans. Amer. Geophys. Union*, 89:13.
SWIPA – Snow, Water, Ice and Permafrost in the Arctic (2011) AMAP Secretariat, Oslo.
Tadjbakhsh, S. (2008) *Human Security: Concept and Implications*, London: Taylor & Francis.
UNDP – Human Development Report (1994) Washington, DC: Woodrow Wilson Center.
Walt, S. (1991) 'The renaissance of security studies', *International Studies Quarterly*, 35(2):211–39.
Weiss, L. (2000) 'Globalization and state power', *Development and Society*, 129(1):1–15.
Wouters, J. and Hachez, N. (2010) *Business and Human Rights in the EU External Relationships: Making the EU a Leader at Home and Internationally*. Online. Available at: http://www.europarl.europa.eu/document/activities/cont/200904/20090430ATT54804/20090430ATT54804EN.pdf (accessed 3 November 2012).
Yalowitz, K. S., Collins, J. F. and Virginia, R. A. (2008) *The Arctic Climate Change and Security Policy Conference: Final Report and Findings*, Dartmouth College, Hanover, New Hampshire, USA. Online. Available at: http://www.carnegieendowment.org/files/arctic_climate_change.pdf (accessed 2 November 2012).
Young, O. (1985) 'The age of the Arctic', *Foreign Policy*, 61:160–79.
—— (1999) *Governance in World Affairs*, Ithaca/London: Cornell University Press.
—— (2002) *The Institutional Dimensions of Environmental Change: Fit, Interplay, Scale*, Cambridge, MA: MIT Press.

Part I

Differing conceptions of security in the Arctic

Chapter 2

Cold War legacies in Russia's Svalbard policy

Kristian Åtland and Torbjørn Pedersen

Introduction

In the past two decades, interstate relations in the European Arctic (see Figure 2.1) have undergone a series of important changes. People-to-people contacts across Russia's northwestern borders have flourished, regional cooperation arrangements have been established and issues such as environmental protection and nuclear safety have been placed on the agenda of all Arctic states, including Russia. Notwithstanding, traditional 'hard' security concerns

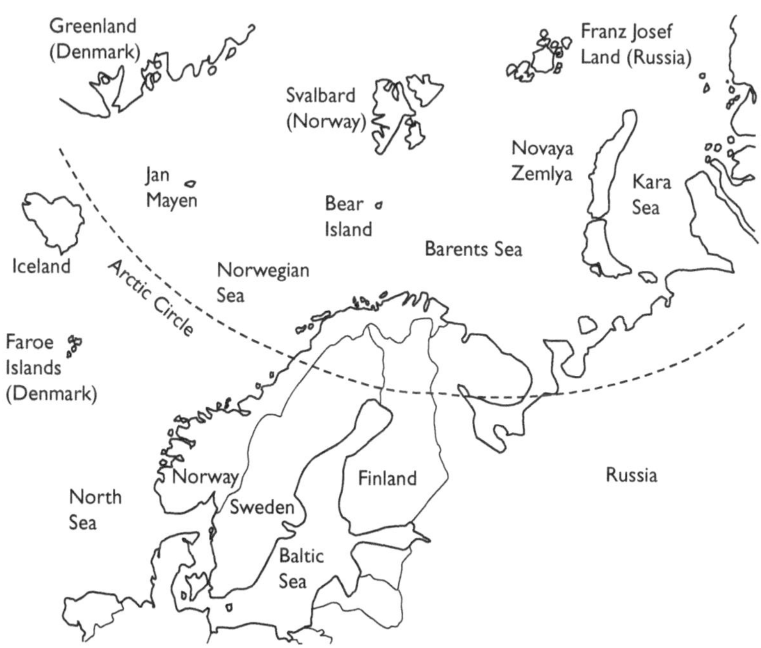

Figure 2.1 The European Arctic

continue to be a factor in Arctic politics, and East–West relations in the High North are still affected by defence-related controversies. As in the past, Russia's perceptions and framing of political developments on and around Svalbard are still shaped by the fact that Norway is a member of the North Atlantic Treaty Organization (NATO) – an organization with whom Russia still has an utterly complicated relationship.

In April 2003, the online-version of *Pravda* published an article titled 'Spitsbergen: NATO's outpost under Russia's nose' (Rivetov 2003).[1] It listed a number of security concerns pertaining to the Svalbard archipelago, which had been frequently voiced in Russian civilian and military journals and newspapers in the late 1990s and early 2000s. The article claimed, first, that the Svalbard Environmental Protection Act – a piece of legislation adopted by the Norwegian Parliament in 2001 – was aimed at obstructing Russian mining operations and could be forcing Russia to abandon the archipelago; second, that Norway was violating the Freedom of the Seas by taking enforcement measures against Russian fishing vessels in the Svalbard Fisheries Protection Zone (FPZ); and, third, that Norway was violating the demilitarization clause of the Svalbard Treaty by allowing the construction and operation of radars and satellite stations that could potentially be used in a US missile defence scheme (Rivetov 2003). Norway's Svalbard policy, allegedly part of a NATO conspiracy, was perceived to constitute a growing threat to Russia's political, economic and military-strategic interests in the European Arctic.

The roots of the Russian–Norwegian controversies can be found in the unique history of Svalbard and the two countries' differing interpretations of the Svalbard Treaty (1920), in which Norway's 'full and absolute sovereignty' over the archipelago was recognized (article 1). Under articles 2 and 3 of the Treaty, the nationals of all parties to the Treaty 'enjoy equally the rights of fishing and hunting' and may engage in 'all maritime, industrial, mining and commercial operations on a footing of absolute equality' (Svalbard Treaty 1920). Norwegian regulatory measures have often been perceived in Russia as infringements on the latter articles.

The provisions of the Svalbard Treaty apply to 'all the islands and rocks located between 74° and 81° northern latitude and 10° and 35° eastern longitude' (article 1), as well as to the 'waters' of these territories (articles 2 and 3; Svalbard Treaty 1920). While Norway maintains that the Treaty applies to the said land and sea territory of Svalbard, Russia and other parties to the Treaty contest Norway's claim to exclusive rights in zones beyond the territorial sea and question the legal status of the FPZ adjacent to Svalbard. The FPZ is a 200-nautical-mile zone of a non-discriminatory Norwegian fisheries jurisdiction, established in 1977.

Article 9 of the Svalbard Treaty (1920), under which Norway undertook not to establish or allow the establishment of naval bases or military fortifications on the archipelago, or use it for 'warlike purposes', has also been a source of debate. Russia interprets the provision as a demilitarization clause. Norway, however, does not interpret the provision as an absolute prohibition against

Norwegian military activity on Svalbard. Nor does it, in the Norwegian view, preclude visits to Svalbard by Norwegian naval or coast guard vessels, landings by military aircraft or temporary presence by Norwegian military personnel in uniform (Ministry of Justice and the Police 1999).

Given the strategic significance of the Barents Sea in the Cold War period, as the Soviet Northern Fleet's only gateway to patrol areas in the North Atlantic and the Arctic Ocean, Norway's handling of its Treaty obligations regarding the strategically located archipelago was a subject of constant Soviet scrutiny. There was widespread fear that Svalbard would be used for military purposes by Norway or its NATO allies, and it was the task of the Soviet Foreign Ministry and the country's consulate in the Soviet mining town of Barentsburg to monitor the situation and report any suspicious activity.

The purpose of this chapter is to examine contemporary Russian–Norwegian security relations in the northernmost part of Europe and to discuss similarities and differences between the old and the new security dynamics on the northern flank of NATO. Specifically, the chapter explores Russian perceptions of, and responses to, Norwegian Svalbard policies in the 1990s and 2000s. The next section of this chapter presents an overview of the theoretical framework, methodology and sources of the study. The three sections after that discuss the three cases mentioned above: Russian reactions to the establishment of Norwegian ground radars and satellite stations on Svalbard (1997–1999), the adoption of the Svalbard Environmental Protection Act (2001) and the enhanced Norwegian efforts to enforce fishery regulations in the FPZ (early 2000s). Findings from the case studies' analysis are summarized and discussed in the concluding section.

Theory, method and sources

Securitization theory, as laid out by the so-called 'Copenhagen School'[2] of security studies in the mid- and late 1990s, offers a conceptual framework for studying how and why certain issues become security issues. In addition to the military sector (military security), the Copenhagen School includes the political sector (political security), the economic sector (economic security), the societal sector (societal security) and the environmental sector (environmental security) in its theoretical framework (Buzan 1991:19; Buzan *et al.* 1998:7–8).

The essence of the theory is that security can be understood as a 'speech act' through which a particular issue is framed by one or more 'securitizing actors' as constituting an existential threat to a designated referent object. In the words of Danish political scientist Ole Wæver, the term 'securitization' denotes 'the intersubjective establishment of an existential threat with a saliency sufficient to have political effects' (Wæver 2003:11). The aim of a 'securitizing move' is typically to enable 'emergency measures' that can secure the survival of a referent object. If and when the content of the security 'speech act' is acknowledged as legitimate by a (significant) 'audience', the issue in question has become successfully 'securitized'. The issue has then been moved out of the sphere of 'normal politics' and into the sphere of 'emergency politics', where it can be dealt with in an urgent manner and with fewer formal and informal restraints. By uttering 'security', a state representative moves a particular development into a particular area and thereby claims a special right to use whatever means are necessary to block it (Wæver 1995:55).

The outcome of security 'speech acts' is determined by 'internal' (linguistic-grammatical) as well as 'external' (social) conditions, which either facilitate or impede the completion of the act (Buzan *et al.* 1998:32).[3] Both conditions are necessary, but not sufficient, to bring about successful securitization of a potentially threatening development. Often, the outcome of 'securitizing moves' hinges on factors associated with the alleged threat, such as its perceived specificity, proximity in time and space and 'whether or not perceptions of the threat are amplified by historical circumstances' (Buzan 1991:134).

This study will devote special attention to the latter dimension. Russia's Svalbard policy in the post-Cold War period features a number of similarities with the Soviet Union's Svalbard policy in the Cold War period. Many of the issues that have figured centrally on Russia's Arctic (military, societal and economic) security agenda in the 1990s and 2000s have historical analogies in Soviet Svalbard policy. These include Svalbard's demilitarized status, Norway's treatment of the Russian population on the archipelago and Russia's mining and fishing rights in the region. Judging from the contemporary Russian political debate with regard to the Svalbard archipelago (*Barents Observer* 2007), and previous articles in Russian defence and security periodicals (Motsak 2000; Smolovskiy 2000, 2006), the perceived threats to Russia's current interests in the region include the danger that: (1) the archipelago may be used for military purposes by Norway, the United States and/or NATO, (2) future Russian mining operations on the archipelago will be obstructed by Norwegian environmental legislation and (3) Russian fishermen will be pushed out of the FPZ by the Norwegian Coast Guard. Therefore, the working hypothesis of this study is that contemporary Russian threat perceptions and security policies with regard to the Svalbard region, perhaps more than elsewhere in the European Arctic, are shaped, or at least heavily influenced, by perceptions and policies formed in the Cold War period.

In order to understand how and why these particular issues were 'securitized' in Russia, and to explore more generally the role of 'historical circumstances' as a 'facilitating condition' for securitization, a case study approach is applied. The selection of cases for this study is based on an overall assessment of the Russian Svalbard debate in the 1990s and 2000s, as well as the content of Svalbard-related diplomatic exchanges between Russia and Norway during the same period. The three threats singled out above cover different dimensions of the concept of security: the first case (the establishment and use of

Norwegian radars and satellite ground stations) is illustrative of Russia's *military* security concerns, the second case (the adoption of the Svalbard Environmental Protection Act) is illustrative of Russian *societal* security concerns and the third case (the Norwegian fishery enforcement measures in the FPZ) is illustrative of Russian *economic* security concerns.

This study is based on data from both primary and secondary sources, such as official policy documents, diplomatic notes and interviews, as well as data from other research publications and articles in (mainly Russian) newspapers and defence periodicals. In each of the three cases, data from one source has been compared with, and interpreted in view of, data from one or more other sources.

Case 1: Radars and satellite ground stations

In the 1990s, space-related activities increased on Svalbard. The Norwegian and international space research community, as well as the aerospace and telecommunications industries, had long been aware of the archipelago's potential as a platform for high-latitude atmospheric research and commercial satellite projects. The extreme northern location allowed for studying geophysical processes in the near-Earth space over the Arctic, not to mention round-the-clock downloading of data from polar-orbiting satellites.

As in the Cold War period, Norwegian authorities were intent on making sure that space-related activities on Svalbard did not violate article 9 of the Svalbard Treaty (1920), which states that the archipelago may not be used 'for warlike purposes'. The Soviet Union, and later the Russian Federation, interpreted this provision as a general prohibition of all military activity (Federation Council of the Russian Duma 2007:4–5). In Norway, the argument was occasionally made that 'Svalbard is only partly demilitarized', in the sense that article 9 of the Treaty related only to 'war operations' (Ulfstein 1995:388). Still, the Norwegian Government had no plans to change its long-standing practice of prohibiting activities on Svalbard that could reasonably be interpreted as being of a military nature, such as the military use of ground radars and/or satellite stations. All space installations on Svalbard are therefore required to document the civilian nature of their operations and are regularly subjected to inspections by Norwegian authorities as well as visits by foreign (including Russian) experts.

So far, Norway's assurances of the non-military nature of the space-related activities on Svalbard appear to have had a limited effect on Russian perceptions. Although Russia has not formally protested against the activities,[4] the country's military and security establishment continues to be suspicious of the radar installations, which are perceived to serve a 'dual' (civilian and military) purpose. A key issue in this regard is the radars' ability to detect and collect data about the flight paths and 'signatures' of Russian inter-continental and submarine-launched ballistic missiles (ICBM and SLBM) during test launches in the Arctic. Should such data be relayed in real time from Svalbard to, say, US military authorities, Russia's nuclear deterrence capability could, according to Russian experts, be jeopardized (Gundarov 2002b:3).

In addition to the Globus II radar in Vardø on the Norwegian mainland close to the Russian border, the installations that caused most concern in Russia in the late 1990s were the EISCAT (European Incoherent Scatter Scientific Association) Svalbard radar, the Svalsat (Svalbard Satellite) station and the Ny-Ålesund rocket range. In Russian media reports and policy documents, the space installations that were established on Svalbard in the mid-1990s (the EISCAT radar, the Svalbard satellite station and the Ny-Ålesund rocket range) were often mentioned in the same sentence as the Globus II radar in Vardø and portrayed as 'proof' of increased Western military activity in the Arctic under the cover of scientific research. The Russians' attitude at the time fully represented the Russian Northern Fleet's former Chief of Staff Rear Admiral Mikhail Motsak's article where he emphasized the following:

> In recent years, a number of dual purpose installations have been deployed in the region. The EISCAT radar, purportedly built to study atmospheric anomalies, has been deployed near Longyearbyen. The station has the capacity to monitor the flight paths of ICBMs and SLBMs in the Arctic. Not far from it ... a missile test site has been set up, including a runway that can be used by heavy transport aircraft, a space communications center, a radar, a missile assembly and storage facility, and a mobile platform for launching various types of geophysical rockets.
>
> (Motsak 2000:8)

Russia's 'securitization' of the Norwegian space projects on Svalbard followed a classic 'speech act' pattern. What was claimed to be at stake was the country's ability to strike the United States and its allies with SLBM launched from Russian submarines operating in northern waters, or ICBM launched from the Russian mainland. Interestingly, few of these security 'speech acts' included a clear identification of countermeasures. Admiral Motsak advocated a strategy of 'indirect action' [*strategiya nepryamikh deysviy*], an expression often used in crisis situations to describe preventive and/or retaliatory measures other than entry into open armed confrontation. Others, like Dmitrii Rogozin, Chairman of the Foreign Affairs Committee, indicated that the Norwegian radar installations, 'according to the logic of war', could be wiped out in the early stage of an armed conflict (*Strazh Baltiki* 2000:6). The 'securitizing actors' were high-ranking, active-duty and retired military officers, civilian members of the military and security establishment, defence journalists and right-wing Duma deputies. The main 'audiences' seem to have been the Russian Government and the Russian Security Council.

Concerning the foreign policy context at the time of the 'securitizing moves', three developments in the mid- and late 1990s stand out as particularly relevant to Russia's securitization of the Svalbard radar issue: the eastward enlargement of NATO, Norway's adaptation of its self-imposed restrictions on allied military activity east of the 24th meridian and the United States' announcement of its intention to withdraw from the 1972 US–Soviet Anti-Ballistic Missile (ABM) Treaty. These three factors are frequently mentioned in Russian news articles and policy documents dealing with the Svalbard issue and Russia's security situation (e.g. Fedorov *et al.* 2001:3).

The eastward expansion of NATO, as outlined in the 1995 Study on Enlargement, was perceived in Russia as a 'betrayal' (Lukin 1995:3) and a 'logical justification for Russian military elites to continue to perceive the West as a potential enemy' (Barany 2007:185). NATO's membership invitation to Poland, Hungary and the Czech Republic, as well as its commitment to further expansion of the Alliance into the Baltic and the Balkans, led a majority of Russians to believe that their country was increasingly being 'encircled' by its former adversary and that political and military countermeasures were necessary in order to maintain security on the Western front.

On NATO's northern flank, the Norwegian Government's decision in 1997 to lift some of Norway's self-imposed restrictions on foreign military activity east of the 24th meridian, that is, in the country's northernmost and easternmost land, sea and air space, was interpreted in Russia as yet another potentially threatening departure from the status quo. The adaptation, which in Norway was seen as a step towards a 'normalization' of the situation in the northern border region, was not well received by the Russian military and security establishment, which feared an increase in the number and scale of NATO exercises close to Russia's northwestern border. Russia's decision to turn down a Norwegian invitation to participate in the 1999 'Barents Peace' exercise, a NATO/Partnership for Peace activity, was regarded as 'a logical political step' (Fedorov *et al.* 2001:3). The 'Barents Peace' exercise took place shortly after the Kosovo war, which apparently made Russian participation even less likely (Smolenskiy 2000). But a major source of concern for the Russians in connection with the developments in northern Norway, on Svalbard and in NATO's new member countries was the signals coming from Washington in the late 1990s about a possible termination of the 1972 ABM Treaty.

To fully understand how and why the radar issue was placed on Russia's post-Cold War security agenda, one needs to take into consideration the foreign policy context at the time, and, perhaps equally important, the prehistory of Soviet–Norwegian disagreements over the interpretation of the Svalbard Treaty and over the status of the strategically located archipelago. Both factors appear to have served as 'facilitating conditions' for securitization.

The fear of being denied access to the open sea, that is, the North Atlantic, and later the Arctic Ocean, had concerned Russians ever since World War II. In 1944, the Soviet Foreign Minister Molotov presented Norwegian Ambassador Trygve Lie with a proposal to revise the Svalbard Treaty (1944–1947) and turn the archipelago into a Soviet–Norwegian condominium (Lie 1958:158).

The proposal was raised again after the war, at the Peace Conference in Paris in August 1946, but was eventually put to rest the following year. Throughout the 1950s and 1960s, however, Norway was frequently faced with Soviet accusations of Treaty violations aimed at undermining the Soviet Northern Fleet's combat potential.

Soviet reactions to the inclusion of Norway's northernmost territories (Svalbard including Bear Island and the island of Jan Mayen) in the command area of the Atlantic Pact in 1950–1951 followed a similar pattern. The Soviet Government maintained that the arrangement was a direct breach of the demilitarization clause of the Svalbard Treaty. Norway responded that no military fortifications or bases had been established on the archipelago and that the country would not allow any state to do so (Østreng 1977:54–5). Thus, in the Norwegian view, the Svalbard Treaty obligations had been upheld. Not reassured by the Norwegian response, the Soviet Government reiterated in a diplomatic note dated 12 November 1951 that it was 'unable to recognize the act as legal' (Østreng 1977:55) and continued to be concerned about the possibility of military use of the archipelago by Norway and/or its NATO allies.

The most striking historic parallel to the securitization of the Svalbard radar issue in the 1990s was the 1965–1969 debate between the Soviet Union and Norway concerning the Norwegian decision to let the European Space Research Organization (ESRO) establish a telemetric station at Kongsfjord, not far from the settlement of Ny-Ålesund.[5] The station, which was set up for civilian research purposes to receive and transmit signals from orbiting satellites, was seen in the Soviet Union as a military-purpose, intelligence-gathering installation:

> The Soviet Government is ... of the opinion that the telemetric station in Svalbard, apart from its purely scientific purposes, could be used for military purposes, in particular for carrying out cosmic, radio-technological, and other forms of intelligence-gathering activity over the territory of the Soviet Union and that its real tendency can only be determined by constant surveillance by Soviet experts.
>
> (Soviet memorandum of 30 April 1969, cited in Østreng 1977:58)

While the statement regarding the Kongsfjord station in 1969 was made by a semi-official group of Russian security policy experts, rather than by an official government source, the Soviet reactions were similar: 'Russia cannot allow itself not to worry that these objects, in the event of a U.S. withdrawal from the ABM Treaty, will become important elements in a national missile defence system' (Fedorov *et al.* 2001:3).

There are many similarities between these issues and the issues that have dominated Russia's post-Cold War Svalbard agenda; the current pattern of securitization is in reality not very different from the Cold War pattern, and it seems fair to assume that the historic baggage of mistrust and suspicion still serves as a 'facilitating condition' for securitization. Throughout most of the Cold War period, Norwegian research and infrastructure projects on Svalbard, such as the establishment of a telemetric station at Kongsfjord in cooperation with ESRO (mid-1960s), and the building of an airstrip at Hotellneset (early 1970s), were initially met with protest notes from the Soviet Foreign Ministry (Østreng 1977:55–8). As late as in the early 2000s, the 'radar issue' had not been removed from Russia's security agenda, and the archipelago Svalbard was still seen in Russia as 'one of the key elements in Norway and NATO's system for widened influence in the Arctic' (Gundarov 2002b:3). The Russian Government and commentators still tend to 'securitize' perceived changes to the status quo, whether it is new research installations, environmental protection measures or Norwegian fishery enforcement measures.

Case 2: The Svalbard Environmental Protection Act

The Svalbard Environmental Protection Act, passed by the Norwegian Parliament on 15 June 2001, aimed to preserve Svalbard as 'one of the best-managed wilderness areas in the world'. In Russia, Norway's viewpoint that environmental concerns should weigh more heavily in the event of a conflict between environmental aims and other interests (Ministry of Justice and the Police, 1999:section 6.2) was seen as a mere pretence for other long-term Norwegian objectives. At the time, provisions that would establish several new environmental protection areas for Svalbard – pursuant to the Svalbard Environmental Protection Act – were in the pipeline, including a 14km^2 biotope protection area in Coles Bay, east of the Russian settlement of Barentsburg. However, apart from rare plants, the Coles Bay area also held the largest and most promising coal reserves of the Russian state-owned mining company Trust Arktikugol. The proposed protection area in Coles Bay coincided conspicuously with the area in which Trust Arktikugol had planned its new coal mine. In 1998, Russia's largest-ever Svalbard settlement, the mining town of Pyramiden, had been abandoned due to depletion of coal reserves. Now, the reserves in the mines outside the only remaining Russian settlement, Barentsburg, were coming to an end. Barentsburg is run like a company town, where the state-owned Trust Arktikugol employed virtually the entire population (Jørgensen 2004:189). Hence, by 2002, a new coal mine in Coles Bay was considered crucial in order to maintain a significant Russian presence on Svalbard. Russian authorities were not convinced by the assurances from the Governor of Svalbard that the protection of the surface biotope in the Coles Bay area could be combined with underground mining activities (Pedersen 2002).

In a diplomatic note, Russia asserted that the protection area would annul their right under article 3 of the Svalbard Treaty (1920) to mine on the archipelago. Hence, the promising coalfields outside Coles Bay would 'in reality be excluded from the regime established by the Svalbard Treaty'.[6] In Russia's view, the proposed provisions would obstruct further mining on the archipelago just as the Svalbard Environmental Protection Act the previous year had introduced formal hindrances to new mines. The Act had imposed restrictions that would make it more difficult to establish new industrial activity on Svalbard if it affected the environment. Some asserted that the strict Norwegian rules aimed at preserving 'a virtually untouched environment on Svalbard'[7] favoured the differentiated Norwegian activity on the archipelago and disfavoured present and future Russian mining activity (e.g. Charodeev 2001:7; Gundarov 2002a:24).

Thus, the Svalbard Environmental Protection Act and its pursuant provisions were perceived in Russia as threats to the only remaining Russian community on Svalbard. Just like coal mining operations had long been the backbone of the Norwegian settlements on the Arctic outpost, Russia's presence there was also based on coal mining. Russia's physical presence on Svalbard enabled the country to maintain its allegedly 'privileged' position among the contracting parties to the Svalbard Treaty (Moe 1983:42), and to closely monitor Norway's activities on the archipelago. If the mineral reserves were exhausted, large settlements could no longer be sustained.

The public debate in Russia, as well as the diplomatic and political responses to the Norwegian legislation, suggest that Norway's environmental protection measures on Svalbard in the early 2000s were seen in Russia as a security issue. The Svalbard Environmental Protection Act and its pursuant provisions could affect the future of the Barentsburg settlement and ultimately Russia's ability to monitor Norway's activities on the archipelago. Some saw the Act and its provisions as part of Norway's security policy, allegedly aimed at ousting Russia from Svalbard. The Russians regarded Norway's environmental legislation as a disguised political instrument, aimed at forcing Russia to leave the archipelago (Jørgensen 2004:184–5).

In a study of Russian perceptions of Norwegian Svalbard policies by Jørgen Holten Jørgensen (2004), several Russian respondents expressed the view that Norway's motives were different from those it had publicly declared. In a 2001 report from the semi-governmental Russian Council on Defence and Foreign Policy (SVOP), Fedorov *et al.* (2001) explicitly stated that Norwegian environmental regulations were mere pretexts for limiting Russia's presence on Svalbard. Similar thoughts were found in Russian security and defence periodicals at the time (*Literaturnaya gazeta* 1997; Smolovskiy 2000; Gundarov 2002a; Pomortsev 2005). 'The Svalbard Environmental Protection Act is more about politics than the environment,' one informant asserted; '[t]he Act is in line with Norway's aim to become the only actor with a presence on Svalbard' (Jørgensen 2004:184). Jørgensen (2004:188) concluded that 'as long as Russia looks at Svalbard from a security policy perspective, Russians will find it hard to believe that Norwegian environmental protection policies are not part of a security policy agenda'.

During his two-day visit to Norway in November 2002, President Vladimir Putin raised the issue of the Svalbard Environmental Protection Act and a pursuant Protection Plan. In the course of his meeting with Prime Minister Kjell Magne Bondevik, Putin expressed worries over the environmental protection regulations for Svalbard (Fyhn and Tjønn 2002:2; Pedersen 2002:3). The Norwegian measures, he asserted, could obstruct Russian coal mining and thus also impede the Russian presence on the archipelago. Implicitly, the issue was considered existential to Russia's only remaining Svalbard settlement and possibly a matter of Russian security. The attention that President Putin devoted to the Svalbard Environmental Protection Act during his visit to Oslo in 2002 indicated that the issue was highly politicized. Still, according to observers, Putin's worries were perceived as 'mild',[8] reflecting that the president hardly had serious security concerns regarding Svalbard (Jørgensen 2004:187–8).

The numerous 'securitizing moves' in Russia in connection with the adoption of the Svalbard Environmental Protection Act in 2001 can perhaps be seen as indications of an underlying value conflict between Norway and Russia. Norway's main objective at the time was to protect Svalbard's fragile Arctic environment (environmental security), whereas Russia was mainly concerned about how the Act would affect ongoing and future Russian mining activities on the archipelago (economic security) and ultimately the future of the Russian settlement in Barentsburg (societal security).

Environmental concerns were not at the top of Russia's regional and global agenda, since the country had 'limited financial possibilities to deal with these issues to the extent they deserve' (Fedorov *et al.* 2001:3). International environmental disputes, or, as the Finnish lawyer and former diplomat Martti Koskenniemi (1991:73) calls them, 'disputes concerning the relationship between environmental and economic values', are often hard to resolve.

In this particular case, the apparent incompatibility of Russia and Norway's value hierarchies led the Russian public as well as decision makers within the country's political and military establishment to believe that Norway had a hidden agenda, that is, that the Environmental Protection Act was part of a Norwegian plan to undermine Trust Arktikugol and force the Russian population to leave Svalbard.[9] Such a turn of events could in Russian eyes jeopardize Russia's military security if Norway at a later stage were to take advantage of the situation and convert the archipelago into a military stronghold, in violation of article 9 of the Svalbard Treaty (1920).

The Russian moves to 'securitize' the Svalbard Environmental Protection Act have many similarities with Soviet efforts to counteract Norwegian conservation measures on Svalbard in the Cold War period. As noted by Clive Archer and David Scrivener (1982:74), '[t]he Soviet Union has often argued that Norwegian regulatory measures, such as the post-1971 introduction of conservation areas covering almost 45 per cent of Svalbard, are subject to prior Soviet agreement on their content'. In the 1970s, as in the late 1990s and early 2000s, Russian concerns were primarily related to the potential for a radical downsizing (and closing down) of Russian mining activities on the archipelago, which, at least officially, sustained the settlement in Barentsburg. Whereas the number of Norwegians on Svalbard has almost doubled from the 1970s to the present day, from approximately 1,000 to approximately 1,800, the number of Russians and Ukrainians on the archipelago has fallen from more than 3,000 at the height of the Cold War to less than 500 in 2007 (Deshayes 2007:12). This development is often seen in Russia as the result of a deliberate Norwegian policy to 'oust' the Svalbard Russians and reduce Russia's influence on Svalbard affairs (Rivetov 2003). At the same time, Russian subsidies to the mining operations in Barentsburg, which in 2000 was estimated at NOK 40 million (USD 7.5 million) (Åtland 2004:3), indicated that Russia, regardless of the costs, found it imperative to maintain at least one Russian settlement on the archipelago.

Despite these many securitizing moves, it is unclear whether the issue was ever successfully securitized. The perception that environmental protection legislation served Norwegian security policy objectives was apparently stronger and more widely felt in the conservative mid-level of the Russian bureaucracy (Jørgensen 2004:187). The environmental protection issue was predominantly handled on a lower level, that is, between the Governor of Svalbard and Trust Arktikugol. Although Trust Arktikugol strongly objected to the Norwegian plans (Pedersen 2003), it did not share the view that they were aimed at driving Russia off the archipelago. For instance, Director Vadim Mardanov of the mining company found that assertion to be far-fetched (Jørgensen 2003:40). Nevertheless, the Governor of Svalbard, Odd Olsen Ingerø, found the responses of the company to be more assertive than expected, noting that the issue was 'considerably more conflict-laden' than he had anticipated.[10] Eventually, in 2003, the Coles Bay area was withdrawn from the Protection Plan on the Governor's recommendation (Pedersen 2002:3). Both of the factors mentioned above, the underlying Norwegian–Russian 'value conflict' and the pre-history of controversies over the legality of Norwegian environmental protection measures on Svalbard, may have served as 'facilitating conditions' for securitization of the 2001 Environmental Protection Act. At the same time, the securitizing moves were, at least to some extent, neutralized by Norway's willingness to address the Russian concerns, as indicated by the somewhat controversial decision to take Coles Bay out of the Protection Plan area.

Case 3: Norwegian fishery enforcement measures in the FPZ

The waters south of Svalbard have long been considered to be of military-strategic importance. The Soviet Union, largely locked in by land and ice, had few outlets to the Atlantic Ocean, and post-Soviet Russia has even fewer. The Northern Fleet, based on the Kola Peninsula, is the country's most powerful grouping of naval strike forces. However, the vessels of the Fleet need to pass through potential naval warfare choke points south and east of Svalbard in order to access forward operation areas. Thus, the waters between the southernmost Svalbard island – Bear Island – and mainland Norway (see Figure 2.1) have traditionally been perceived in Russia as a sensitive 'strait'.

The maritime areas south of Svalbard are also a treasure chest of living marine resources. Productive fishing grounds are located within the 200-nautical-mile FPZ, where commercially important stocks feed, such as the Northeast Atlantic cod and the Norwegian spring-spawning herring. Both fish species are among the migrating stocks that are managed jointly by Norway and Russia. Quotas for the highly profitable cod fisheries are split 50–50 between the two states. The cod caught in the FPZ is more mature and larger in size than the cod caught in the Russian exclusive economic zone further east (Spiridonov and Nikolaeva 2005:44). Thus, the economic returns from the Svalbard waters are substantial, not least to Russia. More than a quarter of Russia's annual catch of Northeast Atlantic cod is taken in the waters off Svalbard.[11]

The Kremlin has consistently opposed Norwegian claims to exclusive rights in the maritime areas outside Svalbard. By 1970 Norway had formed its continental shelf doctrine based on the notion that the shelf area around Svalbard was part of a continuous Norwegian continental shelf to which it had exclusive rights. The Soviet Union strongly objected.[12] Not only did it object to the legal basis for the unilateral actions of Norway that concerned Svalbard waters, it also noted that the area was of special interest to Moscow due to the fact that the archipelago 'lies in the immediate proximity of the northern borders of the USSR'.[13] When Norway in 1977 established a 200nm FPZ around Svalbard by a Royal Decree,[14] the Soviet Union protested again. The Soviet Government found Norway's actions to be an illegal attempt to expand its rights into the Svalbard area. In a harsh diplomatic note, it expressed its right to 'take similar actions to protect the interests of the USSR'.[15]

Despite its generally good cooperation with Norway in the management of common fish stocks, Russia has occasionally called for additional consultations and joint mechanisms or procedures in the exercise of fisheries jurisdiction in the waters outside Svalbard.[16] Russia, viewing itself as a privileged party to the Svalbard Treaty (Moe 1983:42; Pedersen 2006:339–58), initially seemed to comply with Norway's exercise of fisheries jurisdiction in Svalbard waters, which at first seemed to affect only third parties. However, when the Norwegian Coast Guard arrested a Russian trawler ('Novokuybyshevsk') in the FPZ for the first time in 1998, it resulted in hectic diplomatic activity. Norway decided not to prosecute the captain and released the trawler before it reached a Norwegian port. Three years later, the arrest of another Russian trawler ('Chernigov') sparked further debate about the motives behind Norway's enforcement of fishery regulations in the zone. The protest, which set aside the normal courtesy of diplomatic correspondence, seemed to reverse

Russia's former legal views, now placing the Svalbard waters within the High Seas.[17] Russia's representative to the Joint Norwegian–Russian Fisheries Commission warned that Norway's actions 'could hamper the conduct of normal fishery' by Russian fishermen and harm further cooperation as well as the joint fisheries management in the region (Pedersen 2008:249–50). Among Russian actors in the Kola-based fishing industry, distrust of Norwegian motives for seizing Russian vessels became particularly widespread (Jørgensen 2004:186). The Chairman of the State Fisheries Committee, Yevgeniy Nazdratenko, threatened to have the Norwegian Coast Guard vessels sunk by the Northern Fleet if the Norwegians again attempted to arrest Russian fishing vessels in the FPZ (Hønneland 2003:66).

In the summer of 2002, the 163m anti-submarine warfare (ASW) destroyer 'Severomorsk' sailed into the Svalbard FPZ. Initial reports suggested that the warship had been deployed there by the Northern Fleet to protect Russian fishing vessels from the Norwegian Coast Guard, which the previous year had seized the Russian trawler 'Chernigov' in these contested waters. While Norway maintained that it had a coastal state's right under international law to exercise jurisdiction in the maritime areas adjacent to Svalbard, the Kremlin objected to the Norwegian position. Hence, the arrest of the 'Chernigov' in 2001 sparked strong reactions in Russia. The country delivered a formal protest to Norway, asserting that the actions against the 'Chernigov' '[could] not be regarded as legitimate or in line with the norms of international law'.[18]

The deployment of the 'Severomorsk' to the FPZ in 2002 raised new concerns in Norway. The dispatch of a heavy Russian warship was unprecedented, and it triggered the first use of an emergency hotline between the regional defence command in northern Norway and the Russian Northern Fleet. As the Russian destroyer steamed toward Svalbard, the Norwegian fishery enforcement measures in the FPZ seemed to have been perceived in Russia as a threat and as risking peace and stability in the region. Here again, a Svalbard issue of a seemingly legal-technical character was politicized, and to some degree securitized. Russia's show of force in 2002 (the deployment of the 'Severomorsk') suggests that Russia responded to a perceived security concern in that the Norwegian fishery enforcement measures in the FPZ could pose a threat to vital Russian interests. *Izvestiya* reported that the destroyer had been sent to Svalbard at the request of Chairman Nazdratenko of the State Fisheries Committee (Jørgensen 2004:63).

The distrust of Norway's motives behind its exercise of fisheries jurisdiction in the FPZ was apparently widespread in the Russian fishing industry, but less so in the Kremlin. In its diplomatic correspondence with Norway, Russia mostly referred to the differences of opinion in legal terms. Sources in the Russian Ministry of Defence, however, asserted that the Svalbard patrol was part of a long-planned military exercise, and pointed out that 'the sailing routes of Russian naval vessels are not determined by the State Committee for Fisheries' (Getmanskiy *et al.* 2002:2). Still, the public debate in Russia suggests that the controversy was related not only to the legal provisions of the Svalbard Treaty, but more generally that the Norwegian Coast Guard's actions were perceived as an attempt to undermine Russia's economic interests in the Svalbard region.

The debate was revived when the Norwegian Coast Guard tried to seize the Russian trawler 'Elektron' in the FPZ in 2005. With the Norwegian fishery inspectors still onboard, the Russian vessel made a bolt for its home waters while being chased by a number of Norwegian Coast Guard vessels (Åtland and Bruusgaard 2009:339–40). After the trawler's successful escape from Norwegian enforcement, there were numerous claims in Russia that the Northern Fleet should take on a more active role in defending the country's economic interests in the Euro-Arctic region (Solov'ev and Ivanov 2005).

The attempts by certain actors in Russia, particularly within the fishing industry, to play the 'security' card in connection with Norway's arrests of the Russian trawlers 'Chernigov' (2001) and 'Elektron' (2005) represented the culmination of an almost 30-year conflict over the legal status of the FPZ. Since then, things seem to have calmed down. Those in control of Russia's primary military muscle in the Euro-Arctic region – the Northern Fleet – appear to have a genuine interest in maintaining political stability in the region, and, as illustrated during the 'Elektron' incident, they are reluctant to intervene militarily on behalf of Russian fishing vessels in the zone. Despite its reservations, Russia does not want the current management regime to collapse, since this could lead to a radical increase in the number of third-country fishing vessels in the FPZ. The underlying issue is still unresolved and unlikely to become fully resolved, at least in the foreseeable future. But Norway has at least managed to increase Russia's awareness about – and commitment to addressing – the problem of illegal, unregulated and unreported (IUU) fishing in northern waters. Consequently, this has contributed to a softening of the Russian rhetoric and a toning down of the 'national security' dimension of the issue.

Conclusions

The end of the Cold War, the fall of communism and the disintegration of the Soviet empire are often portrayed as watershed events that came to change the security dynamics on both sides of what was once the Iron Curtain. As this study has shown, it takes time and effort to overcome the legacy of fear, mistrust and suspicion. NATO is still seen as Russia's primary security challenge, and East–West relations in the European Arctic are still marked by a relatively high level of 'securitiness'. Attempts to change the situation have often been perceived as representing a threat to Russia's security and economic interests in the region.

Based on the foregoing attempt to explore on a case-by-case basis how and why three Svalbard-related issues were placed, or at least attempted to be placed, on Russia's post-Cold War Arctic security agenda, we can draw four main conclusions.

First, the issues surrounding Svalbard in Russia were framed in security terms by people presumed to be in positions of authority, such as government officials at the federal and regional level, high-ranking naval officers, defence and security analysts and industry representatives (Table 2.1). Many of the 'securitizing actors' held similar positions in the Soviet period and were inclined to analyze the situation in the region with a 'Russia-versus-the-West' mindset. In addition, the 'audiences' that the 'securitizing actors' were courting, such as the Russian Security Council, the Foreign Ministry and the Defence Ministry, shared many of their concerns. The 'audiences' were generally receptive to the calls for extraordinary measures aimed at neutralizing the effects of the developments that were unfolding on and around the archipelago at the time. Wanting to maintain its presence on the archipelago and, to the extent possible, moderate the behaviour of a small and vulnerable neighbour country, Russia had much to win and little to lose by attempting to 'securitize' certain aspects of Norway's Svalbard policy. It cannot be dismissed that Russia's protest notes, as well as the country's occasional display of military force in the FPZ, may have had at least some effect on Norway's behaviour. Norway's decision in 2003 to pull the Coles Bay area from the Protection Plan can be seen as an indication that the Russian protests were duly noted and taken into consideration in Longyearbyen as well as in Oslo.

Second, there seems to have been a high degree of cross-sectoral spillover of securitization dynamics. Hence, when the Russian 'securitizing actors' were

Table 2.1 A 'Copenhagen School' perspective on Russia's Svalbard policy

<table>
<tr><th>Case</th><th>Sector of security</th><th>Perceived subject of threats</th><th>Perceived object of threats</th><th>Historical parallel?</th><th>Securitizing actor(s)</th><th>Audience(s)</th></tr>
<tr><td>1</td><td>Military</td><td>Radars and satellite ground stations</td><td>Russia's nuclear deterrence capability</td><td>ESRO telemetric station, 1960s</td><td rowspan="3">Russian naval officers, security analysts, defence journalists, Murmansk Region Governor, the fishing industry</td><td rowspan="3">The Russian Government (primarily the Foreign and Defence Ministries and the Fisheries Committee), the Russian Security Council</td></tr>
<tr><td>2</td><td>Economic, societal</td><td>The 2001 Svalbard Env. Prot. Act</td><td>Mining activity, Barentsburg settlement</td><td>Norwegian env. prot. measures, 1970s</td></tr>
<tr><td>3</td><td>Economic</td><td>Norwegian fishery enforcement measures</td><td>Russian fisheries in Svalbard waters</td><td>First Soviet protests to the FPZ, 1977</td></tr>
</table>

Source: authors

trying to make their case about the emergence of threats to vital Russian interests in the Svalbard region, they drew on references to multiple cases and multiple sectors, using what might be described as 'replication logic' (Rivetov 2003). Cross-sectoral 'clustering' of security concerns is, however, not unique to Russia. It is largely in line with findings from other case studies, such as Buzan *et al.*'s (1998:188) study of securitization patterns within the European Union.

Third, turning to the question of 'facilitating conditions', the Russian moves to securitize various Svalbard-related issues, particularly in the late 1990s, appear to have been stimulated by a number of simultaneous developments on the Norwegian mainland (the lifting of Norway's self-imposed restrictions on NATO training activities east of the 24th meridian), on the European continent (the eastward enlargement of NATO) and on the other side of the Atlantic (the US withdrawal from the ABM Treaty). At least in the short term, the level of trust in Russian–Norwegian relations seemed to have been adversely affected by these developments. Whereas the three specific policy issues discussed in this study were seen in Norway as separate, they were seen in Russia as interrelated and dealt with as part of a comprehensive Norwegian (or Western/US/NATO) strategy.

Fourth, in addition to the above-mentioned linkages between securitization patterns in different sectors and on different levels, there seems to be a linkage between securitization patterns in different time periods. Many, if not all, of the cases studied in this chapter have historical analogies that can be traced back to the Cold War period (Table 2.1). There are many indications that the historical baggage of more than four decades of East–West confrontation on NATO's northern flank may have served as a 'facilitating condition' for securitization, even as late as in the 1990s and 2000s. More than 40 years ago, the Russian philosopher and natural scientist Valentin Turchin described a phenomenon which he called 'The Inertia of Fear' (Turchin 1981 [1968]). The topic of Turchin's piece was thematically far removed from the topic of this study, but its title captures well some of the main features of Russia's Svalbard policy in the post-Cold War period. The Russian fear of being 'ousted' from the strategically located Arctic archipelago and its surrounding waters has not disappeared. Neither have the accusations that Svalbard is being, or may be, used for military purposes by Norway and its NATO allies, in violation of the Svalbard Treaty (1920). The dynamics at play on the Russian side, including the tendency to securitize perceived changes to the status quo situation in the Svalbard region, may be seen as a reflection of Russia's strained relationship with NATO and the West in the post-Cold War period.

Thus, this study largely confirms the hypothesis that Russia's threat perceptions and security policies in the Svalbard region are still influenced by perceptions and policies formed in the Cold War period. As the world prepares for the 'Age of the Arctic',[19] issues such as the ones discussed in this chapter are likely to figure centrally on the security agenda of the Arctic rim states and will need to be subjected to further studies. Securitization theory, as outlined in the 1990s by the 'Copenhagen School' of security studies, can be a helpful analytical tool in the exploration of security policy-making in and on the Arctic. With its emphasis on the intersubjective dimension of the concept and practice of security, the theory can help us understand the process of security policy agenda-setting on various levels and within various sectors. It represents a good alternative to other approaches, whether they belong to the 'traditional' or 'critical' camps.

Notes

1 'Spitsbergen' is the old term for 'Svalbard'. The former term is still being used in Russia. In contemporary Norwegian, the term 'Spitsbergen' denotes only the largest of the Svalbard islands, previously called 'Vestspitsbergen'.
2 The term 'Copenhagen School' was originally coined by Bill McSweeney (McSweeney 1996) in (critical) reference to the works by a group of researchers associated with the Centre for Peace and Conflict Research, later renamed the Copenhagen Peace Research Institute (COPRI). COPRI was merged into the Danish Institute for International Studies (DIIS) in 2005. Barry Buzan and Jaap de Wilde are considered the School's main contributors.
3 Wæver's concept 'facilitating conditions', understood as 'the conditions under which the speech act works, in contrast to cases in which the act misfires or is abused', is inspired by speech act theorist J. L. Austin's concept 'felicity conditions' (cf. Austin 1962:14–15).
4 According to the Norwegian Foreign Ministry, there were no Russo-Norwegian diplomatic correspondence on the radars and satellite ground stations in the 1990s and the 2000s.
5 The station was in operation until 1974, when the agreements with ESRO elapsed.
6 Russia, diplomatic note 9002/2ED to Norway, 15 October 2002.
7 Svalbard Environmental Protection Act, §1.
8 Kim Traavik, personal communication. Also see Pedersen (2002).
9 This claim was categorically denied by Norwegian diplomats, as were the Russian statements that the regulatory measures had an 'anti-Russian' orientation; see Nordsletten (2001:58).
10 Personal communication. Also see Pedersen (2002).
11 Personal communication with representatives of the Murmansk-based Fishing Industry Union of the North (SRPS), 13 September 2006.
12 The Soviet Union, memorandum to Norway, 27 August 1970.
13 The Soviet Union, memorandum to Norway, 27 August 1970.
14 Royal Decree of 3 June 1977, pursuant to the Norwegian Economic Zone Act, 17 December 1976.
15 The Soviet Union, diplomatic note to Norway, 15 June 1977.
16 See, for instance, diplomatic notes to Norway, 17 July and 18 August 1998. Joint measures were also called for by the Soviet–Norwegian Communiqué of 16 March 1978.
17 Russia, diplomatic note 3695/2ED to Norway, 23 April 2001.
18 Russia, diplomatic note 3695/2ED to Norway, 23 April 2001.
19 This phrase, which seems to have had a renaissance in recent years, was first used by Oran Young in the mid-1980s (Young 1985).

References

Archer, C. and Scrivener, D. (1982) 'Frozen frontiers and resource wrangles: Conflict and cooperation in Northern waters', *International Affairs*, 59(1):59–76.

Åtland, K. (2004) 'Russisk Svalbard-politikk', *FFI-Fokus*, 1.

Åtland, K. and Bruusgaard, K. V. (2009) 'When security speech acts misfire: Russia and the *Elektron* incident', *Security Dialogue*, 40(3):333–54.

Austin, J. L. (1962) *How to do Things with Words*, Oxford: Clarendon Press.

Barany, Z. (2007) *Democratic Breakdown and the Decline of the Russian Military*, Princeton, NJ: Princeton University Press.

Barents Observer (2007) 'Russia under threat at Spitsbergen', 29 June. Online. Available at: http://www.barentsobserver.com (accessed 4 February 2011).

Buzan, B. (1991) *People, States & Fear*, Harlow: Pearson Longman.

Buzan, B., Wæver, O. and de Wilde, J. (1998) *Security: A New Framework for Analysis*, Boulder, CO: Lynne Rienner.

Charodeev, G. (2001) 'Ledyanoi zanaves: my mozhem poteryat' svoi shpitsbergenskie vladeniya' [Ice curtain: We may lose our Spitsbergen possessions], *Izvestiya*, 8 June:7.

Deshayes, P.-H. (2007) 'Northern neighbors', *Moscow Times*, 21 September:12.

Federation Council of the Russian Duma (2007) 'Prisutstvie Rossiyskoy Federatsii na arkhipelage Shpitsbergen: politiko-pravovye, ekonomicheskie i gumanitarnye aspekty' [Russia's presence on the archipelago of Spitsbergen: Legal-political, economic and humanitarian aspects], Protocol from meeting on 19 June 2007, The Federation Council's Committee on the North and Small Indigenous Peoples, Moscow, Russia.

Fedorov, A. V., Voronov, K. V., Deryabin, Yu. S., Karaganov, S. A., Moshes, A. L., Nikonov, V. A., Oznobishchev, S. K., Salmin, A. M. and J. I. Yu. (2001) 'Interesy Rossii na severe Evropy: v chem oni?' [What are Russia's interests in the North of Europe?], *Dipkur'er*, 22 March:1 and 3.

Fyhn, M. and Tjønn, H. (2002) 'Putins besøk: Styrket norsk-russisk naboskap', *Aftenposten*, 13 November:2.

Getmanskiy, K., Charodeyev, G. and Chuyov, A. (2002) 'Voyennye moryaki pomogut rybakam' [Naval servicemen help the fishermen], *Izvestiya*, 22 June:2.

Gundarov, V. (2002a) 'Rossiyskie pozitsii v Arktike' [Russian positions in the Arctic], *Morskoi sbornik*, 4:24–5.

—— (2002b) 'Shpitsbergen v tumane' [Spitsbergen in the fog], *Krasnaia zvezda*, 6 February:3.

Hønneland, G. (2003) *Russia and the West: Environmental Co-operation and Conflict*, London: Routledge.

Jørgensen, J. H. (2003) 'Svalbard og Fiskevernsonen: Russiske persepsjoner etter Den kalde krigen', *FNI Report*, 13.

—— (2004) 'Svalbard: Russiske persepsjoner og politikkutforming', *Internasjonal politikk*, 62(2):177–97.

Koskenniemi, M. (1991) 'Peaceful settlement of environmental disputes', *Nordic Journal of International Law*, 60(1–2):73–92.

Lie, T. (1958) *Hjemover* [Homewards], Oslo: Tiden.

Literaturnaya gazeta (1997) 'Shpitsbergen – eto geopolitika!' [Spitsbergen – that's geopolitics!], 15 October:14.

Lukin, V. (1995) 'Chto stroit'sya v Evrope: noviy dom ili stariy zabor?' [What is being built in Europe: A new home or an old fence?], *Izvestiya*, 12 May:3.

McSweeney, B. (1996) 'Buzan and the Copenhagen School', *Review of International Studies*, 22(1):81–93.

Ministry of Justice and the Police (1999) *Svalbard, Report No. 9 to the Storting (1999–2000)*, Oslo: Ministry of Justice and the Police.
Moe, A. (1983) *Utenrikspolitiske rammebetingelser og norsk Svalbard-politikk*, Lysaker: Fridtjof Nansen-stiftelsen.
Motsak, M. V. (2000) 'O natsional'nykh interesakh Rossii v Arktike' [Russia's national interests in the Arctic], *Voyennaia mysl'*, 6:8–10.
Nordsletten, Ø. (2001) 'Norvergiya i Rossiya: razvitie sotrudnichestva' [Norway and Russia: The development of cooperation], *Voyennaya Mysl'*, 6:55–60.
Østreng, W. (1977) *Politics in High Latitudes: The Svalbard Archipelago*, London: C. Hurst & Company.
Pedersen, T. (2002) 'Verneplanen utsatt', *Svalbardposten*, 15 November:3.
—— (2003) 'Tar Colesdalen ut av verneplanen', *Svalbardposten*, 4 July:1 and 6–9.
—— (2006) 'The Svalbard continental shelf controversy: Legal disputes and political rivalries', *Ocean Development & International Law*, 37(3):339–58.
—— (2008) 'The dynamics of Svalbard diplomacy', *Diplomacy & Statecraft*, 19(2):236–62.
Pomortsev, S. (2005) 'Nezakoniy promysel varyagov' [The illegal catch of the Varyags], *Voyenno-Promyshlenniy Kur'er*, 2 November:2.
Rivetov, P. (2003) 'Shpitsbergen – forpost NATO pod nosom Rossii' [Spitsbergen: NATO's outpost under Russia's nose], *Pravda*, 14 April. Online. Available at: http://news.pravda.ru (accessed 2 May 2008).
Smolenskiy, I. (2000) 'Kholodnaya voyna vozrozhdaetsya na severe Evropy' [The Cold War is reborn in the North of Europe], *Nezavisimoye Voyennoe Obozrenie*, 19 May. Online. Available at: http://nvo.ng.ru/wars/2000-05-19/1_coldwar_revival.html (accessed 26 March 2013).
Smolovskiy, A. (2000) 'Arkhipelag Shpitsbergen i bezopasnost' Rossii' [The Spitsbergen archipelago and Russia's security], *Morskoi sbornik*, 6:38–43.
—— (2006) 'Istochniki ugroz interesam Rossii v Arktike' [Sources of threats to Russia's interests in the Arctic], *Morskoi sbornik*, 11:57–65 and 12:55–64.
Solov'ev, V. and Ivanov, V. (2005) 'VMF samoustranilsya ot vypolneniya ekonomicheskikh zadach' [The Navy has distanced itself from the undertaking of economic tasks], *Nezavisimoye Voyennoe Obozrenie*, 4 November:1.
Spiridonov, V. A. and Nikolaeva, N. G. (2005) *Fisheries in the Russian Barents Sea and the White Sea: Ecological Challenges*, Oslo: WWF Russia and WWF International Arctic Programme.
Strazh Baltiki (2000) 'RLS ryadom s nami' [Radar next to us], 30 May:6.
Svalbard Treaty (1920) 'Treaty between Norway, the United States of America, Denmark, France, Italy, Japan, the Netherlands, Great Britain and Ireland and the British Overseas Dominions and Sweden concerning Spitsbergen signed in Paris 9th February'. Online. Available at: http://www.lovdata.no/traktater/texte/tre-19200209-001.html (accessed 31 March 2011).
Turchin, V. (1981 [1968]) *The Inertia of Fear and the Scientific Worldview*, Oxford: Martin Robertson. [The piece first circulated in the Russian *samizdat* press in the autumn of 1968.]
Ulfstein, G. (1995) *The Svalbard Treaty: From Terra Nullius to Norwegian Sovereignty*, Oslo: Scandinavian University Press.
Wæver, O. (1995) 'Securitization and desecuritization', in R. D. Lipschutz (ed.) *On Security*, New York: Columbia University Press.
—— (2003) *Securitisation: Taking Stock of a Research Programme in Security Studies*, unpublished paper, Copenhagen: University of Copenhagen.
Young, O. (1985) 'The age of the Arctic', *Foreign Policy*, 61:160–79.

Chapter 3

A new northern security

Environmental degradation and risks, climate change, energy security, trans-nationalism and flows of globalization and governance

Lassi Heininen

Introduction

It could be argued that classical geopolitics, meaning both military security based on the technology models of geopolitics, as well as natural resource exploitation based on the resource models of geopolitics, has dominated the relations between the circumpolar North and the outside world (see Heininen 2004a:218, 2010a:231–5). Since the end of the Cold War period there has, however, been a significant shift from the military confrontation between the superpowers, the USA and the Soviet Union, to international and regional cooperation in many fields, e.g. economics, environmental protection, research and higher education. This shift has included a change in the way that security has been defined (see Häyrynen and Heininen 2002).

Part of this shift can be exemplified through, for example, the increased reference to energy security, meaning the guarantee and improvement of both availability to and accessibility of energy resources (mostly oil and gas). At least since the beginning of the twenty-first century, energy security has become a core national interest for many countries and a central goal of several governments. In 2006 alone, energy security was often proclaimed the main theme of many international gatherings, e.g. the European Union Summit of March 2006, the Summit of the Shanghai Treaty Organization in June 2006, the Summit of the Group 8 under the Russian Presidency in July 2006 and the sixth Asia–Europe Meeting in August 2006. This trend has continued and, due to the growing scarcity of oil and continued interest in energy security, important diplomatic efforts such as the EU–Russian dialogue have become critical to strengthening stability, well-being and sustainable development in northern Europe, especially following EU enlargement into eastern Europe. The European North, or the western part of the Eurasian North, has become more important to Europe as a whole, as well as to the European Union and its member states, both economically and strategically. Consequently, in geopolitics, the region's rich energy resources are significantly outweighing any focus on military security on the global stage. Trends towards less militarized understandings of security are increasing in the Arctic, demanding closer examination by analysts.

The primary hypothesis of this chapter is that a holistic picture of Arctic geopolitics and northern security must consider many factors and approaches, including when dealing with security. This is especially the case if security is defined in a more comprehensive way, so as to include the perspectives of human beings, societies and regions, rather than just states. Environmental and energy security are considered to be among the main factors contributing to a comprehensive security in the Arctic context. The empirical focus will be the military and military security, and the ways that we can observe the changes taking place in the concept of security through this traditional security tool, particularly with regard to nuclear safety. Nuclear safety as a part of environmental security thus acts as a metaphor for change in the definition of traditional security, which was redefined by the Arctic states due to nuclear problems within the Barents Sea area. Other factors include environmental degradation and risks, impacts of climate change, competition for hydrocarbons and also trans-nationalism or globalization causing a need for a control of state sovereignty in this 'borderless' space of the Arctic. Regional security, which defines security from the point of view of a region, is a new and alternative approach to traditional security and security discourses.

New and comprehensive approaches are needed when trying to define the security of the post-Cold War circumpolar North. Comprehensive security has already become a part of political vocabularies and included onto foreign policy agendas of some states, e.g. Canada and recently Iceland. Environmental security and human security are also relevant for describing foreign policy approaches. Following from this, the conceptual starting point of this chapter is that security is not, and cannot be, objective, but is always socially constructed, including aspects of culture, society and the environment (Buzan 1991).

This chapter first discusses different concepts of security, linking them through a discourse on regional security. Second, it describes the relevant stages of security in the circumpolar North from the Cold War period into the beginning of the twenty-first century. Third, this chapter goes beyond traditional security and discusses comprehensive security in the North, focusing upon the EU's Northern Dimension policy that aims to increase stability and comprehensive security. Finally, this chapter lists different aspects and factors, hereafter 'key phenomena', that are important when trying to comprehensively define security in the North, i.e. a new northern security.

Discourses on security

Comprehensive security

Academic discourses on defining comprehensive security emphasize environmental and human aspects and their relationship to traditional security (see Buzan 1991; Käkönen 1994). When dealing with security there are terms like 'danger', 'threat' and 'risk', indicating that many aspects of security are taken either as a threat or risk (e.g. cross-border threats and environmental hazards and catastrophes). Consequently, there are also discourses about what is meant by threats and risks and whom they affect, e.g. community/ies, (sub)region(s) or country/ies. Therefore, before discussing different stages of security in the circumpolar North and defining what might be a new type of northern security, different concepts of security need to be briefly discussed, as well as threats/risks.

Traditional security mostly means national security, i.e. security-policy and security from the perspective of the state (generally meaning a nation or unified state), guaranteed if needed by the military. It can also be called a weapon-oriented security or 'a unilateral competitive national military security' (Newcombe 1986). This means that a state's or nation's security and stability, as well as its peace, is guaranteed by the military or a deterrence of military use. Security guaranteed by the power of the state and its military organizations is the core of the unified state system, which is correspondingly the most fundamental part of the 'anarchic' international system, and thus very much consistent with traditional theories of International Relations (see Waltz 1979). Behind this thinking is the idea that a state would like to have hegemony over a geographical region and its natural resources, and, if required, it is prepared to use military power to attain and maintain it. This approach of security has traditionally dominated circumpolar geopolitics and international politics in the North, largely due to the Cold War.

The security concept was changing globally during the second half of the twentieth century. Despite ongoing regional and civil wars and a few annual inter-state wars, there were also significant changes to what the traditional concept of threat and security entailed, including alternative or so-called 'soft' issues of traditional security. These included trans-boundary pollution, nuclear accidents and other environmental catastrophes, infectious diseases (such as HIV/AIDS), illegal immigration as well as organized and international crimes (such as drug and human trade). In the 1980s, several UN commissions started discourses to define security in a different, alternative and more comprehensive way that would incorporate these soft issues. Reports on security also launched new approaches and concepts of security like *common security* (Independent Commission on Disarmament and Security 1982) or *comprehensive security*. The goal of these terms was to have a broader agenda of security by including the traditional military sector and also social, economic and environmental sectors into a security complex (see Huru 2004:42). This was also the case in the European North, the former 'military theatre' of the Cold War period, where one of the aims was to have a new security discourse by introducing 'soft' security parallel to 'hard' security (see Heininen 1999b).

Comprehensive security can emphasize either environmental security or human security, or both together. The concept of environmental security focuses upon the role of the environment and includes elements such as access to natural and renewable resources, quality of renewable resources and human population growth (see Dwivedi *et al.* 2001:263). Thus, not only can environmental degradation be caused by resource scarcity, but also by environmental insecurity. Sustainable development as a part of environmental and comprehensive security relies upon mechanisms that draw upon, among other factors, international cooperation, global governance, regionalization, democracy and cultures, and civil society.

Comprehensive security also refers to human security, or civil security, with a focus on human beings as individuals and citizens; Griffiths (1993) has called this *civility*. This kind of security, or safety, is derived from the point of view of a human being as an individual or citizen within a society, or in the global community. These concepts have been used to emphasize the daily security of people at an individual level with a focus on human beings, and provide an alternative perspective to 'traditional security'. Human security is understood as 'freedom from fear, freedom from want', and is meant to widen the approach to security to include the following categories of security: 'economic, health, food, personal, political, community and environmental'; all this is frequently argued to have relevance to the Arctic setting (Hoogensen 2005a; Hoogensen *et al.* 2009; see also United Nations General Assembly 2005). The concept of human security is much influenced and defined by the work of the United Nations in general, and particularly for human development in the 1990s, like for example 'the threefold theme "Democracy – Development – Human Rights"' (Commission on Human Rights 1993:11).[1] In the Arctic context, a comprehensive security framework includes human security as well as the so-called human ecological security in such a way as it is connected to environmental security – it is holistic and interprets environmental degradation (e.g. pollution, global warming) as a threat or a risk to people and their societies (see Langlais 1995:270–7).

Implementation of comprehensive security requires a consideration of practical issues pertaining to an individual's life, such as ensuring shelter, good health and social and economic well-being, as well as a life in peace without conflict, war and violence. In addition, however, comprehensive security also includes more immaterial values like political freedom, democracy, human rights and freedom from a range of threats and risks, such as disasters, pollution and other environmental problems, hunger and starvation, diseases or other illness, and terrorism. It can also be interpreted to include cultural survival, freedom of expression and security of communication (see *Poverty and Environmental Times* 2004:6).

Broad concepts of security are not just theoretical or academic, but are widely accepted in both national and international contexts and used by many political actors, even governments. For example, the concept of human security has been used in politics worldwide, e.g. the earlier political agenda of the Canadian government (Dwivedi *et al.* 2001) and the strategies of the European

Union, which includes issues of trans-boundary influences and cross-border cooperation (see European Commission 2003).

The approach of comprehensive security has also been used when dealing with internal security, i.e. security within a state and society, such as in the Finnish programme for internal security across sectorial borders. The main challenges for the internal security of the Finns are defined by an increased possibility of re- or displacement (such as arising from poverty and unemployment), maintenance of society's daily security, increase of accidents (e.g. at home), foreign-organized crime and maintenance of border control due to (international) terrorism (Sisäasiainministeriö 2004:1–15). Finally, this extended definition and comprehensive interpretation of security is based on the idea that there is a vast number of threats and risks to national security, besides traditional military threats, trans-border crime and international terrorism (Buzan 1991).

Defining threats and risks

Having mentioned the terms of a threat and a risk here, I should clarify that there is a principal difference between them. Although a threat and a risk are often taken as synonyms, a threat can be physical, economic or social in status, but is mostly subjective and psychological, often rooted in fear and difficult to control. Thus, when dealing with a threat, it is pertinent to ask from whose point of view it is considered a threat (Haila and Heininen 1995:167; Nokkala 1999:48–52).

Conversely, it is possible to qualify, rank and measure a risk; in spite of this, a risk is relative because people measure risks differently. A risk exists in society if it is interpreted as being real, regardless of whether it has materialized (Haila 1989:10; Wahlström 1994:37–40). Risks are also conscious calculations that presuppose an ability to control or at least manage uncertainty, whilst threats may arise without consciously pursuing them. Indeed, risks are present in our modern world; they are even said to be a part of normal human life, which has brought Ulrich Beck (1992) to claim that modern society is a risk society with new kinds of environmental risks emerging, mostly due to modern technology.

New security discourses: critiques and discussion

There have been arguments against discourses on comprehensive security, to a degree for environmental security but especially for human security. In these debates some scholars argue that the concept of security carries such heavy militaristic, nationalistic and ideological burdens that it is dangerous to national security to label non-military matters as security issues (Deudney 1999). The primary concern lies in the possible application of the traditional means of providing security to new kinds of threats. Conventional security institutions, like armies, are not very practical for securing people and states against new threats. To the contrary, armies are likely to produce these threats themselves, as was the case in the accident of the Kursk nuclear submarine in August 2000; this Russian submarine was very modern and new with secret military technology, but was destroyed and sunk when the crew tested a new torpedo (Häyrynen 2003).[2]

Indeed, trying to broaden the mission of the military in the 1990s to include environmental protection partly meant some sort of militarization of the environment or 'militarized environment' rather than 'green security' (Käkönen 1994). The fact that the military tried to control ideas about environmental problems, and the so-called 'environmental rearmament' of an army, greatly strengthened authoritarian tendencies in environmental policy, increasing the role and legitimacy of the military and even giving a new mission to it (Haila and Heininen 1995:162–4). Collectively, environmental security has appeared less fruitful as an alternative approach to traditional security than hoped by many of those who launched and supported it, and, therefore, the academic discourses in the 1980s and 1990s have resulted in some backlash (see Häyrynen and Heininen 2002).

What is, however, important and interesting behind these alternative security concepts is that there are theories and perspectives that consider and include the fields of the so-called 'low' politics (a moniker that is challenged by newer security approaches as these fields have the most meaning to individuals and communities), such as environmental protection, social and cultural issues and civil society. For example, a new understanding of geopolitics, or new geopolitics, supports this wider approach to the concept of power and security. Specifically, these approaches provide alternatives to the challenges of the power and hegemony politics of the Cold War period and the geo-strategic discourse of the unified state system, through new actors, such as indigenous peoples, and the consideration of their autonomy (see Heininen 2010b). Here, the focus is not how to control a region and gain more power through control, but how to reach a socially stable and peaceful situation as well as environmentally sustainable order (Chaturvedi 2000). Thus, instead of power being defined as the traditional type of coercive power by military or money, or both, there are other types of connotations, such as governance and influence. Combined with this we see that the main factors of new geopolitics, i.e. actors, space (including social space) and identity/ies, mirror the priorities being made through human security. Thus not only are geography and political systems important, but also people(s), societies and the environment. The above, broadened and deepened understanding of security is connected to the relationship between people and their spaces, thus integrating the point of view of a region and its inhabitants, linking comprehensive security to regional security.

Regional security

Previously, I have asked 'how and by whom is security to be defined from the point of view of a region, and if it is defined, how can it be implemented?' (Heininen 2004b:38–9). I continued by arguing that 'although a complicated question and challenge, it is possible to define security based on a region, which opens up the opportunity to discuss democracy, identity and culture as a part of citizens' everyday security' (Heininen 2004b). This differs from the regional security approach taken by Buzan and Wæver which relies heavily on a realist-informed, state-centric framework (Buzan and Wæver 2004). Perhaps not surprisingly therefore, they neglect the Arctic as a region of study (Hoogensen 2005b). My approach here complements the work of Adler and Barnett on international security communities where regions could be understood as spaces of security composed of a 'region of states whose people maintain dependable expectations of peaceful change' (Adler and Barnett 1998:30), where trust is high, based on social interaction and (often) democratic processes. Here, I would like to continue the discourse and ask: are local people and regional governments real actors when defining and implementing national security, i.e. are they involved in decision-making for security policy, and even military policy? Furthermore, how is security from the viewpoint of a (sub)region defined and, in particular, is it even possible to define when dealing with an international cooperative region, such as the Barents Euro-Arctic region?

Security from the perspective of a region, or *regional security*, can be understood to mean the security of a regionalized space or a distinctive region, such as the circumpolar North, which is claimed to be defined by regional and local actors (Griffiths 1988; Young 1996), including citizens as well as local and regional authorities. A consideration of both security and a region also raises an opportunity to discuss people(s) and their identity/ies, needs, other interests and culture(s), as well as development of societies' interactions with regions and the environment. Further, as mentioned earlier, regional security provides an opportunity to discuss democracy and citizens' rights, either within a state or within a distinctive, international region transcending national borders. Thus, the focus is not only (traditional) security of a state, or environmental security or human/civil security, but security as a comprehensive way of including all the above-mentioned aspects of security from the perspective of a region and its citizens.

A good example of a regional security issue, and the way it goes beyond 'traditional' security approaches, is the nuclear problem of the Barents Sea area, as well as nuclear safety throughout the entire Arctic in the post-Cold War period. Nuclear safety in the Barents Sea area is a legacy of the Cold War period and its traditional security preoccupations between two central powers, the Soviet Union and the USA. Since environmental degradation and risks of nuclear waste cross national borders, radioactivity was seen to threaten not only the Russian North, but also northern Norway and northern Finland, and even Iceland, due to nuclear submarine accidents (Heininen and Segerståhl 2002:263–4). This was enough to initiate international cooperation for nuclear safety, as a part of Arctic environmental protection under the auspices of the Arctic Environmental Protection Strategy (see 'Rovaniemi Declaration' 1991) between the governments of the Arctic states in the Barents Sea area, as well as in the entire Arctic region. As a result of this close technical cooperation between the former enemies, the Russian Federation, Norway and the USA, the Arctic Military Environmental Cooperation (AMEC) was founded. In this example, nuclear safety was redefined as a security threat to the peoples and communities of the Barents Sea area rather than a military threat from radioactivity, as defined by traditional security.

Based on the mainstream definition of traditional security, it is not surprising that the purpose of regional security, i.e. to define security from the perspective of a region and its citizens, is a sensitive and complicated issue in the context of the unified state system. This is also the main reason that traditional security policy and the military are not a part of the agenda of the new international organizations and forums of the Arctic and entire North,[3] though they are said 'to decrease tension and increase stability and to promote sustainable development' in the region (see Heininen 2004b:32–3). To citizens it is, however, important, even critical in some situations, to know exactly how decisions about their security and security policy are made and by whom. This approach of security closely deals with governance, though it is challenging.

Different stages of security in the North

In the post-Cold War circumpolar North, there has been a significant shift from military tension into international and regional cooperation in many fields such as economics, environmental protection, research and higher education. Governments and other state organizations, in addition to northern indigenous peoples' organizations, sub-national governments, civil organizations and trans-national corporations, are expressing interest in the North in new ways to search for their share of control of space and resources (see Heininen 1999a). Thus, as mentioned earlier, instead of, or parallel to, traditional geopolitics, new approaches of geopolitics have raised human-oriented concerns like human capital, societal responsibility and the question of identity politics (see Chaturvedi 2000). This strongly indicates the importance of new actors in (geo)politics, and economics over politics. Correspondingly, 'critical geopolitics' is interested in this kind of new approach dealing with control of a space, where identity has become relevant (Harle and Moisio 2000).

Summarizing the situation at the beginning of the twenty-first century, the Arctic Human Development Report (AHDR) identified three main themes, or trends, of international relations and geopolitics in the circumpolar North: increased circumpolar cooperation by indigenous peoples' organizations and sub-national governments; region-building with nations as major actors; and the relationship between the Arctic and the outside world, including traditional security policy, since the North is still highly strategic to the USA and Russia (Heininen 2004a). Since then, another significant environmental, geopolitical and geoeconomic change with growing global interests has occurred to the region (see Heininen 2010b).

Looking back at the state of security and its development in the second half of the twentieth century until now, three, or even four, stages and interpretations of security can be identified (see also Heininen 2010a). First, the militarization of the circumpolar North in the first half of the twentieth century included two world wars with activities and influences in the entire North such as the convoys by the Allies to Murmansk to bring supplies for the Soviet Union in World War II. This stage emphasizes the importance of traditional security. The second stage, from the 1960s until the end of the 1980s, can be called the 'military theatre' of two superpowers, the Soviet Union and the USA. This stage included the deployment of nuclear-weapon systems into the region, and collective security in North America and the western part of North Europe through the 'security community' of NATO. At this stage, traditional geopolitics, especially the technology models of geopolitics, dominated the utilization of natural resources. National borders became barriers, which meant that most of international cooperation in the North was frozen. The third stage is the so-called transition period due to the end of the Cold War and the collapse of the Soviet Union. This stage could be, however, interpreted to have already begun by the end of the 1980s, inspired by the Murmansk Speech of the Soviet President Gorbachev (1987). This is due to the fact that the speech's initiatives declared the Soviet North open for international cooperation. Indeed, international and multilateral cooperation in economic, scientific and environmental protection in the Arctic started at the end of the 1980s and flourished by the beginning of the twenty-first century. This stage also meant de-militarization and the rationalization of the military, i.e. 'from quantity into quality', and thus it can be described by the slogan 'cooperation instead of confrontation'. From the point of view of comprehensive security as well as that of human security, the third stage is both relevant and very interesting. For example, based on the above-mentioned nuclear problem of the Barents Sea area, it is possible to conclude that there was a change in the narrow definition of traditional security in the 1990s, by the states of the region, dealing with threats and risks of nuclear waste and accidents, and, more broadly, with respect to the relationship between the military and the environment (Heininen 2010a). First, this deals with how state actors and intergovernmental organizations (IGOs) have changed their attitudes toward environmental degradation by the military or, perhaps more accurately, how they have taken the issue of environmental protection more seriously. As a result, nuclear safety in the Arctic has come onto the political agenda of many states and IGOs, but also other environmental issues dealing with the military, and thus traditional security, have become items on these political agendas.

In addition, one key result of the changing definition of security is the European Union's Northern Dimension (the EU's ND), which has responded to the new comprehensive human security agenda. From the point of view of the European Union, the ND is a framework and process for continuous dialogue on cooperation between the EU and its neighbours in North Europe (the Russian Federation, Norway, Iceland and Greenland) for coordination, and even management, of cross-border cooperation across the EU borders (European Union 2003). In keeping with its new concern for comprehensive and human security, the 'areas for cooperation' under the EU's ND include, among others, the environment, nuclear safety and energy cooperation. The ND ensures that the EU's environmental requirements are met, as well as ensuring the necessary actions are taken to monitor relevant and acute environmental threats. This is evident in a heightened interest by the ND towards: i) building the capacity for cooperation in nuclear safety and environmental issues between Russia and the EU, ii) focusing upon 'sustainable development' in terms of resource utilization, and iii) 'securing the border' while harmonizing legislation, standards and procedures in the interests of protecting and promoting civil society and environmental security.[4]

Lastly, if we agree that the early twenty-first century marked the end of the 'post-Cold War period' in the Arctic, then it is possible to define the fourth stage of security in the North. It can be seen to incorporate new kinds of security threats, such as global environmental problems and risks of environmental (long-range) degradation and impacts of climate change, and emphasizes the strategic interests of world powers through economic rather than military security interests in the region.

All this indicates that Arctic geopolitics, as well as northern security, is in a dynamic state of transition, and there are new factors influencing security and affecting perceptions of and changes in the definition of security. One indicator of this is that in the circumpolar North there have already been concrete and severe consequences of globalization, particularly due to long-range air and water pollution, and impacts by climate change (see Heininen 2010b:95–6). Many of these complicated environmental problems are mostly due to the increased utilization of, and competition over, natural resources. One result of this is that in the European North, as well in the broader circumpolar North, there have been, and are, asymmetric environmental conflicts mostly on the use of land, water and resources among different actors and their interests (see Osherenko and Young 1989; Riipinen 2005). More environmental catastrophes are expected and calculated to occur due to off-shore oil drilling and intensive sea traffic and growing transportation, particularly of exported Russian heavy oil and liquid gas (LNG) from new Russian oil terminals to central and southern Europe and North America. Examples of these new risky hotspots are the new Russian terminals at the bottom of the Gulf of Finland and the Kola Peninsula in the Barents Sea area, and the consequent growing transportation of oil and liquid gas (see Frantzen and Bambulyak 2003; Flamini 2005). Increased drilling and transportation of oil and natural gas and potential nuclear accidents will affect other aspects of security, e.g. cultural survival and food security (see Paci *et al.* 2004).

Although the Cold War is over, and there are discourses of comprehensive security and even changes in problem definition on security premises, the reality is that for Russia and the USA, as well as partly for the UK, the circumpolar North still has high strategic importance both politically and militarily. This means that a holistic picture of a (new) northern security includes traditional security dimensions in combination with the influences of newer threats and risks. Thus military security is still very relevant, as is regional security due to impacts of climate change, energy security meaning both access to, and import and export of, oil and natural gas, and also environmental security due to oil transportation, nuclear accidents and impacts of climate change. Overall, there are both traditional and alternative security dimensions that all deal with the security of the circumpolar North in the early twenty-first century.

Comprehensive security in the North

There are many security dimensions present in the circumpolar North, but there is no common definition for circumpolar or Arctic regional security. Instead of that, comprehensive security, either emphasizing environmental or human aspects, gives another, alternative and partly contrasting point of view to the traditional security of the Cold War period defined in the southern capitals.

In the 1980s, people in northern Europe started to become concerned about trans-boundary pollution in general and especially about radioactive contamination due to nuclear tests in Novaya Zemlya, the Chernobyl accident and accidents of nuclear submarines (Heininen 1994). The risk of radioactive contamination due to nuclear submarine accidents in the northern Atlantic was seen as a real threat among the Icelanders and the Norwegians, and the anxiety it created gave the Icelandic government a good reason to restate its position concerning growing risks of nuclear accidents (see Pa'lsson 1988). Behind this kind of statement is the fact that in the circumstances of a nuclear accident the relationship between the environment and the military will easily become concretized, since as a result there might be real environmental damage, meaning radioactive contamination. Even rumours of a possible radioactive leakage can damage, for example, the fishing industry, which is of great economic importance to Iceland and Norway (Heininen 1999a).

Environmental organizations such as Greenpeace International (see Greenpeace 1987), civil organizations such as the (Nordic) Saami Council (see Samiraddi 1990) and the first Nordic researchers became active in studying environmental degradation and risks caused by the military because of the popular concerns of the 1980s. Following this, Nordic citizens and the Nordic countries considered themselves as stakeholders in the international nuclear negotiations process, although, in most cases, they were, and are, outside the formal negotiations. In actuality, this can be interpreted to mean an effort to redefine security from the point of view of a region.

To the northern indigenous peoples, the alternative viewpoint of comprehensive security has been important, first to decrease the influence, and the dependence, of the region on non-regional actors and outside forces, including those who often interpret the region either as a potential military arena, such as the Thule US air and radar base (e.g. Brösted and Fægteborg 1985), or as a reserve of natural resources, and, second, to promote sustainable development in the region (Heininen *et al.* 1995). Indeed, northern indigenous peoples have recently become concerned about, and active in addressing, both toxic threats by the military and impacts of climate change, especially with respect to their traditional livelihoods such as hunting. This activity has been successful at pushing governments to sign the Stockholm Convention on persistent organic pollutants (POPs). It can be taken as an interesting success story on fruitful cooperation between northern indigenous peoples and Arctic epistemic community (AMAP 2002:36; Downie and Fenge 2003).

In contrast, and as a continuity of the debate, there are also counter-arguments to environmental concerns that there is enough space, land and sea in the North for both human activities, like the traditional livelihoods of the northern indigenous peoples, as well as the testing, training and other needs of armies and military presence. Behind this argument is the fact that armies have brought development to northern peripheries, especially for the northern indigenous peoples and others who live in the North, through employment, needed services, tax revenues, flight routes and infrastructure. There are also indigenous groups who have supported military defence and think that they should have at least some representation in national decision-making on security policy and defence (see Gaup 1990; Jull 1990).

Further counter-arguments might be that it is difficult, or even impossible, to show any dramatic evidence of pollution that would affect the Arctic ecosystem and that the ownership of some areas by armies has prevented environmental degradation. Additionally, other arguments are that there are much larger and serious sources of pollution than the military, like oil-drilling, long-distance air and water pollution, traffic, heavy industry and impacts of climate change which are both global and rapid. Finally, the environmental effects of the military may no longer be a relevant issue in the circumpolar North, because military presence and activities have been decreased and, correspondingly, clean-up has already started to reduce their past impacts as, for example, the international cooperation for nuclear safety in the Barents Region, the DEW Line Clean Up Protocol and the statements to clean up severe contamination by the military in the Russian Arctic indicate.

Indeed, there is ongoing cooperation for cleaning up the environmental damages of the military at national and international levels, such as the development of new technology for cleaning up by the Norwegian–Russian–US Arctic Military Environmental Cooperation programme as mentioned earlier. AMEC is not an official part of the international cooperation for Arctic environmental protection under the auspices of the Arctic Council, as more general traditional security and security-policy issues were excluded from the political agendas of organizations such as the Arctic Council and the Barents Euro-Arctic Council (see Heininen 2004b). It can be, however, taken as evidence that the environmental impact caused by the military has been acknowledged to be a problem.

All this indicates that, in spite of the above-mentioned counter-arguments, the concept of comprehensive security is presently widely accepted and used when issues of trans-boundary influences and cross-border cooperation are discussed, and in the context of states and society when dealing with internal security or defining security from the point of view of citizens. Due to the fact that there is both stability in the Arctic region and more confidence among northern actors, and hopefully less nationalist and ideological aspects, it would be possible to broaden the discussion on security to include factors previously considered taboo. It is not honest, but artificial, to exclude some relevant aspects of an issue when trying to build a comprehensive, holistic picture. For example, it is important to include all possible pollutants and potential polluters when trying to define all sources of pollution of a region.

However, the whole viewpoint of the military as a polluter is complicated, sensitive and in a way 'hidden' and, consequently, at the same time it is very political and relevant. This is also the case in the circumpolar North, particularly as it is strategically important to both the major nuclear powers and other Arctic states, as previously mentioned. Furthermore, the nuclear weapons system, which is still the backbone of major powers, as well as nuclear power and facilities, always include a risk of a nuclear accident, which cannot be overlooked. Finally, environmental pollution has been seen as a real problem only for a relatively short time and, therefore, more monitoring and evaluation of the environmental impacts and socio-economic impacts of military activities is needed.

Key phenomena of a northern security in the early twenty-first century

When applying the technology models of (classical) geopolitics, the end of the Cold War has not meant an exclusion of the military because, in many parts of the circumpolar North, heavy military structures are still present.

Among them are the nuclear weapons system and the command, control, communication and intelligence systems, as well as military activities such as the strategic nuclear submarines patrolling in the Arctic Ocean (see also Heininen 2010a; Wezeman 2012). Thus, as mentioned earlier, the region has a high strategic importance politically and militarily, especially for the USA and Russia.

When applying the resource models of geopolitics, the fact remains that the circumpolar North is rich in natural resources. Indeed, according to the AHDR, the annual gross product of the circumpolar North was about $US 230 billion in 2001.[5] This figure mostly consists of oil and natural gas drilling, for which production is foremost intended for southern latitudes of the northern hemisphere to meet the energy needs of the centres of developed countries (Duhaime 2004). Further, if Eurasia plays a big role in the global game, then the western part of the Eurasian North has a special role due to its rich natural resources, especially the fossil minerals of West Siberia and the shelves of northern seas. It has even been predicted that the Eurasian North might become a new 'Heartland' of energy, and the Barents Region as its sub-area will develop into a 'big oil playground' or a 'geopolitical hub' (see Brunstad *et al.* 2004; Flamini 2005). Indeed, this process has already started.

Following from this and indicating the continuing relevance of classical geopolitics and political power based on new realism at the beginning of the twenty-first century, northern or Arctic geopolitics includes two main phenomena: first, (mega) projects that utilize natural resources, consequently increasing transportation and plans for new navigation routes, and, second, state-centrism emphasizing the importance of state sovereignty and national security, including military capability and presence in the region. This is, of course, a new version of state politics relevant to globalization and globalized world economics. For example, if earlier governments mostly controlled the utilization of natural resources in the North, today trans-national corporations together with state-owned and controlled corporations are active and mobile in transferring investments and labour into the region for larger profits. Correspondingly, national and sub-national governments try to attract these corporations and their investments to the region for jobs. Further, if the above-mentioned actions are interpreted to describe traditional security and the importance of economics in national security, then there are also other relevant aspects and factors, such as environmental degradation and climate change and the uncertainty related to them, influencing security in general that emphasize the relevance of a comprehensive security in the Arctic. The following are among the relevant and interesting aspects of this more human-oriented northern security:

1 Pollution and environmental degradation: Many existing, as well as expected, environmental conflicts of, and in, the North are due to the increased utilization of, and competition over, natural resources.

Environmental degradation in general, and especially environmental risks, indicate not only conflict between man and nature, but mostly between socio-economic and political forces (see Käkönen 1995). In the North, these include radioactive waste and nuclear accidents (see the nuclear problem of the Barents Sea area), long-range air and sea pollution from the South to the North (especially radioactivity, toxics, mercury and POPs) and pollution from the military (see the DEW Line). Further, land claims by, and environmental conflicts between, the northern indigenous peoples and the Arctic states might also continue in the circumpolar North, even in the context of the Arctic Council, when trying to implement 'sustainable development' as a pillar of the Council and maintain it as one of the main goals (see Heininen 2004b:39–41).

2 Utilization and transportation of energy resources: Existing oil and natural gas drilling and exploration of new fields of hydrocarbons represent competition and some sort of 'hunt' for new and more natural resources that indicate the strategic importance of the northernmost regions of the globe. Following from this, there are plans for transportation of raw materials and other traffic, e.g. transportation of heavy oil and LNG gas from the Russian Arctic to Europe and North America, and even to Asia via the trans-Arctic sea routes. Transportation is becoming a relevant issue for comprehensive security by affecting environmental security in the sea and coastal areas, and human security in small communities and among indigenous peoples.

3 Impacts of climate change: Climate change has multiple impacts, e.g. melting of sea ice and glaciers and rising sea levels. By creating insecurity in northern regions and communities, especially in coastal areas, threatening towns and settlements and other man-made infrastructure as well as traditional livelihoods based on hunting and fishing, climate change has a relevant security dimension, too. Dealing with the above-mentioned fourth stage of northern security, the dramatic changes in the sea ice conditions of the Arctic Ocean and other northern seas in the near future might be particularly important and challenging. Current and rapidly increased thinning of the sea ice of the Arctic Ocean and many of the glaciers in Greenland, which are much influenced by climate change, is estimated to continue (ACIA 2004), and even melt faster than previously expected. This will increase the ice-free areas of the Arctic, which, correspondingly, creates new possibilities to increase both the utilization of natural resources and the transportation of them, e.g. by oil tankers in the trans-Arctic sea routes through the Arctic Ocean. Correspondingly, this will all easily increase other activities and other traffic like, for example, tourism by passenger yachts, and military patrolling and manoeuvres.

Climate change means new challenges and probable dramatic changes in the northernmost regions of the globe, and it brings uncertainty that is keenly related to climate change there. For example, one impact of climate change is to threaten traditional diets, which has raised concerns about food security. The traditional diet of many northern indigenous peoples is endangered due to the new climate conditions, because 'the continued and predictable availability and access to food, derived from northern environments through Indigenous cultural practices', is no more (Paci *et al.* 2004:1). Other possible impacts include smuggling and other associated uncontrolled traffic in northern seas, which is defined as a challenge, or even a threat, to the state sovereignty and national security of the Arctic states such as Canada and its sovereignty claim over the Canadian Arctic archipelago (see Huebert 2001; CARC *et al.* 2002). Thus, in the conditions of climate change, a growing and partly uncontrolled traffic might also become a new kind of security issue to the sovereignty of coastal states.

4 Trans-nationalism and globalization: Growing international cooperation and region-building and trans-nationalism reflect the influence of the globalized world and global economy in the North. These all influence human security in the North, especially because most northern inhabitants live in cities and no northern settlement remains isolated from the rest of the world (Bogoyavlenskiy and Siggner 2004). It is not, however, fair or even possible to claim that globalization only brings problems to the North, and is thus a threat, because there are both negative and positive impacts (see Heininen and Southcott 2010). Among the negative ones are impacts of climate change, privatization and liberal global economy, and weakening of nation-states. Correspondingly, among the positive impacts are decolonization and growth of regional autonomy, recognition of indigenous peoples' rights, the rule of law and multilateral environmental initiatives. The problem is often that 'local actors resist the effects of globalization, and they adapt to globalization by using the same tool kit that globalization uses or is constructed of' (Helander-Renvall 2010:207).

Trans-nationalism and globalization also indicate the importance of a borderless space, where cooperation occurs across national borders, e.g. building an international region like the Barents Euro-Arctic region. Trans-nationalism and borderless spaces certainly indicate the importance of regional cooperation and regionalization, but they also create new possibilities of threat and, thus, there is also a need, or a feeling of need, to control national borders.

5 Innovations in governance: In the circumpolar North there are also influences and examples of innovations in governance, i.e. 'the development of innovative political and legal arrangements that meet the needs of the residents' of the region (Young and Einarsson 2004:287–8). This is on the one hand due to devolution, giving self-determination, which is not against the larger, mostly national, political systems that are influential in the region (Bankes 2004), regionalization, indicating mostly a bottom-up activity, and finally region-building, which is mostly led by state

actors. On the other hand, this is due to a high capability in applying advanced technologies to address social problems. Following from this, sustainable development indicates human security and reminds us of the importance of implementing accepted policies on sustainability and environmental and human security by international organizations and other external political structures of, and for, the North.

Conclusions

In the early twenty-first century there are both aspects and factors of traditional security and those of comprehensive security in the circumpolar North. Among the main aspects and factors of the current security-political situation, traditional security is primarily indicated by state sovereignty and national security guaranteed by the military. This includes international cooperation for stability and peace in the context of some intergovernmental organizations, such as the European Union and NATO, and either common or national military activities like patrolling of strategic nuclear submarines in northern seas. Further, there are national interests, like, for example, economic welfare, prosperity and competitiveness, access to natural resources and, finally, energy security. Correspondingly, among the aspects and factors of comprehensive security emphasized by peoples and regional and local actors, there are, for example, environmental degradation, impacts of climate change, regional development and sustainability, human health and well-being, and devolution of power and innovations of governance.

If oil and natural gas drilling mostly deal with energy security and traditional security, then the transportation of heavy oil and liquid gas mostly deal with environmental security. Correspondingly, the impacts and uncertainty of climate change indicate new aspects of global security, either naming it environmental or human security. The claim that climate change has a relevant and important security dimension is also proved by the fact that climate change might also threaten national security and state sovereignty, and thus appear as a traditional security-political conflict. The issue is, however, neither defined nor interpreted in a traditional way and thus needs a new approach to security and an understanding of comprehensive security, including aspects of both environmental security and human security. This will also raise the above-mentioned question of how to define, or redefine, security from the point of view of a region, which would be a relevant part of a new northern security. Finally, this approach needs a holistic, geopolitical picture and requires not only multi-disciplinary research, international cooperation and global governance, but also implementation of comprehensive security and a perception of a global, 'borderless' space.

Notes

1 The work to discuss and define human freedom and security by the United Nations was continued for example by Resolution 1325 (United Nations Security Council 2000) on gender equality; among others the resolution declares 'to protect women and girls from gender-based violence in situations of armed conflict'.
2 Though the Russian military did not admit that the test was too risky, the Russian authorities accepted, or had to accept, international assistance when trying to rescue the crew.
3 For example, in the founding document of the Arctic Council there is a footnote that the cooperation does, and should, not include security policy or the military ('Ottawa Declaration' 1996).
4 The latter is of particular interest to the countries of the shallow Baltic Sea due to intensive sea traffic and growing sea transportation of oil from the new Russian terminals at the bottom of the Gulf of Finland to the Atlantic Ocean and the planned new oil pipeline from Russia to Germany at the bottom of the Baltic Sea.
5 About 67 per cent of this production comes from the Russian North.

References

ACIA (2004) *Impacts of a Warming Arctic: Arctic Climate Impact Assessment*, New York: Cambridge University Press.

Adler, E. and Barnett, M. (1998) 'A framework for the study of security communities', in E. Adler and M. Barnett (eds) *Security Communities*, Cambridge: Cambridge University Press.

AMAP (2002) *Arctic Pollution, Arctic Monitoring and Assessment Program*, Oslo: AMAP.

Bankes, N. (2004) 'Legal systems', in N. Einarsson, J. N. Larsen, A. Nilsson and O. R. Young (eds) *Arctic Human Development Report*, Akureyri: Stefansson Arctic Institute.

Beck, U. (1992) 'From industrial society to risk society: Questions of survival, social structure and ecological enlightenment', in M. Featherstone (ed.) *Cultural Theory and Cultural Change*, London: Sage.

Bogoyavlenskiy, D. and Siggner, A. (2004) 'Arctic demography', in N. Einarsson, J. N. Larsen, A. Nilsson and O. R. Young (eds) *Arctic Human Development Report*, Akureyri: Stefansson Arctic Institute.

Brösted, J. and Fægteborg, M. (1985) 'Expulsion of the great people when U.S. Air Force came to Thule: An analysis of colonial myth and actual incidents', in H. Kleivan and J. Brösted (eds) *Native Power*, Bergen: Universitetsforlaget.

Brunstad, B., Eivind, M., Swanson, P., Honneland, G. and Överland, I. (2004) *Big Oil Playground, Russian Bear Preserve or European Periphery? The Russian Barents Sea Region towards 2015*, Delft: Eburon.

Buzan, B. (1991) *People, States and Fear: An Agenda for International Security Studies in the Post-Cold War Era*, 2nd edn, Hertfordshire: Harvester Wheatsheaf.

Buzan, B. and Wæver, O. (2004) *Regions and Powers: The Structure of International Security*, Cambridge: Cambridge University Press.

CARC, CMSS and CPC (2002) *On Thinning Ice: Climate Change and New Ideas about Sovereignty and Security in the Canadian Arctic. Presentation Abstracts*, Ottawa: Canadian Arctic Resources Commission.

Chaturvedi, S. (2000) 'Arctic geopolitics then and now', in M. Nuttall and T. V. Callaghan (eds) *The Arctic: Environment, People, Policy*, Amsterdam: Harwood Academic Publishers.

Commission on Human Rights (1993) 'Question of the realization of the right to development'. Online. Available at: http://www.unhchr.ch/Huridocda/Huridoca.nsf/(Symbol)/E.CN.4.1994.21.En?Opendocument (accessed 4 July 2013).

Deudney, D. (1999) 'Environmental security: A critique', in D. H. Deudney and R. A. Matthew (eds) *Contested Grounds: Security and Conflict in the New Environmental Politics*, Albany, NY: State University of New York Press.

Downie, D. and Fenge, T. (eds) (2003) *Northern Lights against POPs: Combatting Toxic Threats in the Arctic*, Montreal: McGill University Press.

Duhaime, G. (2004) 'Economic systems', in N. Einarsson, J. N. Larsen, A. Nilsson and O. R. Young (eds) *Arctic Human Development Report*, Akureyri: Stefansson Arctic Institute.

Dwivedi, O. P., Kyba, P., Stoett, P. J. and Tiessen, R. (2001) *Sustainable Development and Canada: National & International Perspectives*, Peterborough: Broadview Press.

European Commission (2003) *A Secure Europe in a Better World: European Security Strategy*, Paris: The European Union Institute for Security Studies.

European Union (2003) 'The Second Northern Dimension Action Plan (2004–2006)'. Online. Available at: http://eur-lex.europa.eu/LexUriServ/LexUriServ.do?uri=COM:2003:0343:FIN:EN:PDF (accessed 8 August 2013).

Flamini, R. (2005) 'Analysis: Oil rich Norway's troubles', *UPI News*. Online. Available at: http://www.upi.com/Business_News/Security-Industry/2005/03/09/Analysis-Oil-rich-Norways-troubles/UPI-32301110384890/ (accessed 16 February 2011).

Frantzen, B. and Bambulyak, A. (2003) *Oil Transport from the Russian Part of the Barents Region*, Barentssekretariat, Svanhovd miljösenter. Online. Available at: http://www.crrc.unh.edu/workshops/arctic_spill_summit/oil_transport_ru_no_2007.pdf (accessed 23 March 2013).

Gaup, O. (1990) 'Indigenous peoples and the militarisation of the Arctic', unpublished conference paper. *Conference on Security and Cooperation in Europe (CSCE): 'Parallel Activities'*, Copenhagen.

Gorbachev, M. (1987) 'The speech of President Mikhail Gorbachev on October 2, 1987 in Murmansk', *Pravda*, 2 October.

Greenpeace (1987) 'Nuclear free seas', *The Greenpeace Campaign for Nuclear Free Seas*, Campaign Bulletin.

Griffiths, F. (1988) '"Introduction": The Arctic as an international political region', in K. Möttölä (ed.) *The Arctic Challenge: Nordic and Canadian Approaches to Security and Cooperation in an Emerging International Region*, Boulder, CO: Westview.

—— (1993) 'Defence, security and civility in the Arctic region', *The Nordic Council's Arctic Conference in Reykjavik, Iceland*, Copenhagen: The Nordic Council.

Haila, Y. (1989) 'Ekologinen tasapaino vai luonnon palautuvuus – mistä voimme johtaa normeja luonnon käsittelylle?', in L. Heininen (ed.) *Arktiksen Ristiriitoja – Turvallisuus, Kehitys, Ympäristö?*, Helsinki: Teoksessa.

Haila, Y. and Heininen, L. (1995) 'Ecology: A new discipline for disciplining?', *Social Text*, 42:153–71.

Harle, V. and Moisio, S. (2000) *Missä on Suomi? Kansallisen Identiteettipolitiikan Historia ja Geopolitiikka*, Jyväskylä: Vastapaino.

Häyrynen, N. (2003) 'Environmental security: The case of the Kursk', *Environmental Politics*, 12(3):65–82.

Häyrynen, N. and Heininen, L. (2002) 'Changes in problem definition: A case study of nuclear risks in the Barents sea region', in P. Strand, T. Jölle and Å Sand (eds) *Proceedings from the 5th International Conference on Environmental Radioactivity in the Arctic and Antarctic, Extended Abstracts*, Oslo: Arctic Monitoring and Assessment Program.

Heininen, L. (1994) 'The military and the environment: An Arctic case', in J. Käkönen (ed.) *Green Security or Militarized Environment*, Aldershot: Dartmouth.

—— (1999a) *Euroopan pohjoinen 1990-luvulla. Moniulotteisten ja ristiriitaisten intressien alue* [The European North in the 1990s: A Region of Multifunctional and Conflicting Interests], Rovaniemi: University of Lapland and the Arctic Centre.

—— (1999b) 'The nuclear problem and different actors: The problem onto a political agenda', in L. Heininen and G. Lassinantti (eds) *Security in the European North: From 'Hard' to 'Soft'*, Rovaniemi: Arctic Centre and Olof Palme International Centre.

—— (2004a) 'Circumpolar international relations and geopolitics', in N. Einarsson, J. N. Larsen, A. Nilsson and O. R. Young (eds) *Arctic Human Development Report*, Akureyri: Stefansson Arctic Institute.

—— (2004b) 'New external political structures in northern cooperation and northern governance: From quantity to quality', in T. Koivurova, T. Joona, R. Shnoro and J. Lapponica (eds) *The Northern Institute for Environmental and Minority Law*, Rovaniemi: University of Lapland.

—— (2010a) 'Globalization and security in the circumpolar North', in L. Heininen and C. Southcott (eds) *Globalization and the Circumpolar North*, Fairbanks, Alaska: University of Alaska Press.

—— (2010b) 'Post-Cold War Arctic geopolitics: Where are the peoples and the environment?', in M. Bravo and N. Triscott (eds) *Arctic Geopolitics and Autonomy, Arctic Perspective Cahier*, Ostfildern: Hatje Cantz Verlag.

Heininen, L. and Segerståhl, B. (2002) 'International negotiations aiming at a reduction of nuclear risks in the Barents Sea region', in R. Avenhaus, V. Kremenyuk and G. Sjöstedt (eds) *Containing the Atom: International Negotiations on Nuclear Security and Safety*, New York: Lexington Books.

Heininen, L. and Southcott, C. (eds) (2010) *Globalization and the Circumpolar North*, Fairbanks, Alaska: University of Alaska Press.

Heininen, L., Käkönen, J. and Jalonen, O.-P. (1995) *Expanding the Northern Dimension: The Final Report of the International Arctic Project of TAPRI*. Tampere Peace Research Institute, Research Report 61, Tampere: University of Tampere.

Helander-Renvall, E. (2010) 'Globalization and traditional livelihoods', in L. Heininen and C. Southcott (eds) *Globalization and the Circumpolar North*, Fairbanks, Alaska: University of Alaska Press.

Hoogensen, G. (2005a) 'Human security in the Arctic', unpublished presentation to ICARPII, 11 November, Copenhagen.

—— (2005b) 'Bottoms up! A toast to regional security?', *International Studies Review*, 7:269–74.

Hoogensen, G., Bazely, D. R., Christensen, J., Tanentzap, A. and Bojko, E. (2009) 'Human security in the Arctic: Yes, it is relevant!', *Journal of Human Security*, 5:1–10.

Huebert, R. (2001) 'Climate change and Canadian sovereignty in the Northwest Passage', *Isuma*, 2(4):86–94.

Huru, J. (2004) 'Mitä tekemistä identiteetillä on tirvallisuuden kanssa? McSweeney vastaan Buzan ja Waever', *Kosmopolis*, 34(3):29–50.

Independent Commission on Disarmament and Security (1982) *Common Security: A Blueprint for Survival*, New York: Simon and Schuster.

Jull, P. (1990) 'Social change in NATO's Far North', *NATO Review*, 38(2):23–8.

Käkönen, J. (ed.) (1994) *Green Security or Militarized Environment*, Aldershot: Dartmouth.

—— (1995) *Konfliktit, Turvallisuus ja Ympäristö*, Saarijärvi: Atena Kustannus Oy.

Langlais, R. (1995) *Reformulating Security: A Case Study from Arctic Canada*, Göteborg: Göteborg University.
Newcombe, H. (1986) 'Collective security, common security and alternative security: A conceptual comparison. Peace', *Research Reviews*, 10(3):1–8 and 95–9.
Nokkala, A. (1999) 'Kylmä sota päättyi – miten Suomen uhkakuvat muuttuivat?', *Kosmopolis*, 29(4):47–66.
Osherenko, G. and Young, O. R. (1989) *The Age of the Arctic: Hot Conflicts and Cold Realities*, Cambridge: Cambridge University Press.
'Ottawa Declaration' on the Establishment of the Arctic Council. Ottawa, Canada, 19 September 1996.
Paci, J., Dickson, C., Nikels, S., Chan, L. and Furgal, C. (2004) 'Food security of northern indigenous peoples in a time of uncertainty', position paper for the 3rd NRF Open Meeting, Yellowknife, North West Territories, Canada, 15–18 September 2004. Online. Available at: http://www.nrf.is/Publications/The%20Resilient%20North/Plenary%204/3rd%20NRF_Plenary%204_PP_Paci%20et%20al.pdf (accessed 16 February 2011).
Pa'lsson, T. (1988) 'Islannin ulko- ja turvallisuuspolitiikka', *Paasikivi-Seuran monistesarja*, 77.
Poverty and Environmental Times (2004) A periodic publication by UNEP/GRID-Arendal, No. 2, March 2004.
Riipinen, M. (2005) 'Local community as a stage for land use discourse: Before and after a global actor', unpublished presentation in the Calotte Academy, 26–28 May 2005, Inari.
'Rovaniemi Declaration' (1991) Online. Available at: http://arcticcircle.uconn.edu/NatResources/Policy/rovaniemi.html (accessed 23 March 2013).
Samiraddi (1990) Nordiska samerådet och World Council of Indigenous Peoples (WCIP) samlat till ett gemensamt mote I Tromsö den 7 och 8 mars 1990 får harmed tillkännage följande uttande.
Sisäasiainministeriö (2004) Arjen turvaa. Sisäisen turvallisuuden ohjelma. Sisäasiainministeriön julkaisuja, Finland.
United Nations General Assembly (2005) 'In larger freedom: Towards development, security and human rights for all'. Report of the Security-General. A/59/2005 Distr.: General, 21 March 2005. Online. Available at: http://www.un.org/largerfreedom/contents.htm (accessed 4 July 2013).
United Nations Security Council (2000) Resolution 1325, S/RES/1325 (2000). New York: United Nations Security Council.
Wahlström, E. (1994) *Ympäristöriskit – Kokonaiskuvaa Etsimässä*, Jyväskylä: Schildts.
Waltz, K. N. (1979) *Theory of International Politics*, New York: McGraw-Hill.
Wezeman, S. T. (2012) *Military Capabilities in the Arctic*, SIPRI Background Paper, March 2012, SIPRI: Solna.
Young, O. R. (1996) *The Arctic Council: Marking a New Era in International Relations*, New York: Twentieth Century Fund.
Young, O. R. and Einarsson, N. (2004) 'A human development agenda for the Arctic: Major findings and emerging issues', in N. Einarsson, J. N. Larsen, A. Nilsson and O. R. Young (eds) *Arctic Human Development Report*, Akureyri: Stefansson Arctic Institute.

Chapter 4

Virtuous imperialism or a shared global objective?

The relevance of human security in the global North[1]

Gunhild Hoogensen Gjørv

Introduction

Until the end of the Cold War, security was largely confined to the militarized and elite notion of state security bound within an anarchic international system regulated by superpowers.[2] Establishing and maintaining state security did not ensure individual or community security, particularly with regard to poverty, access to basic resources and care, and general human well-being (UNDP 1994). The security referent, or target, has since expanded from a narrow focus on the state and state system to include other levels of analysis and security perspectives such as those articulated by individuals or groups. The 'human security agenda' was proposed in the early 1990s as a global initiative focusing on the security of individuals, but seemed to be largely adopted as a foreign policy approach by a number of global North countries, practising a 'new' way of understanding and providing assistance to the global South.[3] It was intended to update and complement a 'traditional' (narrow) notion of security, making more visible issues of security that traditional security failed to address, such as continuing poverty and human strife within failing or weak states, or in so-called secure, but non-democratic, states (ibid.). However, an important critique of the human security agenda, in addition to questions of definition, has been its perpetuation of the superior–subordinate relationship, such that the concept can be understood to entrench linear and elitist thinking with regard to security (Chandler 2008). As it has been practised, it is a concept that is assumed to be mostly relevant for states and peoples living in the global South. Although the human security concept appeared to break the stranglehold that a narrow state-based conception managed to co-opt since WWII and particularly through the Cold War, taking into consideration a broader range of security referents than that of traditional security, the prevailing assumption that it is a concept only relevant for particular regions of the world perpetuates ahistorical claims about the state and sovereignty, claims that are strongly rooted in the realist tradition that assumes that 'strong states provide better security' (Smith 1991 in Wibben 2011:70). The global North is assumed to be composed of 'strong' states, thus having eradicated its own human security issues, and is well placed to assist the South in eradicating the same. It is also a very state-based concept assuming that security is addressed by state actors alone, rather than a process that involves multiple actors, state and non-state (Hoogensen Gjørv 2012). The result becomes an imbalance in perceptions and explanations of what occurs within and across regions and the globe, as well as disguising the contributions and competencies of different actors in providing security at different levels (Stuvøy 2009; Hoogensen Gjørv 2012). Assuming that human security perspectives have little relevance in the global North serves to mute voices of individuals and communities who recognize threats to their existence as they perceive it (Deiter and Rude 2005). Thus while rendering passive global South actors, it also renders invisible any human insecurities and vulnerabilities that may be occurring in northern states. It assumes that northern voices are fully represented and attended to by a state actor, and it disguises and prevents any possible shared human security concerns and experiences between peoples across regions, between the global North and South.

Instead of broadly brushing one area of the world as 'secure' and another as 'insecure', we can look at securities of non-dominance (Hoogensen 2006; Hoogensen and Stuvøy 2006). Relations of dominance/non-dominance exist in all parts of the world, albeit distributed in different ways. In the global North, where it is broadly understood that the entire area is 'secure', a securities approach recognizing a dominance/non-dominance relationship would recognize non-dominant insecurities otherwise invisible to the dominant discourse, such as those experienced by women or indigenous communities, for example. Equally it can be recognized who is secure in the global South by acknowledging the dominance/non-dominance relationship. Despite the generally more pronounced insecurities (and in some cases extreme insecurities) present in global South communities, there are nevertheless dominant groups living in relative broad security. These relationships need to be made visible in the securities context, and securities of non-dominance might be recognized to be broadly shared across constructed divides such as the North and South.

Broadened notions of security are becoming more prevalent (Buzan *et al.* 1998; Thomas, 2001; Buzan and Wæver 2003; Fierke 2007; Glasius 2008; Newman 2010). In this chapter I argue in favour of security defined from the 'bottom up', which can dig deeper into the security perspectives of communities, states and regions but also transcend North/South divisions, by identifying security articulated from the position of those who are most insecure; in other words, those often on the margins. This is not a call to focus on the global North at the expense of the global South, nor a competition of insecurity between regions and peoples. Security and insecurity from the bottom up implies that security is understood within the context from which it is identified and expressed. In this chapter I am interested in understanding security from the 'bottom up', and will look more specifically at positions of gender and indigenous perspectives. I will begin however with a focus on the perceptions of human security as they developed within Canada since the popularization of the human security concept in the 1990s. Canada had been one of the most prominent proponents of human security, although with a strictly global South focus (Government of Canada 2004).[4]

A deeper reflection by some of the previously enthusiastic human security proponents upon how there might be human insecurity in one's own state, or shared human security concerns with the global South, has not been in evidence. Drawing upon some of the concerns articulated by women and indigenous peoples at the same time as the Canadian government trumpeted its human security platform, I hope to illustrate some of the problems with a 'one-way' approach to thinking about human security, and encourage debate on how different understandings of human security from the bottom up (from people in their different contexts) might lead to a more shared sensitivity to insecurity across the globe. This argument is inevitably linked to a broader understanding of what security is, or rather its meaning. Ideally, the debate could significantly move away from a central problem that plagues assistance-focused concepts and programmes like human security – the 'us–them' polarization.

Challenges of definition

Mainstream security definitions that isolate security as a matter for high politics, 'securitizing' as little as possible except obvious threats to national security, are created by elite interests, or those in power (the state). Preserving the state and maintaining a focus on the military caters to ensuring and maintaining security for those interests that are most secure in the first place, largely the state apparatus and elites within. As much as this aspect of security has validity for state security, it does not articulate all security interests. Nevertheless, its preeminence has remained due to a certain logic connected to the definition – the state itself is expected to provide security to 'the people', to individuals.

State-based security requires a level of state mobilization that would otherwise not be called upon to address a particular issue (Buzan *et al.* 1998). Such mobilization is relegated to the level of 'high politics' or the top priority of the state, and does not include action at the political level (such as social security or economic policy) or 'low politics' (Robertson 1997). As such, the state has a central role, and addresses threats by eliminating the chance that such threats would successfully overthrow the state and its apparatus. It is largely due to this special relationship of security to the state and to military or 'high politics' concerns that many in the traditional security community argue vehemently against a widening of the concept. However, state security, even in the most allegedly secure state, can and does threaten individual security. Military build up and arms accumulation often play a central role in the design of state security, but these same features have been frequently identified as threats to human beings, within and across state affiliations.

'Human security' was popularized in the 1994 United Nations Development Programme (UNDP) Human Development Report, expanding the notion of security to include dimensions of food, health, community, environmental, economic, personal and political security, with the intention to, in part, address some of the glaring weaknesses of security policy, but also spurred on theoretical debates. Human security focuses on the individual as the security referent. In the UNDP report, human security is defined as 'freedom from fear, freedom from want', and consists of four essential characteristics: universal, interdependent, easier to ensure through early prevention, and people-centred (UNDP 1994). Almost 20 years later, both academic and policy literature are still ambivalent about a definition. At the turn of the twenty-first century when Canada was developing its own position on the concept, Rob McRae claimed the Canadian position on human security was solely 'freedom from fear', which was understood as prior to development (McRae and Hubert 2001). Human security applied to migration required a broader definition, using the definition parameters of the UNDP report which went beyond the Canadian application (Poku *et al.* 2000). The human security concept has, since then, been subject to much discussion and debate, both arguing how the concept is difficult for policy makers to operationalize when it is defined too broadly, how it is meaningless to focus upon individuals per se, and how it is a necessary concept that finally brings non-state actor voices into the security debate (Hansen 2000; Paris 2001; Burgess and Owen 2004; Buzan 2004; Hoogensen and Vigeland Rottem 2004; Burgess *et al.* 2007; Hoogensen Gjørv 2012). Although not supportive of a human security concept, the Copenhagen school's securitization process opens up the potential for non-state actors to have a voice in the security dynamic (see Åtland and Pedersen, this book, for a more thorough discussion about the Copenhagen school). Through a speech act or an articulation of security, security becomes 'what we make of it' (Buzan *et al.* 1998). Human security can indeed be a platform to allow for more voices to decide what security is made of.

As noted in Chapter 1 of this book, the concept of security is not a neutral concept but a political concept that is subject to struggles of power. A problematic dynamic of the speech act process described above is the relationship between the speech actor and the respondent or audience, or those who 'hear' and respond to the security speech act. Whose expressions of security are actually 'heard' by those who have the power to act on these articulations (Hansen 2000)? And if security speech acts of security are made but not recognized by those with the choice and power to respond, does that mean these are no longer articulations of security or insecurity? Another difficulty lies in the assumption about who provides security. Narrow conceptions of security assign the role of security provider to the state (often through the tool of the military). This assumption makes it problematic if one is going to recognize a broader and more complex understanding of security. The state (particularly the military) is not capable of addressing all 'on the ground' human security-related issues, and nor is that desirable. This assumption also implies that other, non-state actors are passive recipients of security, and do nothing to contribute to establishing and maintaining security for individuals and communities at sub-state levels or beyond. A deeper analysis of processes of security, speech acts as well as security practices, demonstrates that security is about both the identification of threats as well as the capability to address such threats, by both state as well as non-state actors (Hoogensen *et al.* 2009; Hoogensen Gjørv 2012).

Canada and the development of a human security concept

Canada was a champion for human security at the beginning of the millennium, playing a leading role in human security promotion, and as such its policies serve as a good illustration to see how human security has been employed or operationalized in practice. In this section I want to further develop an analysis of Canada's position with regard to human security that I started a few years ago when I referred to the Canadian *policy level* leadership in an earlier article addressing non-state actor security and terrorism (Hoogensen 2005). I drew upon the following quotations to illustrate the Canadian position at the time, the first quote coming directly from the then Minister of Foreign Affairs, Lloyd Axworthy, and the second quote following, reflecting the influence of the Canadian position in supporting *scholarly analyses* of human security in different contexts:

> The present discourse of security reflects this change in the global reality and the change in perspective that goes with it. No longer are we limited to discussions of states' rights and national sovereignty. Protecting civilians, addressing the plight of war-affected children and the threat of terrorism and drugs, managing open borders, and combating infectious diseases are now part of a dialogue.
>
> (Axworthy 2001:19)

Axworthy played an important role in stressing the importance of security at the individual level, as human security would likely not have managed to capture the interest it did without such significant political recognition. Axworthy's role as human security champion was acknowledged in academic circles as well, where the concept was gaining ground: 'Canadian Foreign Minister Lloyd Axworthy, perhaps the most conspicuous exponent of the concept, has listed "safety for people from both violent and nonviolent threats" as core preconditions [of human security]' (Brower and Chalk 2003:5). The Canadian government had thus been one of the more active world leaders in the promotion of human security (MacLean 2000). The Canadian government championed human security as a feature of its foreign relations policy programme (Government of Canada 2004) and had supported leading policy-relevant research institutes, which continued to explore and develop the concept.[5]

This leadership was celebrated in May 2013 when a number of the leading policy makers who fronted the human security agenda 15 years earlier met in Bergen, Norway, to re-visit the concept and agenda and look back on how it has developed. The original signatories to the Lysøen Declaration, a declaration of intent signed by the then foreign ministers Lloyd Axworthy (Canada) and Knut Vollebæk (Norway), played a lead role again by opening a conference that for all intents and purposes demonstrated that despite all the controversies surrounding the concept, human security is still relevant. But it still remains to be asked – in what ways is it relevant? Who is this concept really for?

Although the current Canadian government has significantly downplayed (if not rejected) an explicit role in the human security agenda since it came into power in 2006, the rationale that was behind the development of this concept in the 1990s and early 2000s, and its further implicit operationalization, provides a good example of how the human security concept has been burdened with what I call a 'virtuous imperialism' reputation. This term expresses the way in which human security has often been perceived, to put it bluntly, as a one-way-feel-good concept for the global North to attend to insecurity in the global South. It also served as a way to protect the North (northern states) from the South (southern peoples/southern threats).

At that time, human security appeared to have provided Canada with its *raison d'être* in the international community. Human security was a concept and policy application that was consistent with Canada's soft power approaches, in Canada but also amongst other middle power states such as Norway. The 'complementary' nature of human security, where it does not replace state security but provides an added dimension that should be addressed, suited middle power politics. Thus states like Canada and Norway emphasized, and tried to contribute to, this complementary role by promoting human security (particularly the protection of civilians) at the Security Council (MacLean 2000:275). Canada and Norway assumed much of the leading roles in the drafting of the 1997 Anti-Personnel Landmines Treaty, and the creation of the International Criminal Court in 1998 (ibid.). Canada took the human security approach on board in a more vigorous manner however, designing a whole foreign policy programme around the concept, and the choices Canada was making was influential in the human security debate (Mack 2004; Human Security Centre 2005). This included the debates regarding definition. The definition offered by the UNDP Human Security Report in 1994 was vague and rather inelegant. On one level the statement 'freedom from fear, freedom from want' might generate intuitive agreement, but finding out what this meant concretely was decidedly more difficult. The UNDP report focused on the categories of food, health, environmental, community, personal, economic and political security (UNDP 1994), but the Canadian government considered this 'unwieldy as a policy instrument' (MacLean 2000). Although the UNDP report represented a policy-based platform rather than an academic and theoretical deliberation, it did little to provide guidance for concrete measures taken by states, if any, and under what contexts.

Canada thus focused on an interpretation of 'freedom from fear', emphasizing trade in small arms, people, drugs, as well as other transnational threats such as terrorism (Government of Canada 2005). This approach facilitated policy making (narrow) and ensured a separation of resources where 'fear' is associated with the notion of security which is further linked to violence, and 'want' is linked with development which is considered something other than security, as violence is not a key feature of development issues. Human security seems to stretch the boundaries of security as understood by the Canadian government, by including issues of development as well. The Canadian government's approach reflected a way of thinking of security where their focus was on negative security (fear-based) that focuses on issues of violence and its containment (often by policing or military means) and not allowing for understanding security issues not directly related to violence issues either in theme or in treatment, known as positive security (Hoogensen Gjørv 2012). Focusing on issues of negative security or fear and violence facilitated policy making. This Canadian priority was illustrated in a 2002 report by the Liu Centre for the Study of Global Issues where the linkage of security and development was explicitly addressed, but it was equally stressed that development must not become an 'arm of traditional national security policy' (Liu Centre 2002:15–16). The state apparatus could focus on traditional security tools (military or police) to address these narrow human security needs without straining or overstepping state responsibilities. Thus the Canadian approach to human security did not appear far off from so-called traditional national security policy. George MacLean (2000:274) noted then that 'Canada's position has been that the UNDP criteria does not pay sufficient attention to the threats encountered by people as a result of violent conflict'. But it was more than this. Canada's threat-based orientation equated threats to people with threats to states seemed to step only very cautiously into human security terrain.

Narrowing the focus did not reduce problems, particularly with regard to implementation. Canada was not really willing to risk its own people for those in 'far-flung places' (Hataley and Nossal 2004:17). Keeble and Smith (2001:136), who were also following the Canadian policy closely at that time, stated that the approach was a 'state-based philosophy embedded in a foreign policy that is concerned, first and foremost, with Canadian economic growth and competitiveness'. The human security concept opened a door for non-state actor security concerns, for giving a voice to those whose voices were not considered relevant in a state-based security approach, but the Canadian operationalization of the concept did little to find ways to make this happen.

Human security was supposed to be transformative, but Canada's view of human security was so limited that there was nothing transformative about it, no commitment to addressing inequalities (ibid.). Hataley and Nossal noticed Canada's hesitance to really engage when human security seemed to be threatened. Keeble and Smith (2001:136) summed up the problem with the question: 'How can a country promote human security while participating in activities that perpetuate so many sources of insecurity?'

Dominant security discourses perpetuate the 'us–them' polarization

Dominant human security definitions, and more such practices (as illustrated by the Canadian example), have contributed to, I would argue, an 'us–them' polarization. It is assumed that freedom from fear and want is not relevant for a particular experience – more specifically for those living in a stable state. For example, regarding fear and want in the average Canadian or North American's life it is assumed that the state is well placed to address these fears. As noted above, human security was about people in 'far-flung places'. Human security was fronted as part of a foreign policy regime for northern states like Canada and Norway, not as a lens to assess insecurities within communities at home. Thus the approach assumed that all human insecurity must lie outside of their borders, largely in the global South, without recognizing the complex relationships between different security practices and actors at home and abroad. A broader definition of human security calls for greater recognition of context, non-state actors (particularly individuals and communities) and security practices, which in turn will say more about security/insecurity origins and solutions, and would not exclude or isolate the North or South. Human security issues like the sex trade, the proliferation of disease, the arms trade, poverty and famine occur within the global North, as well as between the North and South. Insofar as the Canadian approach acknowledged an increased insecurity amongst its own civilians, it looked only in one direction, where threats did not occur internally but usually emanated from the outside or the 'other': 'Whether it is a terrorist bomb exploding and killing or injuring Canadians travelling or working abroad, or a sexual predator half way around the world talking to a child who is using the Internet in Charlottetown' (McRae and Hubert 2001:11).

Beyond identifying and perhaps 'solving' human security crises abroad as a foreign policy platform, human security also seemed to become a way of identifying the 'other' as a source of insecurity for Canadians, both as individuals and communities but also state security. Human security was thus still about state security. Insecurity was perceived to originate *from* the outside *to* Canada, not the other way around: 'The tentacles of terrorism or crime can stretch from the Middle East or the Balkans to the backyards of Toronto' (ibid.). Human security was not a problem *inside* Canada because insecurity on the individual level was assumed to be addressed by and through the state apparatus, at least for stable and secure states in the North. State-centric security definitions are considered sufficient in democratic, economically complex developed settings because it is assumed that the people can dictate their needs/securities to government and these will be/are executed at the people's behest. As a foreign policy tool therefore, human security protected northern state security while at the same time offering assistance and support to the 'other', at least where northern states chose to identify relevant and worthy human security threats.

The danger of the 'us–them/we–other' polarization can be further illustrated in the tendency towards humanitarian intervention, where the virtuous imperialism critique can be most strongly felt, and we see this critique often in debates about 'Responsibility to Protect' (R2P), which is a policy platform that has strong roots in human security (ICISS 2001; Bellamy 2009; Kofi Abiew 2010). A common criticism of human security and its related concept of R2P is that it can easily lead to humanitarian intervention if the right 'conditions' apply. In the interest of ceasing conflict, protecting individuals and establishing security, humanitarian intervention can be a logical consequence of human security initiatives. As Michael Pugh states, 'rational, civilized "humanitarian intervention" may be a part of the packaging in which Western security culture, self-perception and self-interest are wrapped' (Pugh 2002). This view is well exemplified in Canadian policy in the early 2000s:

> For Canada, human security is an approach to foreign policy that puts people – their rights, their safety and their lives – first. Our objective is to build a world where universal humanitarian standards and the rule of law protect all people; where those who violate these standards are held accountable; and where our international institutions are equipped to defend and enforce those standards. In short, a world where people can live in freedom from fear.
>
> (Government of Canada 2004)

For some critics of human security, approaches such as those found within American security policy could logically follow from the previous Canadian position:

> the ultimate goal of U.S. strategy must be to spread democracy everywhere. The United States must finish the job that Woodrow Wilson started. The world, quite literally, must be made safe for democracy, even those parts of it, like the Middle East, that have so far resisted that tendency. Terrorism – and by implication the authoritarianism that breeds it – must become as obsolete as slavery, piracy, or genocide: 'behaviour that no respectable government can condone or support and that all must oppose'.
>
> (Gaddis 2002)

Although the USA did not buy into the human security agenda, it manipulated the same priorities that are dominant in human security foreign policies and practices – 'or it claims to be pursuing values that, as the National Security Strategy puts it, are "true for every person, in every society"' (ibid.) – and therefore could be equated with the motives behind such human security initiatives. As succinctly expressed in *The Economist*, 'rich countries must sometimes be prepared to kill' (*The Economist* 2003) in the interest of facilitating development and security. These positions are top-down expressions of security, applied by secure elites judging their own societies to be free of insecurity but threatened from the outside. This vision of human security has not disappeared or died down in the past 15 years, but as noted above can be argued to have instead found new life in the R2P doctrine. Without a deeper understanding of what human security can and should entail, how it might be a way towards reducing inequalities, incidences of violence and poverty, both in the global South and North, a one-way virtuous imperialism can result (Chandler 2012).

The need for more voices in security discourses: learning from indigenous perspectives

> To discover that the food which for generations has nourished them and kept them whole physically and spiritually is now poisoning them is profoundly disturbing and threatens Indigenous Peoples' cultural survival ... The Indigenous Peoples of the Arctic continue to see contaminant and climate change variability as a major threat to our collective survival as Peoples.
>
> (Statement prepared by the Indigenous Peoples Secretariat on behalf of Arctic Council Permanent Participants: A Call for Further Action, in Nilsson and Huntington 2002:v)

Human security perceptions in and from the North can be seen to be influenced by a number of voices, not least indigenous (highlighting security of identity and the environment, as well as economy; see chapters by Slowey and Irlbacher-Fox *et al.*, this book) and women (highlighting personal security; see chapters by Lvova and Stuvøy, this book). In the next two sections I wish to draw attention to both indigenous and women's positions. The relevance of indigenous voices should be significant to the security discourse (Stern 2006). These are voices that span the world over, transcend the constructed boundaries of North/South, East/West, West/'other', as these voices represent and highlight insecurities from ethnic and cultural genocide, disease epidemics, colonization, oppression and discrimination by race, gender, class and ethnicity, to economic and cultural destruction of key natural resources and, for some, the planet itself. Indigenous issues and human security are linked, as exemplified by a statement from the United Nations Guide for Indigenous Peoples:

> There are an estimated 370 million indigenous people in more than 70 countries worldwide ... Despite their cultural differences, the various groups of indigenous peoples around the world share common problems related to the protection of their rights as distinct peoples ... Indigenous people are arguably among the most disadvantaged and vulnerable groups of people in the world today.
>
> (Office of the United Nations High Commissioner for Human Rights n.d.)

Insofar as indigenous communities have experienced war, these experiences with their attendant needs have not been the target for human security policy, with the effect that they had been some of the least noticed, acknowledged, supported or attended-to vulnerable groups (and it can be argued that this situation has largely remained):

> Before September 11, 2001 indigenous nations in many places in the world had been engaged in low-intensity conflicts either in defence of their peoples and territories or in movements of self-determination (as many as 80 different conflicts). These conflicts had been described in the press and by foreign policy spokespersons as 'internal conflicts' or as regional conflicts.
>
> ('Program Synopsis' 2002)

In the above respect, indigenous experiences with violence and violent conflict play a role insofar as these experiences are a part of traditional security logic and interests (state security and the threats to state security). However, they have also been positioned as the 'other', either as a source of the threat defined by dominant voices or, on the other hand, are included amongst the many unheard voices in regional conflicts that did not yet warrant the attention of global North or Western attention on the basis of human security or otherwise.

Understanding the nature of security in the indigenous context goes even further when it is taken into consideration that natural resources, a significant element of many indigenous cultures and livelihoods, are often exploited to the detriment of indigenous cultures, threatening cultural, community and economic securities, in addition to personal and political securities if these resources are extracted under threat or action of violence (ibid.), or economic isolation. An extensive International Polar Year (2006–2007) project mapping and examining the relationships between northern Russian indigenous groups in the Nenets Autonomous Okrug in northwest Russia and the oil and gas industries revealed very specific tensions between economic security (the benefits of the oil and gas industry) and community/identity security of the Nenets people (Dallmann *et al.* 2010). Although some (but not all) of the communities surveyed during the project welcomed the economic development that oil and gas activity would bring to their region, many expressed a fear about not having their concerns heard about the impacts of such activity on their communities and livelihoods (ibid.). Poor, unrepresentative legislation, a lack of participation and consultation, a fear for the impact on the environment upon which many Nenets rely for other economic activities (reindeer herding) and a fear for their very identity and existence as Nenets or Komi communities were all raised as primary concerns (ibid.). These concerns, not least the perceived impacts from local actors regarding negative changes that could have detrimental effects upon the existence of their communities as indigenous groups, as well as the knowledge and capacities of indigenous peoples to mitigate, reduce or eliminate these fears, could be better included. Power devolving to people who have a better, or at least more complex, on-the-ground knowledge about security issues for their own communities, but also for communities and regions beyond local spaces, where knowledge can be shared more widely with others, is crucial to understanding human security (in both the global South and North). Human security policies cannot be determined from above, from elites or state actors who cannot identify human security concerns alone. As argued in Oran Young's chapter in this book, room must be made for integrating knowledge from and between different systems, ecological and social. Knowledge about the interconnections of these systems is further demonstrated to be insufficient amongst dominant knowledge institutions (see Young, this book). Thus seeking out bodies of knowledge that have longer traditions of such integration is relevant to today's policy making regarding a broader operationalization of security.

Feminist approaches to human security

> The traditional goal, and perceived outcome, of the narrative selection process has been the achievement of order, unity, and coherence in support of a particular modern social, symbolic, political order. To achieve this goal, some narratives are relegated to the margins, and others silenced ... every instance of selection is also a political moment.
>
> (Wibben 2011:64)

Feminist approaches draw the focus to the voices of women, whose perspectives have also not been traditionally included in the security debate, but which have the ability to transcend and integrate many of the levels and sectors of security that scholars have otherwise chosen to analyse separately: 'security narratives limit how we can think about security, whose security matters, and how it might be achieved' (ibid.:65). Instead of playing into the dominant approaches to security studies that focus on a very small portion of the security grid and from the top down, gender analysis takes its starting point from the bottom up; it reaches all the way down to the individual, as gender analysis acknowledges that even the personal is political, and therefore the individual's experience is relevant, not just at the level of the individual however, but also in relation to the community, the state and the global order. At the same time it is recognized that individuals are part of communities, and that gender is a significant feature of individual identity in relation to others and is therefore a part of societal security (Hoogensen and Vigeland Rottem 2004). The social constructions of gender come in to play in the analysis and the ways in which humans have constructed their societies on the basis of gender roles, who has the 'right' to play which roles in society (as decision-makers or as the subjects/objects of those decisions) and how people are supposed to relate to one another. Gender analysis is inherently social, as it relies on the construction of relationships in society as the point of analysis. Even further, however, globalized gender analysis has demonstrated not only the dominance of male or patriarchal-based societies, but culturally dominant societies, where the gendered demands (for example, Western feminists) of one society are imposed upon other, less dominant societies. As such, gender analysis integrates the individual, national and global levels through acknowledgement of the social power interactions between peoples and constructions of societies.

By identifying the articulation of security needs by those who are least secure or in positions of non-dominance, security is reoriented away from elite interests. Feminist analysis has much to offer in this respect, and contributes to the identification of security needs from the margins. Feminist approaches have had a long and established tradition of highlighting marginalized realities faced on a daily basis by, arguably, the majority of the world's population, from economic insecurity and domestic violence, to rape as an institutionalized strategy of large-scale warfare, as well as from the gendered roots of war itself. Such realities are open to, and open up, the meaning of what it is to be secure. Additionally, it is a theoretical tradition that has a logical place in the human security discussion, bringing the political 'down' to the level of the individual, to bring a voice (and power) to the personal. The personal is political, and human security, with its focus on the individual, has the potential as a concept and a theoretical platform for policy to support these personal voices.

It is often women who find themselves at the forefront of security challenges, but also they who find themselves immediately marginalized when attempting to arrive at solutions for these same challenges (Karamé 2001; Hoogensen and Stuvøy 2006; Youngs 2008). International relations, and security in particular, has sustained and maintained a militarized, elite actor focus. 'An IR lens focused exclusively on elite interstate actors and narrow definitions of security keeps us from seeing many other important realities' (Peterson and Runyan 1999:32). Often the dichotomy between the narrow conception of security and a broader conception can be illustrated by the differences between militarization and structural violence; the former focused on military defence and the removal of a physical, often institutional threat such as war or large-scale violent conflict, and the latter representing 'reduced life expectancy as a consequence of oppressive political and economic structures ... that especially affects the lives of women and other subordinated groups' (ibid.). If the Canadian approach is any example, state elites operating in isolation from community voices tend to still understand security in narrow, state-based, militarized and violence-focused terms.

Feminist and gender security studies have been active for decades, as long as, if not longer than, the human security agenda. There are some important, principled overlaps between human security and feminist and gender security perspectives, though dominant human security theorizing and policy making have failed to adequately draw on these connections. While the human security concept was still nascent, Christine Sylvester's work in 1994 was an already well-developed demonstration of the complexity of a gender security perspective that accounted for individual or human security. In that work she stated that a gender perspective, and feminist perspective more particularly, 'paints security and cooperation in multihued multistandpoints that may be discredited as monstrous by modernists but that will not go away' (Sylvester 1994:169). Sylvester's multidimensional approach recognized that there 'can be no coherent statement, inscription, or morality that makes the definitive home, covers all the contingencies of contingent existence, and thereby ends insecurity' (ibid.:183). We nevertheless struggle with these contingencies, working with the 'elusiveness' that is security:

> To work with this elusiveness requires a recognition that [security] is intensified in the late or postmodern era by the number and types of insecuring actions that rivet our attention. This is a time of simultaneous struggles, of storms with many centers unfolding on many fronts at once ... Some simultaneous struggles are relatively easy to see, as in South Africa, where efforts to homestead the acrid terrain of apartheid move in cross-cutting directions; there are similar struggles, it seems, in Peru, erstwhile Yugoslavia, Liberia, Canada, Angola. Other types of struggles are more difficult to follow, as in the see-sawing efforts to 'secure' the international environment or to secure reproductive rights or religious identities.
>
> (ibid.)

Sylvester thus acknowledged that we face obvious insecurities (those widely recognized by dominant discourses) and less obvious ones (such as reproductive rights) that reflect the securities of the unheard or marginalized. The problem of isolating only one dimension of security is further illustrated in the divide and discourses of non-Western feminisms, highlighting the 'us–them' polarization inherent within Western thinking overall, and Western feminisms themselves. Western feminisms may aptly identify the lack of acknowledgement of diverse insecurities by the traditional security framework, but non-Western feminisms wage similar arguments against Western feminisms themselves: 'Western feminist scholarship cannot avoid the challenge of situating itself and examining its role in ... a global economic and political framework' (Witt 1999:17). Gayatri Spivak writes that indigenous and non-Western feminism increasingly becomes marginalized within mainstream feminism (Spivak 1999). Thus feminist analysis, from both the global North and South, highlights features of the security dynamic that have been isolated, ignored and made invisible because the realities of gender and 'other' have not been acknowledged. This has been very much the case in relation to indigenous approaches, and especially gendered indigenous approaches.

The 1995 Beijing Declaration of Indigenous Women exemplifies this well. Self-described as those who continue to suffer multiple oppressions 'as Indigenous peoples, as citizens of colonized and neo-colonial countries, as women, and as members of the poorer classes of society' (Beijing Declaration of Indigenous Women 1995), this group of women were well placed to give voice to the concerns of the marginalized and least secure. Personal, economic, environmental, community and physical security become inevitably intertwined in their vision of what poses a threat, to them, to their children and to their children's children. In response to the increasingly dominant pressures of international trade, neo-colonialization and 'science', the Declaration explicitly acknowledged the potential and likelihood for 'ethnocide and genocide' by progressively eliminating biological and cultural resources, while continuing to exacerbate conflict over lands and communities (ibid.). The threats identified in the Declaration include those that would be addressed by narrower conceptions of security, but then go well beyond traditional security parameters, recognizing and acknowledging the sources and origins of threats that are shared. Joyce Green and Cora Voyageur demonstrate the dynamics of the non-dominant and insecure position of indigenous women in the Canadian context, noting the multiple insecurities such as poverty, hunger, social and identity marginalization, political isolation and domestic and societal violence faced by many aboriginal Canadian women (Green and Voyageur 1999). Aboriginal women have not been recognized by Canadian government platforms on human security.

Insecurity originates within democratic states as well as non-democratic, North and South, although the degree and nature of these threats often differ. Security defined by elites is not necessarily security as defined by those who are insecure, regardless of location. But we respond to security differently depending on the location, and determine its human security implications on the same basis. In 2001–2002, for example, during Canada's heyday of human security, over 28,000 people fled violence, over half of them children, with less than 11,000 able to find safe shelter or refuge, forcing many to return to the violence they fled from. This was a case of structural violence in a society with limited opportunity to flee certain types of violence, both societally as well as institutionally. State mechanisms to respond to this sort of violence were and are weak, and the seeking of refuge by certain groups of people would in principle qualify as a human security issue. Except that this case referred to women and children fleeing violent homes in Alberta, one of the richest provinces of Canada (Scott 2003). It is assumed that social, economic and political development and an advanced Canadian infrastructure ensures that a legitimate state apparatus is in place to address the insecurities of women experiencing violence. However, the above case is an illustration of insecurity from the margins of society in Canada, of women and children who are not part of the dominant political apparatus, and many of whom live in poverty. The number of homeless women and children continued to escalate during Canada's human security years, as did the violence against women and their children, but it never became a human security issue politically (Alberta Council of Women's Shelters). The problem remains especially acute on First Nation (indigenous) reserves (ibid.).

Highlighting certain insecurities in otherwise affluent and 'secure' parts of the world might be considered a devaluation or marginalization of other insecurities, in other parts of the world, that involve immediate physical threats such as war. It is not uncommon to see a ranking of security issues, usually against the threat of large-scale violence or war, as the assumption is that war is the activity that most likely threatens the existence of the state (which is assumed to protect people and their human security). The security of women has been weighed against this measure as well; when large-scale conflict defines security and particularly security of the state, individual security issues fall by the wayside, where women, for example, are responsible for their own security (prevent your own rape by not walking alone at night, drinking or not wearing provocative clothing; be sure to keep your husband/partner happy so that he will not resort to domestic violence; do not commit adultery for fear of being punished by family or society on religious/cultural grounds), and their insecurity will not be acknowledged in the public, political realm. And when it has been acknowledged, it often falls at a lower priority than does open armed conflict as a security issue. Is it necessary, however, to place these securities at cross-purposes? As 'enemies' of each other, so to speak? The insecurity of women, for example, subject to domestic violence or sexualized violence, is a problem that is shared across regions, states and communities, during war as well as peace time, but little is being done to address this global threat to women. Knowing security as a part of but also beyond immediate physical threat or war-torn contexts can provide a much deeper and relevant understanding of security – demonstrating its complexity and dynamism. As well, it can provide insights into securities that are otherwise ignored or overlooked, whether relevant to the global South or North. It also de-powers the human security agenda defined by elites (without any proper engagement and recognition for other actors), and reduces the tendency towards a virtuous imperialism imposed by those in power.

It is important to continue to illustrate the hypocrisies of narrow definitions of security, and the assumption that the state, especially the 'developed and democratic' state, is the purveyor of security to the people (such as Canada, home to many communities that live without many of the benefits extolled by human security; Green and Voyageur 1999; Deiter and Rude 2005; Slowey, and Irlbacher-Fox *et al.*, this book). Most importantly, recognizing diverse, multi-levelled and multi-actor security perspectives increases awareness about the origins and effects of such insecurities, allowing for early treatment, if not outright prevention, of escalated conflict. Security has to mean more than 'treating' conflict, genocide and other large-scale violence that are allowed to happen precisely because other securities were ignored at earlier stages or in other contexts.

In the discussion above I present an argument for the importance of perceiving security multi-dimensionally, reflecting the fact that individuals feel insecurity in a variety of ways, allowing for the immediate treatment of emergency threats, and not ignoring other features of security, thereby working on prevention simultaneously. The end result is a holistic view of security, moving towards the vision of those like indigenous women that recognizes threat in circumstances assumed by elites to be threat-free or relatively benign (such as the environment and identity). Security can be seen as a dynamic relationship contributing to a process, rather than a static condition. Within a dynamic relationship it is possible to visualize many factors that influence security, from the social relationship between the individual/collectivities to the state and global networks/organizations, to other influential dynamics such as hegemony and sovereignty, 'developed' and 'undeveloped' or 'underdeveloped', secular and religious, and so on. These are all dynamics contributing to the security conversation. This is a process because the dynamic is continuous and always in motion. State security can and should not be ignored because it is a part of this conversation, but other dynamics and realities cannot be excluded.

Conclusion

If human security only means security from open armed conflict or war, there is, at least in recent history, admittedly little shared experience between the global North and South (although war in the global North does occur, albeit less frequently and taking place in locations of the 'other' such as Iraq, Afghanistan, Libya and Mali). There is no denying that violent conflict and war poses an enormous security threat upon individuals and communities and requires the productive attention of the global community. However, there are a great many insecurities that can be identified that occur well before open conflict becomes the result, largely due to poverty and inequalities. Particularly when seen through the lenses of gender and indigenous experiences, there are many shared human security concerns between the North and South that are rooted in identity, poverty and other security threats relevant to local contexts, local environments. A broadened definition of human security includes immediate, emergency efforts directed towards war and conflict-torn regions, but also has a view towards prevention. Without recognizing the diversity of human security, measures to promote it will remain insufficient, and security voids will continue to exist, leading to increased violence and conflict and war which is then 'rescued' (or at least attempted to do so) by human security programmes. In sum, a number of key points can be drawn out:

- Security defined by those who are least secure or in non-dominant positions, not only elites, should be included in security discourses.
- Feminisms, including Western, non-Western and indigenous feminisms, offer powerful arguments articulating the voices of the insecure, and deserve to be heard and responded to by mainstream sources.
- Security is a multi-dimensional, dynamic, fluid process, always in flux.
- A broadened security agenda increases awareness with a sight to prevention, instead of attending to insecurity at its most extreme.
- The 'us–them' polarization creates inequality, exacerbates insecurity and ignores the diverse sources and features of insecurity and security.
- The human security agenda can be a shared orientation/attitude by a variety of policy groups and networks, implemented across and within policy frameworks by both state and non-state actors, cooperating and merging shared foci while retaining independence specific to each individual policy framework.

Current definitions of security force us to render many insecurities as secure because we judge them in relation to severe or extreme insecurities that clearly demand and deserve attention. Ignoring the diversity of insecurities, however, often results in the very extreme insecurities we must rush to stop because they have escalated beyond our 'acceptable' levels. To refer to a very recent example, the issue of rape appeared to be on the brink of finally being recognized globally for the gross danger that it presents to women, families and society. Early in 2013 when brutal incidences of rape, from India to the USA and Canada, were being reported, there were calls for changes in social attitudes about rape, as well as tighter legislation.[6] The calls have died down, raising the question about how severe or extreme insecurity is defined (only by machine guns that threaten men?) and how far it must go before local, national and global responses are really mobilized. But human security is not just about meeting threats early (or better late than never as in the case of rape). It is also recognizing when processes of security production are working, at least on some levels. Cooperation, dialogue and designing open and inclusive structures of governance including state and non-state actors that reflect the priorities and values of their constituent communities go a long way towards creating environments of security where individuals and communities can thrive. Human security and insecurity therefore are relevant across the globe, in the South and the North, and need responses that are based on local understandings of insecurity and that are contextually determined, which may or may not mean intervention by state-based actors. It is a process that is constantly a work-in-progress; and understanding this security from a multi-actor (individuals and communities, organizations, policy makers, etc.), multi-level (state and non-state actor) perspective is essential.

Notes

1 An earlier version of the chapter was presented in 2006: Hoogensen, G. 'Virtuous imperialism: Human security and the exporting of the "wisdom" of the North to the South', International Studies Association, San Diego, 21–27 March 2006. This earlier version is available at AllAcademic research. Online. Available at: http://citation.allacademic.com/meta/p_mla_apa_research_citation/0/9/9/3/7/pages99374/p99374-1.php and http://www.humansecuritygateway.com/showRecord_fr.php?RecordId=15274 (accessed 23 March 2013). Referred to by W. Greaves (2012) in 'For whom, from what? Canada's Arctic policy and the narrowing of human security', *International Journal*, 67(1).

2 The history of the security concept is certainly more nuanced when one examines it more closely. For the purposes of this chapter I refer to the paradigm that dominated the past 50–100 years.

3 Canada and Norway were both leaders in supporting a human security foreign policy approach, and led the way for the 'Human Security Network' composed of 13 member states (mostly from the global North, but not solely). Canada and Norway's leading role in fronting this perspective was celebrated in May 2013 with 'Lysøen Revisited – 15 Years of Human Security' (Lysøen was the island location where Canada and Norway signed an agreement on human security in 1998). The Human Security Network is now largely defunct, its previous website no longer functioning (www.humansecuritynetwork.org), though it is still included in links through other foreign ministry or advocacy websites. See, for example, Institute for Environmental Security, 'Human Security Network': http://www.envirosecurity.org/actionguide/view.php?r=27&m=initiatives; Embassy of Austria, Washington DC, 'Human Security Network': http://www.austria.org/humansecurity-network (both accessed 5 July 2013). The members are/were: 'Austria, Canada, Chile, Costa Rica, Greece, Ireland, Jordan, Mali, Norway, Slovenia, Switzerland and Thailand, with South Africa participating as an observer' (stated on Embassy of Austria link above). There is not much evidence of activity since 2007.

4 Canada's enthusiasm for the human security concept and agenda waned considerably with a government shift in the mid-2000s. However, it still acknowledges a place in this debate, most recently evidenced in its joint participation and organization of a high-profile conference on human security held in Bergen, Norway (Norway being another northern state that has been an eager proponent of the concept on occasion), in May 2013.

5 Up until at least 2007 there was an active Canadian academic and policy network called the 'Canadian Consortium on Human Security' (CCHS). See the Department of Foreign Affairs, Trade and Development Canada for the archive on the CCHS: http://www.international.gc.ca/about-a_propos/oig-big/2007/evaluation/cchs-ccsh07.aspx?lang=eng (accessed 5 July 2013). Academic or research programmes on human security continue, such as at the Liu Centre for Global Studies (http://www.ligi.ubc.ca/?p2=/modules/liu/researches/category.jsp&id=6, accessed 5 July 2013).

6 See, for example, 'It's not her fault: A look at victim-blaming and rape culture', at http://www.newbrunswickbeacon.ca/30981/fault-victim-blaming-rape-culture (accessed 4 July 2013); 'Two teens found guilty in Steubenville rape case', at http://edition.cnn.com/2013/03/17/justice/ohio-steubenville-case (accessed 4 July 2013); and 'Political philosophy behind rape?', at http://kaalratri.com/2013/04/20/political-philosophy-behind-rape (accessed 4 July 2013).

References

Alberta Council of Women's Shelters (2003). Online. Available at: https://www.acws.ca (accessed 4 July 2013).

Axworthy, L. (2001) 'Human security and global governance: Putting people first', *Global Governance*, 7(1):19–23.

Beijing Declaration of Indigenous Women (1995) Online. Available at: http://www.ipcb.org/resolutions/htmls/dec_beijing.html (accessed 26 March 2013).

Bellamy, A. (2009) *Responsibility to Protect: The Global Effort to End Mass Atrocities*, London: Polity Press.

Brower, J. and Chalk, P. (2003) *The Global Threat of New and Reemerging Infectious Diseases: Reconciling U.S. National Security and Public Health Policy*, Santa Monica, CA: Rand Publishing.

Burgess, P. and Owen, T. (2004) 'Editors' note', *Security Dialogue*, 35(3):345–6.

Burgess, P., Amicelle, A., Bartels, E., Bellanova, R., Cerami, A., Eggum, E., Hoogensen, G., Kittelsen, S., Knibbe, K., de Koning, M., Koser, K., Krause, K. and Salemink, O. (2007) *Promoting Human Security: Ethical, Normative and Educational Frameworks in Western Europe*, Paris: UNESCO.

Buzan, B. (2004) 'A reductionist, idealistic notion that adds little analytical value', *Security Dialogue*, 35(3):369–70.

Buzan, B. and Wæver, O. (2003) *Regions and Powers: The Structure of International Security*, Cambridge: Cambridge University Press.

Buzan, B., Wæver, O. and de Wilde, J. (1998) *Security: A New Framework for Analysis*, Boulder, CO, and London: Lynne Rienner Publishers.

Chandler, D. (2008) 'The human security paradox: How nation states grew to love cosmopolitan ethics', *Globalization, Difference, and Human Securities Conference*, Graduate School of Human Sciences, Osaka. Online. Available at: http://www.davidchandler.org/pdf/papers/Osaka%20conference%20paper%20(Chandler).pdf (accessed 25 March 2013).

—— (2012) 'Resilience and human security: The post-interventionist paradigm', *Security Dialogue*, 43(3):213–29.

Dallmann, W. K., Peskov, V. V. and Murashko, O. A. (2010) *Monitoring of Development of Traditional Indigenous Land Use Areas in the Nenets Autonomous Okrug, Northwest Russia*, Tromsø: Norwegian Polar Institute.

Deiter, C. and Rude, D. (2005) *Human Security and Aboriginal Women in Canada*, Ottawa: Status of Women Canada.

Fierke, K. M. (2007) *Critical Approaches to International Security*, Cambridge: Polity Press.

Gaddis, J. L. (2002) 'A grand strategy of transformation', *Foreign Policy*. Online. Available at: http://www.questia.com (accessed 23 March 2013).

Glasius, M. (2008) 'Human security from paradigm shift to operationalization: Job description for a human security worker', *Security Dialogue*, 39(1):31–54.

Green, J. and Voyageur, C. (1999) 'Globalization and development at the bottom', in M. Porter and E. Judd (eds) *Feminist Doing Development*, London and New York: Zed Books.

Government of Canada (2004) 'Audit of the Human Security Program (HSP)'. Department of Foreign Affairs and International Trade. Online. Available at: http://www.international.gc.ca/about-a_propos/oig-big/2002/human_security_program-programme_securite_humaine.aspx?lang=eng (accessed 12 August 2013).

—— (2005) 'Human Security Program'. Online. Available at: http://www.humansecurity.gc.ca/psh-en.asp (accessed 22 June 2005).

Hansen, L. (2000) 'The Little Mermaid's silent security dilemma and the absence of gender in the Copenhagen School', *Millennium: Journal of International Studies*, 29(2):285–306.
Hataley, T. S. and Nossal, K. R. (2004) 'The limits of the human security agenda: The case of Canada's response to the Timor crisis', *Global Change, Peace and Security*, 16(1):5–17.
Hoogensen, G. (2005) 'Gender, identity, and human security: Can we learn something from the case of women terrorists?', *Canadian Foreign Policy*, 12(1):119–40.
—— (2006) 'Virtuous imperialism: Human security and the exporting of the "wisdom" of the North to the South', *International Studies Association*, San Diego, 21–27 March 2006.
Hoogensen, G. and Stuvøy, K. (2006) 'Human security, gender and resistance', *Security Dialogue*, 37(2):207–28.
Hoogensen, G. and Vigeland Rottem, S. (2004) 'Gender identity and the subject of security', *Security Dialogue*, 35(2):155–71.
Hoogensen, G., Bazely, D., Christiansen, J., Tanentzap, A. and Bojko, E. (2009) 'Human security in the Arctic: Yes, it is relevant!', *Journal of Human Security*, 5(2):1–10.
Hoogensen Gjørv, G. (2012) 'Security by any other name: Negative security, positive security and a multi-actor security approach', *Review of International Studies*, 38(4):835–59.
Human Security Centre (2005) *The Human Security Report 2005: War and Peace in the 21st Century*, H. S. Centre, Lieu Institute of Global Issues, University of British Columbia, Canada.
ICISS (2001) *The Responsibility to Protect: Report of the International Commission on Intervention and State Sovereignty (ICISS)*, Ottawa: International Development Research Centre.
Karamé, K. (2001) *Gendering Human Security: From Marginalization to the Integration of Women in Peace-Building.* Fafo-report 252/Nupi-report 261, Oslo: NUPI-FAFO Forum.
Keeble, E. and Smith, H. (2001) 'Institutions, ideas, women and gender: New directions in canadian foreign policy', *Journal of Canadian Studies*, 35(4):130–41.
Kofi Abiew, F. (2010) 'Humanitarian intervention and the responsibility to protect: Redefining a role for "kind-hearted gunmen"', *Criminal Justice Ethics*, 29(2):93–109.
Liu Centre (2002) *Development, Conflict and Peacebuilding: Responses from Canada*. Report on a Roundtable co-sponsored by the Liu Centre for the Study of Global Issues and the Canadian International Development Agency, Vancouver: Liu Centre for the Study of Global Issues.
Mack, A. (2004) 'A signifier of shared values', *Security Dialogue*, 35(3):366–7.
MacLean, G. (2000) 'Instituting and projecting human security: A Canadian perspective', *Australian Journal of International Affairs*, 54(3):269–76.
McRae, R. and Hubert, D. (eds) (2001) *Human Security and the New Diplomacy: Protecting People, Promoting Peace*, Montreal: McGill-Queen's University Press.
Newman, E. (2010) 'Critical human security studies', *Review of International Studies*, 36:77–94.
Nilsson, A. and Huntington, H. (2002) *Arctic Pollution*, Oslo: Arctic Monitoring and Assessment Programme.
Office of the United Nations High Commissioner for Human Rights (n.d.) *Leaflet No. 1 – Indigenous Peoples and the United Nations System: An Overview*. Online. Available at: http://www2.ohchr.org/english/issues/indigenous/guide.htm (accessed 4 July 2013).
Paris, R. (2001) 'Human security: Paradigm shift or hot air?', *International Security*, 26(2):87–102.
Peterson, V. S. and Runyan, A. S. (1999) *Global Gender Issues: Dilemmas in World Politics*, Colorado: Westview Press, Inc.
Poku, N. K., Renwick, N. and Glenn, J. (2000) 'Human security in a globalising world', in *Migration, Globalization, and Human Security*, London: Routledge.

'Program Synopsis' (2002) *World War and the Fourth World* (Centre for World Indigenous Studies – Fourth World Dialogue: A WEBCast and Radio Broadcast Production, 11 March 2002), 1. Online. Available at: http://cwis.org/GML/?post=1125 (accessed 12 August 2013).

Pugh, M. (2002) 'Maintaining peace and security', in D. Held (ed.) *Governing Globalization: Power, Authority, and Global Governance*, Cambridge: Polity Press.

Robertson, C. L. (1997) *International Politics since World War II: A Short History*, Armonk, NY: M. E. Sharpe.

Scott, S. (2003) 'It's a long, tiresome journey for women to escape abuse: Nowhere to run', *Alberta Parent Quarterly*, Spring: 8–11.

Smith, S. (1991) 'Mature anarchy, strong states and security', *Arms Control*, 12(2):325–39.

Spivak, G. (1999) *A Critique of Post-Colonial Reason: Toward a History of the Vanishing Present*, Cambridge, MA: Harvard University Press.

Stern, M. (2006) *Naming Security – Constructing Identity: 'Mayan-Women' in Guatemala on the Eve of 'Peace'*, Manchester: Manchester University Press.

Stuvøy, K. (2009) 'Security under construction: A Bourdieusian approach to non-state crisis centres in Northwest Russia', PhD Doctoral Thesis, Tromsø: University of Tromsø.

Sylvester, C. (1994) *Feminist Theory and International Relations in a Postmodern Era*, Cambridge: Cambridge University Press.

The Economist (2003) 'The poor man's curse: Some things the world can do to prevent wars', 22 May.

Thomas, C. (2001) *Global Governance, Development & Human Security: The Challenge of Poverty & Inequality*, London: Pluto Press.

UNDP (1994) *Human Development Report 1994: New Dimensions of Human Security*, New York: United Nations Development Programme.

Wibben, A. T. R. (2011) *Feminist Security Studies: A Narrative Approach*, London and New York: Routledge.

Witt, D. (1999) *Black Hunger: Food and the Politics of U.S. Identity*, New York: Oxford University Press.

Youngs, G. (2008) 'From practice to theory: Feminist international relations and "gender mainstreaming"', *International Politics*, 45:688–702.

Part II

Environmental security

Chapter 5

The sustainability transition

Governing coupled human/natural systems[1]

Oran R. Young

Introduction

We live in a world of dynamic, often turbulent, and increasingly human-dominated ecosystems. The social systems that generate anthropogenic impacts on ecosystems are similarly dynamic and turbulent. The key to sustainability in this setting lies in developing techniques that allow us to steer or govern these coupled systems in such a way as to avoid outcomes that are unintended, nasty and largely irreversible. Yet many familiar and well-entrenched scientific methods and models are not conducive to this effort; some familiar models actually impede progress in these terms. It follows that we need to rethink our analytic tools and alter our approach to policymaking for sustainability accordingly.

The body of this chapter develops this argument in three steps. The first section focuses on rethinking our analytic tools; it explains the reasons why such an effort is needed and points to some of the cutting-edge questions we should bear in mind in dealing with this need. The second section sets forth three challenges for science that must be addressed in order to produce knowledge that can play an influential and constructive role in moving society towards sustainability or, in other words, making a 'sustainability transition'. The third section spells out six lessons for policymaking that flow from the preceding discussion of the production of knowledge relevant to making a sustainability transition. An underlying theme throughout the argument set forth here is the need to improve the relationship between science and policy in order to make science more policy relevant and to avoid pitfalls arising from efforts to manipulate science for partisan purposes.

Rethinking our analytic tools

Much current thinking about human/nature interactions has its roots in an earlier era in which it seemed reasonable to assume that (i) human actions are not major drivers of ecosystem dynamics and (ii) a reductionist approach to the analysis of ecosystems constitutes an appropriate point of departure for generating policy-relevant knowledge. Those who make these assumptions find it natural to treat anthropogenic forces as exogenous and generally minor perturbations and to construct models focusing on specific components of ecosystems through a liberal use of ceteris paribus assumptions. A number of observers have argued that this analytic strategy had serious drawbacks from the start. Whatever our view of this history, however, it is clear now that such a strategy cannot provide the knowledge needed to make progress towards the sustainability transition and that we require new tools to achieve sustainable ecosystems.

Ecosystems

Whereas members of earlier generations of scientists were satisfied to focus on subsystems and typically to concentrate on efforts to understand the population dynamics of individual species, recent decades have witnessed a sharp turn towards holistic studies that rest on the premise that the interdependencies among the elements or components of ecosystems are too great to ignore or to abstract away through the introduction of assumptions. Certainly, there are those who cling to the development of reductionist models in the interests of preserving the logical power of these constructs; they take the view that the analytic price of moving towards integrated models will exceed any benefits likely to accrue from a more holistic mode of thought. Even so, it is clear that the introduction of whole ecosystem perspectives is a major development that is not likely to go away during the foreseeable future (Golley 1993).

Current thinking about ecosystems also features a heightened awareness of dynamics, including non-linear and chaotic processes, rapid change events, cascades and regime shifts (Ludwig *et al.* 1993). Even in the absence of human interventions, ecosystems are prone to more or less dramatic changes, and there is no basis for assuming that these systems will tend to return to some prior equilibrium in the wake of significant disturbances. Far-reaching changes, often described as regime shifts, are by no means uncommon; less dramatic but still significant changes occur as a matter of course. Under the circumstances, arguments based on the premise that biophysical systems exhibit strong equilibrating tendencies are doomed to failure.

What is the significance of these trends for the development of tools needed to support the sustainability transition? Many familiar models that are widely used in efforts to manage ecosystems of interest to humans are of limited value; their application may actually prove counterproductive in some domains. A prominent example involves the widespread reliance on maximum sustainable yield (MSY) models in fisheries management (Larkin 1977). Because MSY models focus on single (or at most a few) species and do not address the dynamics of the larger ecosystems to which targeted species belong, these models can and often do produce results that are misleading as a basis for establishing allowable harvest levels. Beyond this lies the challenge of modelling complex systems without an undue loss of analytic rigour. Partly, this is a matter of making deliberate choices regarding appropriate trade-offs between tractability and complexity. In part, however, it has stimulated a growing interest in new analytic procedures, such as computerized simulations, in contrast to more traditional exercises in mathematical modelling.

Social systems

The preceding observations about the analysis of ecosystems apply almost equally well to studies of social systems. Growing interdependencies manifested in such processes as globalization and global social change give rise to an unavoidable need for holistic perspectives. Even small-scale systems are impacted routinely by global processes (e.g. worldwide markets for agricultural products), a fact that makes it impossible to explain or predict developments at the local level without paying careful attention to driving forces launched in distant realms. Like ecosystems, social systems seldom revert to the status quo ante following major disturbances or perturbations (e.g. wars, economic cycles, electoral changes). What is particularly notable in this context is the juxtaposition of periods of stasis with abrupt and fundamental changes. The resultant processes of punctuated equilibrium regularly lead to regime shifts that take most observers by surprise (Gunderson and Holling 2002).

It will come as no surprise, under the circumstances, that many familiar models in the social sciences are of limited value to observers endeavouring to understand patterns of change through time. This is not to say that there is no place, for example, for equilibrium models of competitive markets and spatial models of voting. But there is a clear need to transcend these familiar models in efforts to grasp the dynamics of social systems in holistic terms. Among the cutting-edge issues now arising in this context are matters of interplay or the interaction of distinct social systems with one another and matters of scale or the effort to generalize propositions about social dynamics across levels of social organization (Young 2002a).

Coupled socio-ecological systems

Increasingly, ecosystems and social systems interact with one another. In many realms, these interactions have produced human-dominated ecosystems (Vitousek *et al.* 1997), moving us in the process into an era that many scientists have begun to call the Anthropocene (Crutzen and Stoermer 2000). The evidence for these observations is easy to find (Schellnhuber and Held 2002). Humans have appropriated 50 per cent of available freshwater and converted 30–50 per cent of the Earth's land surface for their own use. We have increased emissions of sulphur dioxide 200 per cent over background rates and raised concentrations of carbon dioxide in the Earth's atmosphere well beyond anything that has occurred during the last 400,000 years. We have caused species to go extinct at 1,000 to 10,000 times the normal rates.

It follows that the real challenge today is to increase our understanding of the complex dynamics – including the feedback loops – characteristic of coupled systems (Palmer *et al.* 2004). This is much easier said than done. How can we integrate human actions into models of the carbon cycle, the nitrogen cycle or the hydrological cycle (Global Carbon Project 2003)? How should we address the fact that human actions can give rise not only to self-fulfilling prophesies (e.g. the amplification of business cycles) but also to self-denying prophesies (e.g. forecasts of diseases likely to spread in the absence of preventive measures)? The difficulty in addressing questions of this sort will lead some scientists to avoid them on the grounds that they are not ripe for scientific analysis. But this is an unsatisfactory reaction to the situation in which we now find ourselves. To the extent that problems like climate change and variability are emerging as undeniable realities, a refusal to confront the causal mechanisms underlying such problems is asking for trouble in terms of the need to link science and policy.

Fundamental implications

The argument spelled out in the preceding subsections has a number of implications regarding the need to rethink our analytic tools. First and foremost is the proposition that we can no longer focus our attention on biophysical systems, treating the impacts of human actions as minor disturbances or perturbations. Whatever the merits of earlier analyses that proceeded by setting the impact of anthropogenic forces aside by assumption, such procedures are no longer viable. In the growing category of cases in which human actions constitute major drivers, there is no alternative to integrating human actions. This is not to say that there is no room for studies in which human actions are marginal or even irrelevant. Analyses of fluctuations in the sun's radiation or even the succession of ice ages during the course of the Holocene can proceed without an emphasis on human actions. But as soon as we come to large-scale concerns like understanding the future of the Earth's climate system, the omission of anthropogenic forces is no longer an acceptable strategy.

Understanding human-dominated ecosystems will certainly require the development of a new generation of analytic models and methods. Despite the hopes of some members of the research community, there is no way to graft simple and tractable terms covering human actions onto familiar models of biophysical processes. Here, too, the case of climate change and variability is striking. The present 'no analogue' state of the Earth's climate system reflects the impacts of human actions, mostly in the realms of land use change and industrialization (Steffen *et al.* 2002). Human responses to actual climate change – not to mention anticipatory responses to the prospect of climate change – will be determined by a wide range of social, economic, political and legal factors. The way forward in dealing with these forces is anything but clear. However, refusing to tackle matters of this sort is not an option.

One element of an effective programme to address these issues will centre on the roles that institutions play in guiding human/nature interactions (Young *et al.* 1999). Treated as clusters of rights, rules and decision-making procedures that govern human actions in various issue areas, institutions often loom large as causes of problems in human/nature interactions (e.g. situations that lead to what is known as the tragedy of the commons) as well as responses to such problems (e.g. the creation of the ozone regime as a means of addressing the seasonal depletion of stratospheric ozone). In a sense, institutions are transmitters that channel the impacts of human actions towards ecosystems and provide opportunities for human actors to respond to relevant changes (or predicted changes) in major biophysical systems (Berkes and Folke 1998). This linking role of institutions coupled with the perception that institutions are social artefacts subject to purposive adjustment or restructuring has given rise to a surge of interest in improving our understanding of environmental institutions (Ostrom *et al.* 2002).

Three challenges for science

Rethinking our analytical tools is easier said than done. Although acknowledgement of the logic underlying the argument sketched in the preceding section is spreading rapidly, old ways die hard. The potential contributions of reductionist procedures have not been exhausted; mature scientists socialized into familiar modes of analysis typically find it hard to shift to new and unfamiliar research agendas, and finding tractable approaches to analysing the dynamics of coupled systems is genuinely hard. Yet there is no alternative to tackling this agenda if we want to manage coupled human/natural systems in order to avoid the onset of massive problems as we move progressively deeper into the Anthropocene. Three scientific challenges loom especially large in this context: (i) melding diverse forms of intellectual capital, (ii) exploring interactive causal clusters and (iii) designing governance systems that are well matched to the key attributes of the relevant socio-ecological systems.

Melding diverse forms of intellectual capital

The analysis of coupled systems requires a sustained effort to integrate models and methods derived from both the natural sciences and the social sciences. This is not simply a matter of constructing good biophysical models (e.g. MSY models focusing on the population dynamics of fish) together with models of related social phenomena (e.g. models of markets in fish products) and joining the two together by introducing simple linking devices (e.g. stimulus/response mechanisms). Not only are such models based on fundamentally different premises, they also yield results that are difficult to translate into common metrics of the sort needed to generate meaningful conclusions. We know, of course, that population dynamics for specific stocks of fish are impacted heavily by human actions both in the form of intentional harvesting and in the form of unintended by-products of a wide range of other activities (e.g. agricultural and industrial activities generating marine pollution). We also know that human enterprises, such as the performance of specific markets, are impacted heavily by fluctuations in the condition of targeted fish stocks. But this is not enough to allow for the construction of integrated models of socio-ecological systems in which anthropogenic and biophysical forces are operating at the same time.

What is to be done? For starters, it is essential to frame questions that cannot possibly be answered without input from both natural scientists and social scientists. Obvious examples include questions pertaining to the future of coastal communities dependent on harvesting marine living resources or to projections regarding concentrations of carbon dioxide in the Earth's atmosphere in 2020, 2040 or 2060. But beyond this, we are likely to confront a trade-off between resorting to highly simplified models or working with models that are not as tractable as we would like in analytic – and especially mathematical – terms. Simplified models might make use of reaction/process relationships in which various biophysical and social subsystems react to each other's behaviour on the basis of well-defined coefficients (Rapoport 1960). Models of this type require sacrifices in terms of methodological individualism and granularity in order to represent complex systems in crude terms. Models that sacrifice analytical tractability, on the other hand, produce more complex systems that are not amenable to mathematical or other forms of logical analysis. In the end, some combination of these strategies may prove effective. The simplified models can, and often will, turn up phenomena (e.g. emergent properties of coupled systems) that call for much more detailed consideration. These efforts, in turn, may lead to models that can be used to support computer simulations, even when they do not lend themselves to more traditional forms of analysis (Meadows *et al.* 1992). The results, while not predictive, can turn up numerous insights about the behaviour of coupled systems that identify issues that are clearly relevant to policy, even though they do not take the form of predictions in the normal sense (Bolin 1997).

Exploring interactive causal clusters

The causal mechanisms that drive socio-ecological systems do not lend themselves to well-defined forms of deductive and inductive reasoning that are standard in the natural sciences and that are increasingly common in the social sciences. The problem is twofold: (i) any of a number of combinations of driving forces can lead to the same outcome (e.g. the degradation of marine ecosystems, the loss of biological diversity) and (ii) the individual elements in these clusters of drivers are highly interactive. Situations of this sort are difficult to model in a parsimonious fashion, a situation that leads either to the development of tractable models dealing with special cases or to the creation of complex wiring diagrams that are far less tractable. Familiar statistical procedures (e.g. multiple regression) are hard to apply successfully to phenomena of this sort. Measures of association that seek to attribute a proportion of the variance in a universe of cases to individual variables often produce unsatisfactory results in situations of this sort. What we find instead are distinct combinations of circumstances or, as some would say, syndromes, any one of which can lead to the outcomes of interest, such as patterns of land use change or collapses of harvested stocks of renewable resources (Young 2002b; Lambin *et al*. 2003).

What are the options for sharpening our understanding of these complex systems? Thoughtful observers have proposed a variety of methods for analysing such situations, including causal narratives, genetic explanations, counterfactuals, qualitative comparative analysis, agent-based modelling and computer simulations. What is unclear are the relative merits of these methods with particular reference to the subject matter under consideration here. Preliminary research suggests that there are no silver bullets in this domain; we are unlikely to find one or even two or three methods that can produce the desired results in analysing socio-ecological systems. What is more likely is a messier process in which a range of methods are employed on the part of those concerned with the sustainability transition. Under the circumstances, we can measure our confidence in the conclusions flowing from these procedures in terms of the extent to which analyses using different procedures converge on the same conclusions.

Designing governance systems

Institutions or governance systems created to address specific problems arising from the dynamics of socio-ecological systems occupy a critical role in the analytic framework outlined in the preceding section. Not only do they regulate human actions (e.g. the harvesting of living resources, the treatment of wastes or residuals) likely to prove disruptive to such systems, but they also influence both the timing and the nature of human responses to growing evidence of dysfunctions in socio-ecological systems (e.g. degradation of large ecosystems, changes in the Earth's climate system). A structure of rights that forces industrial facilities to engage in full-cost accounting, for instance, may prevent problems like the build-up of greenhouse gases in the Earth's atmosphere from occurring. Similarly, arrangements that monitor biophysical processes closely and include flexible response mechanisms may be able to detect emerging problems early and take effective steps to deal with them before they burgeon into issues like the contemporary problems of climate change and the loss of biological diversity.

The fact that governance systems are social artefacts suggests that there is more scope for (re)structuring them than there is for altering biophysical processes. Yet designing, adopting and implementing institutional arrangements that turn out to be well adapted to the attributes of specific problems is easier said than done (Young 2002a). Misfits are common in this realm. Among those that crop up again and again are regimes whose jurisdiction fails to encompass the relevant biophysical systems, arrangements that emphasize the creation of incentive mechanisms when the real problems centre on matters of equity, and regulatory systems that lack adequate mechanisms to monitor the behaviour of those subject to specific rules. Partly, this is attributable to the facts that many problems giving rise to a demand for governance are extremely complex and that governance systems tend to become path dependent and difficult to adjust even when their shortcomings are widely known. In part, however, these shortcomings are testimony to the fact that our understanding of governance systems – environmental or resource regimes as they are often called in discussions of these issues – is crude. A major scientific challenge during the immediate future for those concerned with the sustainability transition will be to foster substantial gains in our understanding of governance. We cannot expect to come up with simple recipes for crafting regimes to govern the dynamics of complex socio-ecological systems, but major improvements in this area are well within the realm of possibility.

Six lessons for policymaking

What can we learn from this discussion of the sustainability transition that is relevant for policymaking? Two seemingly contradictory responses to this question emerge at the outset. One emphasizes the absence of any basis for assuming we can solve problems of the sort under consideration here through the application of purely scientific procedures and new technologies. Not only do ethical considerations and rent-seeking behaviour intrude on a regular basis, but also the socio-ecological systems involved are far too complex and turbulent to allow for technocratic solutions. At the same time, however, the argument presented in the previous sections of this chapter gives rise to a number of important lessons for policymaking. Even though the sustainability transition does not lend itself to purely scientific or technical treatment, there is great scope for improving the quality of policymaking in this realm. This section draws attention to six lessons that seem particularly pertinent to this effort.

Use multiple sources and types of knowledge to good advantage

There is a marked tendency today to treat Western scientific knowledge as superior to other types of knowledge, a fact that gives those whose thinking is rooted in other approaches to knowledge a strong incentive to cloak their ideas in the terminology of science. But this is not a helpful way to proceed. Not only is it the case that the capacity of science to understand turbulent socio-ecological systems is severely limited, there is also ample reason to conclude that other types of knowledge have much to contribute to our efforts to navigate the sustainability transition. Consider the case of indigenous or traditional ecological knowledge (TEK) as an illustration (Berkes 1999). TEK is largely experiential, typically place-based and generally long on diachronic or longitudinal observations. Those who possess such knowledge may not be able to generalize across a range of situations involving coupled human/natural systems, but they are often able to say a lot about the behaviour of specific socio-ecological systems over comparatively long periods of time. Knowledge of this type is often holistic, despite the fact that it does not rest on the sort of systematic observations that constitute a hallmark of Western science. Grafting TEK onto scientific analyses is apt to be a losing proposition. At best, such efforts provide certain data that are useful from a scientific point of view. Often, the result is a highly unequal encounter in which the most important insights of TEK are lost.

From the perspective of policymaking, the lesson here is that the contributions of different types of knowledge should be embraced on their own terms and that there is much to be said for establishing multiple channels rather than endeavouring to assimilate other types of knowledge into a single stream labelled 'science'. Our ability to forecast the behaviour of dynamic socio-ecological systems is limited; policymakers will have to cope with relatively high levels of uncertainty under the best of circumstances. In such a setting, it makes sense to take advantage of insights arising from different sorts of knowledge in weighing the pros and cons of a range of policy options. When these insights converge in the sense of producing similar assessments of the relative merits of individual options, policymakers can have greater confidence in proceeding to make their choices. In contrast, when the assessments flowing from different types of knowledge diverge, policymakers will want to be particularly careful in weighing the relative merits of the options available to them. The cautionary implications of divergence may prove just as valuable to the policy process as the confidence-building implications of convergence.

Devise effective responses to uncertainty

Despite our best efforts, high levels of uncertainty constitute a fact of life in efforts to manage coupled human/natural systems. Recasting our thinking about ecosystems to consider turbulent and often non-linear processes is hard enough. But, when we add in the central role of human actions as driving forces in socio-ecological systems, an appreciation of the critical role of uncertainty is unavoidable. Of course, science strives to reduce uncertainty, and there is a convincing case for providing strong support for scientific work that can reduce the impact of uncertainty in this realm. But it is pointless to cling to naive hopes in this context. Despite our best efforts to reduce uncertainty, there is no escaping the fact that decision making under uncertainty is, and will continue to be, a central feature of the policymaking process (Kahneman and Tversky 2000).

What are the practical implications of this conclusion? As uncertainty rises, the value of adopting a precautionary approach to matters of policy becomes more apparent. Partly, this is a matter of taking error bars seriously, considering outcomes that may seem relatively unlikely but that would have highly disruptive or costly consequences should they occur. Thinking carefully about the consequences of rapid climate change events is a case in point. It is not necessary to attach a high probability to their occurrence to take such events seriously. The argument here is much like the rationale underlying the use of worst-case analysis in the realm of national security. In part, this lesson has to do with maintaining a clear understanding of the gap between policies on paper and policies in practice (Ostrom 1990). Policies are often implemented selectively, if at all. Subjects frequently misinterpret rules or simply refuse to comply with inconvenient requirements. Unforeseen and unintended side effects occur with great regularity. Under the circumstances, it is essential to think carefully about uncertainties relating to the anthropogenic side as well as to the biophysical side of socio-ecological systems.

Take long-term consequences seriously

It is easy – some would say perfectly defensible – to devote little attention to or even to ignore the longer-term consequences of individual policy options. The familiar procedure of discounting in benefit/cost calculations exemplifies this proposition. Even moderate discount rates suggest that it makes sense to dismiss consequences occurring beyond a couple of decades into the future. And the incentives to focus on the short term arising from the pressures of electoral cycles and other political processes are well known. The need to win the next election is a political imperative that can and often does have profound impacts on the policymaking process. Investments today that may produce enormous benefits in the long run are often avoided on the grounds that voters have a low tolerance for accepting burdens that will not produce short-term gains and that may, in some cases, produce benefits accruing largely to members of future generations.

Yet socio-ecological systems are path dependent. Not only is it difficult – sometimes impossible – to reverse course once we start down a particular path, but also what might have been easy and inexpensive to alter at an earlier stage can become prohibitively expensive to change later on. Residential patterns and associated transportation systems in the United States provide a dramatic example of the power of path dependence. A disbursed pattern of single-family homes coupled with a reliance on private automobiles as the preferred mode of transportation has long seemed attractive to many Americans. But no one foresaw the environmental implications of this developmental path during the middle decades of the twentieth century. And the costs of changing course today regarding a socioeconomic system of this scope would be enormous. Needless to say, it is generally difficult to foresee the longer-term impacts of current choices. But the power of path dependence should provide us with ample reason to pay attention to the future, whatever the attractions of discounting or the pressures to adopt short-term perspectives arising from the imperatives of electoral politics.

Create sensitive monitoring systems

The combination of uncertainty about the behaviour of turbulent systems and the power of path dependence makes it essential to monitor the results of policies closely. If we could forecast future developments confidently or if there was no reason to care about long-term consequences, we could dispense with monitoring. But neither of these conditions holds. We must expect surprises in the form of unforeseen occurrences that can have far-reaching impacts on human welfare as well as on the welfare of other species and ecosystems. And given the normal volatility of socio-ecological systems, early warning is highly desirable. Although early warning offers no guarantee that we will be able to intervene effectively to prevent disruptive occurrences, warning of this sort is clearly a necessary condition for successful adaptation.

In dealing with socio-ecological systems, there is a need to monitor both biophysical processes and human actions. Consider the case of climate change in this light. We need to track not only rates of change in the concentrations of carbon dioxide and other greenhouse gases in the Earth's atmosphere but also rates of change in emissions of such gases resulting from human actions. Increases in anthropogenic emissions are a critical force changing atmospheric concentrations of greenhouse gases, and the resulting impacts on the Earth's climate system constitute the fundamental concern regarding human interferences in the climate system. To determine the scope of the problem we need to monitor the relevant biophysical processes. But to develop appropriate responses, we need to track the human actions involved so that any measures we devise to address the problem can be designed in such a way as to target the most important and most malleable of these actions.

Encourage social learning

Adaptive management, an idea that is currently popular, is undoubtedly a useful addition to our analytic toolkit (Lee 1993). It recognizes the dynamism of coupled systems and stresses the importance of creating governance systems that are designed to be sensitive to and adjustable in the light of changing circumstances. The regime created to protect the stratospheric ozone layer is a case in point. It allows the parties to agree on accelerated phase-out schedules for ozone-depleting chemicals already subject to regulation (e.g. most chlorofluorocarbons or CFCs) without requiring a laborious process of ratification on the part of the member states. Useful as it is, however, adaptive management emphasizes the value of making adjustments at the margin without calling into question key elements of prevailing governance systems. In dealing with socio-ecological systems that are turbulent and subject to regime shifts that occur over relatively short periods of time, we need to go beyond adaptive management to emphasize social learning as a mechanism for making more extensive changes in the character of the institutions we rely on to manage interactions between biophysical systems and social systems (Social Learning Group 2001). Wherever rates of change in biophysical systems exceed the normal capacity of social systems to respond in a timely manner, the need for social learning is apparent. What is called for in such cases is an enhanced capacity to make major changes in institutional arrangements to ensure that the gap between biophysical changes and social responses does not become too great.

Although the need for social learning is easy to document, it is often difficult to achieve success in this realm. What exactly is social learning and how does it work? Simple examples of this type of learning involve collective but unguided shifts in word usage and vocabulary more generally as well as well-defined changes in norms regarding such matters as smoking in public places. As these examples suggest, social learning can – but need not – occur in the absence of conscious efforts. The need for social learning in conjunction with socio-ecological systems arises, in large part, from the growing human domination of these systems. Whereas it once made sense to focus on human adaptations to more or less uncontrollable biophysical processes, we now find ourselves in a situation in which human actions are major driving forces in these systems. Although there is much to be learned about the biophysical processes involved, the key to success at this stage lies in learning how to manage human actions that pose problems for socio-ecological systems, e.g. emissions of greenhouse gases or activities that degrade habitat critical to endangered species. A primary focus of attention in this realm is the importance of learning how to design and implement institutions or governance systems to regulate human actions that disrupt biophysical processes.

Treat crises as periods of opportunity

Crises are periods of danger. Not only do they call into question systems that had been considered immutable previously, they can also compress into short time periods changes that would normally take decades to occur. Sometimes crises are largely socio-political in character. The fall of the Berlin Wall and the collapse of the Soviet Union during the period 1989–1991 are striking cases in point. Similar bursts of change, triggering what observers often call regime shifts, are common in biophysical systems as well. A contemporary example involves rapid changes in the Bering Sea ecosystem in recent years (National Academy of Sciences 1996). Scientists examining the turbulence of this system talk about cascading changes and turn to the idea of regime shifts in their efforts to understand what is happening in this large marine ecosystem. Both examples feature changes unfolding over short periods of time. Yet our understanding of these crises is limited. We regularly fail to anticipate fundamental and fast-paced changes in both social systems and biophysical systems. Even after the fact, we are often left with more questions than answers regarding the causal mechanisms that trigger crises and produce regime shifts.

Yet crises are also periods of opportunity. They can break logjams that arise from path dependence built up over long periods of time, creating the conditions needed to restructure prevailing governance systems in fundamental ways in contrast to making small adjustments at the margin. The resultant windows offer opportunities to tackle major problems, but they seldom remain open for long. The lesson here centres on the importance of devoting time and energy in advance to thinking through the pros and cons of alternative governance systems. As the case of the former Soviet Union makes clear, a failure to understand important features of changes (e.g. the introduction of Western-style markets) can prove immensely costly. Actors playing important roles in socio-ecological systems often find their resources stretched thin in the effort to cope with day-to-day challenges. But the advantages of thinking about the pros and cons of alternative arrangements before crises erupt will become increasingly evident as experience with the dynamics of socio-ecological systems grows.

From science to solutions

The argument sketched in this chapter has several implications for the science–policy interface. To achieve relevance in dealing with socio-ecological systems science must accept certain compromises, reducing its reliance on reductionist techniques of analysis and devoting more attention to complex systems that are turbulent, prone to regime shifts and likely to give rise to irreversible changes. Success in this endeavour will almost certainly go to teams of natural scientists and social scientists who find ways to overcome prejudices and conceptual barriers that often thwart efforts to foster the collaboration needed to deepen our knowledge of socio-ecological systems. This is a huge challenge; many efforts to come to grips with it are likely to fail. Even so, there are grounds for optimism. The work of the Earth System Science Partnership in launching crosscutting projects dealing with large-scale, socio-ecological systems is promising in this regard. To take a single example, the Global Carbon Project, guided by a steering committee including a mix of individuals with expertise in atmospheric, oceanic, social and terrestrial systems, has launched a number of promising initiatives (Global Carbon Project 2003). If this project makes breakthroughs in efforts to understand the carbon cycle as a coupled human/natural system, it will immediately assume a prominent role in the quest for progress towards the sustainability transition.

If the analysis of coupled systems in the preceding sections is persuasive, it will also strengthen the hand of those who argue that we must devote more energy to understanding the nature of environmental regimes or governance systems. These institutional arrangements sit at the intersection of ecological and social systems, endeavouring to steer the interactions between the elements of these coupled systems in the interests of avoiding destructive outcomes and enhancing the prospect of managing socio-ecological systems on a sustainable basis. Recent years have witnessed a substantial growth of interest in analyses of the determinants of success and failure regarding these institutional arrangements. But this field of study is still in its infancy; there is much to be done to amass a body of knowledge about environmental regimes capable of offering real help to those responsible for creating and implementing a wide range of governance systems intended to contribute to the sustainability transition (Underdal and Young 2004).

Finally, it is important to bear in mind common pitfalls of the science–policy interface. Nothing is more common than the manipulation of scientific findings to suit the needs of influential actors in the policy process. Scientists themselves often lack both the time and the inclination to enter the arena of policymaking. This suggests that there is a critical role for 'knowledge brokers', individuals who are comfortable both in the world of science and in the world of policymaking and who are able to explain the implications of scientific advances in a manner that is accessible to a lay audience (Litfin 1994). Of course, the presence of such individuals offers no guarantee that the integrity of scientific knowledge will remain intact as it enters the arena of policymaking and politics. But efforts to integrate science and policymaking will surely be critical during the foreseeable future if we are to succeed in navigating the sustainability transition.

Note

1 This chapter is based on a presentation at the National Capital Colloquium on the Governance of Sustainable Development, Ottawa, April 2004.

References

Berkes, F. (1999) *Sacred Ecology: Traditional Ecological Knowledge and Resource Management*, Philadelphia, PA: Taylor and Francis.

Berkes, F. and Folke, C. (eds) (1998) *Linking Social and Ecological Systems*, Cambridge: Cambridge University Press.

Bolin, B. (1997) 'Scientific assessment of climate change', in G. Fermann (ed.) *International Politics of Climate Change*, Oslo: Scandinavian University Press.

Crutzen, P. and Stoermer, E. F. (2000) 'The Anthropocene', *IGBP Newsletter*, 41:17–18.

Global Carbon Project (2003) *Science Framework and Implementation, Earth System Science Partnership (IGBP, IHDP, WCRP, Diversitas), Report No. 1*, Canberra: Global Carbon Project.

Golley, B. (1993) *A History of the Ecosystem Concept in Ecology*, New Haven, CT: Yale University Press.

Gunderson, L. H. and Holling, C. S. (2002) *Panarchy: Understanding Transformation in Human and Natural Systems*, Washington, DC: Island Press.

Kahneman, D. and Tversky, A. (eds) (2000) *Choices, Values, and Frames*, Cambridge: Cambridge University Press.

Lambin, E. F., Geist, H. J. and Lepers, E. (2003) 'Dynamics of land-use and land-cover change in tropical regions', *Annual Review of Environmental Resources*, 28:205–41.

Larkin, P. A. (1977) 'An epitaph for the concept of maximum sustainable yield', *Transactions of the American Fisheries Society*, 106:1–11.

Lee, K. N. (1993) *Compass and Gyroscope*, Washington, DC: Island Press.

Ludwig, D., Hilborn, R. and Walters, C. (1993) 'Uncertainty, resource exploitation, and conservation: Lessons from history', *Science*, 260:17 and 36.

Meadows, D. H., Meadows, D. L. and Randers, J. (1992) *Beyond the Limits: Confronting Global Collapse, Envisioning a Sustainable Future*, Post Mills, VT: Chelsea Green Publishing Company.

National Academy of Sciences (1996) *The Bering Sea Ecosystem*, Washington, DC: National Academy Press.

Ostrom, E. (1990) *Governing the Commons: The Evolution of Institutions for Collective Action*, Cambridge: Cambridge University Press.

Ostrom, E., Dietz, T., Dolšak, N., Stern, P. C., Stonich, S. and Weber, E. U. (eds) (2002) *The Drama of the Commons*, Washington, DC: National Academy Press.

Palmer, M., Bernhardt, E., Chornesky, E., Collins, S., Dobson, A., Duke, C., Gold, B., Jacobson, R., Kingsland, S., Kranz, R., Mappin, M., Martinez, M. L., Micheli, F., Morse, J., Pace, M., Pascual, M., Palumbi, S., Reichman, O. J., Simons, A., Townsend, A. and Turner, M. (2004) 'Ecology for a crowded planet', *Science*, 304:1251–2.

Rapoport, A. (1960) *Fights, Games, and Debates*, Ann Arbor, MI: University of Michigan Press.

Schellnhuber, H.-J. and Held, H. (2002) 'How fragile is the Earth system?', in J. C. Briden and T. E. Downing (eds) *Managing the Earth*, Oxford: Oxford University Press.

Social Learning Group (2001) *Learning to Manage Global Environmental Risks*, Cambridge, MA: MIT Press.

Steffen, W., Jäger, J., Carson, D. J. and Bradshaw, C. (2002) *Challenges of a Changing Earth*, Berlin: Springer-Verlag.

Underdal, A. and Young, O. R. (eds) (2004) *Regime Consequences: Methodological Challenges and Research Strategies*, Dordrecht: Kluwer Academic.

Vitousek, P. M., Mooney, H. A., Lubchenco, J. and Melillo, J. M. (1997) 'Human domination of Earth's ecosystems', *Science*, 277:494–9.

Young, O. R. (2002a) *The Institutional Dimensions of Environmental Change: Fit, Interplay and Scale*, Cambridge, MA: MIT Press.

—— (2002b) 'Are institutions intervening variables or basic causal forces? Causal clusters vs. causal chains in international society', in M. Brecher and F. Harvey (eds) *Millennial Reflections on International Studies*, Ann Arbor, MI: University of Michigan Press.

Young, O. R., Agrawal, A., King, L. A., Sand, P. H., Underdal, A. and Wasson, M. (1999) *Institutional Dimensions of Global Environmental Change (IDGEC) Science Plan: IHDP Report No. 9*, Bonn: International Human Dimensions Programme.

Chapter 6

Arctic environmental security and abrupt climate change

Chad Michael Briggs

Introduction

The past two decades have witnessed increasing concern over new security threats and risks, from environmental issues to terrorism and economic instability. The traditional definitions of national and international security, based upon narrowly defined and easily measured metrics of violent interstate conflict, have proven inadequate for describing the spectrum of risks that the international community now faces. The shifts in security discourse and policy since the end of the Cold War have affected academic, policy and intelligence communities, while increasingly engaging certain scientific communities that had previously little contact with or interest in issues of international security. Policy interest in environmental security since 1989 has rested upon both a need to define new operational missions for existing security forces and growing realization that environmental changes may bring overwhelming pressures to bear upon critical systems and vulnerable regions.

Arctic security had long been defined by the Cold War as the naval and aerial proving ground between NATO and Warsaw Pact forces. Yet even during those times, concerns about the environment persisted and grew. Sweden was the host of the first Earth Summit in 1972, and the Nordic contributions to environmental policy (both at the European and international levels) have long been recognized (Hey 1998). Security and the environment has also long been intertwined, whether in the work of Belonna and its concerns regarding nuclear waste in Arctic waters, to Sweden's detection of radiation from the 1986 Chernobyl nuclear accident in the Soviet Union. The continued focus of Nordic states on environmental questions related to human well-being and international development concerns have resulted in perspectives distinct from traditional American definitions, as exemplified by the 2008 Swedish report on climate and security (SIDA 2008). Definitions may have shifted since the 1980s, but still the Arctic remains a key region of interest for environmental security. Much attention has been paid lately to the geopolitical ramifications of Arctic sea ice melt, particularly in light of record sea ice melt in 2012 and accelerated prospects of an ice-free Arctic (Borgerson 2008; Zhang *et al.* 2013). Yet more crucial may be the role of the Arctic (and Antarctic) in potentially abrupt climate system changes. The purpose of this chapter is to explore the key role the Arctic plays as a geographic locus for catastrophic environmental risks, and how these present new and untraditional security risks. Exploration of both topics has implications not only for the Arctic states, but for how cascading effects of climate change may have severe security impacts on the rest of the world.

Security and climate change

Connections between environment and security began with medical concerns over nuclear weapons tests in the early 1960s, but the debate over environmental determinants for security concerns dates to the mid-1980s. Dominant among earlier 1990s studies were those postulating that increased resource scarcities would lead directly to violent conflict between states, often as a result of population increases in less-developed countries (Homer-Dixon 1991, 1994). Such scarcity-conflict models relied upon traditional models of security as interstate conflict, and largely assumed linear relationships in terms of both causality and decision-making. Academic criticisms highlighted that scarcity does not necessarily result in conflict, that such causal relationships were nearly impossible to substantiate even post-facto and that a focus on the state level misleadingly ignored interstate economic relationships that exploited natural resources from afar (Gleditsch 1998; Hauge and Ellingsen 1998; Buhaug *et al.* 2008). From a policy perspective, the scarcity-conflict theses could result in a form of paralysis, as conflicts were blamed on natural conditions and population levels in foreign countries, with little if any direct connection to wealthier, western states (Kaplan 1994). Although general development may help in the long term, the logical response to scarcity-conflict explanations was to bolster border defences and justify the issue as an external problem. Even those concerned with cross-border environmental issues in environmental security tended to downplay the potential role of climate change (Homer-Dixon 1991).

An alternative set of views was promoted by Canadian and Nordic scholars and governments, as exemplified by the Global Environmental Change and Human Security (GECHS) project based in Oslo, the work on resilience and the environment in Stockholm, or International Polar Year projects examining impacts of industry upon the environment and human security (GAPS and MODIL-NAO). Research in this field (Bazely *et al.*, this book) emphasized the human dimensions of environmental change, whether in terms of economic relations, human geography or the interconnections between global shifts and human existence. Such studies attempted to shift the focus from the state level and traditional measures of interstate violence, illuminating the importance of interrelationships at more local levels. These approaches to security, whether associated with Oslo, Stockholm, Copenhagen, Tromsø, Helsinki or Ottawa, have allowed a space in security studies to examine climate change impacts without the inevitable contradictions that state-level security would bring to analyses.

Despite popular intuition that climate change issues affect human and international security, it remains difficult to conceptualize what these connections may be, particularly for a global process so rife with uncertainty. Without a useable framework for analysis, policy action and discussion will remain difficult, and a form of policy paralysis may result (Barnett 2003). The following sections explain briefly the non-intuitive ways that climate systems act and how such complex networks also relate to alternative definitions of vulnerability and risk assessment.

Systems theory

In contrast to the linear systems assumed in the environmental security literature of the 1990s, ecological systems (including the global climate) are better understood as complex emergent systems. The units of environmental systems, however defined, are not nearly as important as the relationships and networks among a system's components. Properties of the system cannot be determined simply by reference to its components, nor can the future state of a system be understood in reductionist terms. Change can occur while maintaining the integrity of the system, but the system may shift to multiple points of stability. Such shifts, as with eutrophication of lakes (when nitrogen added to water, often from fertilizers, results in sudden algae blooms), may occur quite suddenly and with little indication that conditions may suddenly 'worsen' (Walker and Salt 2006). Likewise, palaeo-climatological studies have indicated that atmospheric temperatures can shift very suddenly, perhaps as much as 10–20 degrees Celsius within a few years. Sudden temperature shifts similarly occurred 11,500 years ago at the end of the Younger Dryas Period, when temperatures in Greenland dropped 19 degrees Celsius in less than 20 years (National Research Council 2002; Alley *et al.* 2009).

What are less well understood are the 'tipping points' in such systems, technically known as *catastrophe sets*. How far can a system be pushed before it shifts to a new level of stability? What are the most relevant relationships? In climate terms, the most pressing question is concerned with what amount of greenhouse gases (GHGs) the global climate system can absorb before a large-scale shift in climate stability occurs (Lenton 2011). Atmospheric temperatures may rise gradually over the years, as per the standard Intergovernmental Panel on Climate Change (IPCC) projections, but then suddenly rise (or fall) precipitously. The National Research Council's (2002) report on abrupt climate change described several qualities of a system that creates 'abruptness'. First, the system is non-linear, and shifts from one condition (often measured as temperature) to another occur rapidly, perhaps within a number of years. Second, this change is irreversible, as measured by human time scales, a condition made even more likely by the increases in atmospheric irradiance caused by GHG emissions and atmospheric carbon dioxide (CO_2) lifetimes. Third, the changes may occur due to second-order effects of the global system, as positive feedback loops originally unrelated to the more obvious forcing (e.g. Arctic methane gas release caused by melting permafrost) (Marshall 2013). Finally, abrupt changes are often categorized as such due to the inability of related systems to adapt, leading to the possible description 'dangerous' climate change. The danger does not necessarily stem from threats of violent conflict between people. Climate change is very unlikely to lead directly to conflict (Gleditsch 1998; Floyd 2008), but may adversely affect social, political and economic systems at varying levels, and these overlapping systems may contain feedback loops that accelerate the stability shifts.

Most issues concerning resources and conflict stem not from changes in the environment as a root cause, but rather, the failure of political and economic systems to provide adequate resources or adaptation measures. At times such failure can be deliberate, as with resilience and livelihood targeting during violent conflicts (Brown 2004; Briggs *et al.* 2009). At other times these failures can be second- or third-order effects, such as the breakdown in food and health security experienced in Zimbabwe. The effects of climate change in such situations have been conceptualized as 'threat multipliers', conditions that exacerbate risks and make adaptation more difficult (CNA Corp. 2007), but not conditions that could be understood as the root causes of conflict. The concept of abrupt climate change is both quantitatively and qualitatively distinct, in that such sudden shifts in environmental conditions will not merely emphasize existing inequalities and conflicts. Vulnerable systems, be they ecological, political or economic, may 'fail' completely should environmental conditions shift much more quickly than adaptation allows (Gallopín 2007). Failures of systems need not be understood in simple, binary terms often used to describe failed states in political science. Rather, unstable systems will find a new, often lower, level of stability and functioning, often resulting in severely negative consequences for certain components or populations.

Environmental issues are not necessarily more complex than the socio-political conditions that give rise to violent conflict, but it is far more difficult to attribute intentions or divination of rational thought to how conditions change. Global environmental conditions are also inextricably linked to social, political and economic systems, which can further compound uncertainty when analysts would prefer to study systems as discrete collections of isolated variables. Thus, climate change is not merely the interaction of a few variables, where a simple, linear relationship exists between levels of atmospheric greenhouse gases and average atmospheric temperatures. Rather, the global climate is a complex system, exhibiting emergent properties and influenced by numerous feedback effects, none of which can easily be predicted in advance. Just as with ecological systems, the climate may rest upon multiple points of stability, but these stable levels may be unsuitable for both existing human and environmental adaptation.

The concept of 'abrupt' therefore applies not only to the sudden shifts in average air temperature, but also adaptive capacity and our ability to anticipate such changes. The ability of people to respond to environmental change depends, at least in part, on the congruence between the rate of expected change and the actual physical changes that affect a particular group (Barnett 2003; SIDA 2008). Ecosystem resilience does not depend upon perception of the problem, but is very much tied to both rate and extent of change. Overall, there are four primary components to the concept of vulnerability, and to understanding what makes a particular system more or less able to adapt to changing circumstances. Responses to environmental changes can be both positive and negative, and the ability of a system to react positively or adequately (i.e. the system does not suffer an overall decline) does not merely depend upon the most visible characteristic of the system in question. Resilient ecosystems may have high or low numbers of species, economic systems may be more vulnerable as wealth increases, and adaptation to climate change may depend upon factors not yet well understood.

Vulnerability: risk, sensitivity, resilience and fragility

Policymakers and analysts perhaps assume that regions like Africa are more vulnerable to climate changes, and recent reports have focused on lesser-developed regions to create risk scenarios for climate change and security (Brown *et al.* 2007). This estimation of vulnerability is true to an extent, but only by one aspect of the components of vulnerability. As security implications are based upon notions of vulnerability to climate change, it is notable that often a full definition of vulnerability is omitted from analyses, and what is meant by these terms makes little reference to previous work in risk and ecology. Vulnerability is a general term for risk from environmental change, but is constituted from various traditions of research, such as the biophysical risk/hazard approach, political economy and ecological resilience. Vulnerability definitions for policy can usefully use hybrid models (see Eakin and Luers 2006) to account for geophysical impacts on ecological systems and related social networks, by breaking vulnerability into four main components of risk/hazard, sensitivity, resilience and fragility measures. These categories reflect recent efforts by the defence community in the US and UK, and related work at the Stockholm Institute, Swedish Defence Research Institute (FOI) and others (see also Adger 2000; Brooks 2003; Wisner *et al.* 2005; Haines *et al.* 2006; Lantze and Raven-Roberts 2006; deFur *et al.* 2007; Füssel 2007; Gallopín 2007; Briggs and Bath 2009).

The first component of vulnerability is risk, or the probabilistic measure of adverse outcomes to which a particular group is exposed. Often the most traditional measure of vulnerability, the extent of risk is a function of probability of hazard and the exposure level (R=[H, E]). In some iterations, this function is modified by the level of 'vulnerability', meaning the extent to which the effects of hazards can be mitigated and withstood. In such a construction, however, the hazard becomes something external to the group in question, a force to be withstood rather than part of the complex, adaptive capacity of the population. Rather than make such vulnerability an external modifier of risk, for purposes of clarity we define this variable instead as sensitivity, the second component of vulnerability.

Sensitivity is the degree to which a group is affected by the occurrence of a hazard, or how far from a baseline livelihood one is pushed. Sensitivity can be a result of particular geography, such as a community placed on sandy hillsides where landslides are common, and it may be affected by factors such as physical characteristics, cultural practices, age or past negative pressures. Climate changes can interact with sensitivity measures along multiple pathways, and can create situations of positive feedback loops that create even greater vulnerability. Certain groups, for reasons of geography, age, livelihood or cultural practices, can be much more sensitive to environmental changes. The degree of exposure to climate changes alone cannot be used to predict impacts (i.e. temperature changes alone do not indicate harm), but past exposure to disease can increase the sensitivity of populations to future exposures of similar or related pathogens. Sensitivity depends not merely on location and exposure, but often is affected by numerous small changes in the underlying support structure of the group. Dynamic systems such as social groups can be highly sensitive to small pressures, amplifying dynamic pressures into generating large effects. Low sensitivity can be the result of robust systems that dampen such pressures, providing negative feedback loops that actually strengthen collective response to risks.

The mitigating factor in vulnerability is known as resilience, or the ability to return to a baseline condition in a reasonable time period following exposure to a hazard risk. Resilience cannot be measured in terms of material wealth of aggregate individuals, but is rather a qualitative and emergent measure of networks and social capital, such as the ability and willingness to reallocate resources to assist those in need. Resilience is therefore often not measured at the individual or state level, but rather at intermediate levels of analysis that include families, neighbourhoods, cities or larger collectives within society. Health studies have demonstrated that groups with the largest stocks of social capital are most likely to recover from outbreaks of illness, while recurring illnesses leave neighbourhoods and cities vulnerable to any number of additional stresses, often creating a downward spiral of vulnerability.

The last factor in vulnerability is fragility, a measure that represents the variable nature of the factors above. Rather than conceive of resilience as a 'stock' that merely exists and is drawn down over time, fragility represents the extent to which a group can be stressed before its underlying resilience and support networks are permanently weakened. Resilience is an emergent property of social networks and interactions, and a society's ability to respond from taxing events may be permanently damaged if pushed beyond a particular limit state. Once this limit state is exceeded, the ability of groups to respond and adapt is undermined in crucial ways, possibly resulting in the society falling to a lower level of stability. Fragility is not a matter, as some political analyses would suggest, of a binary stable/unstable categorization, but rather describes how vulnerable a group is to a stability shift (catastrophe set) to an entirely different set of relationships and adaptive capacity. When combined, these factors of vulnerability help to describe those system components or conditions that need to be assessed if we are to understand potential security risks of abrupt climate change. One can begin by considering the geographic region where the climate system is most sensitive and potentially fragile, the Arctic.

Abrupt climate change risks in the Arctic

The Arctic may provide a useful illustration of how foresight of environmental conditions and knowledge of non-linear effects can contribute to a greater understanding of potential impacts. Rather than examine global models of average climate change, a number of potential tipping points exist in Arctic ecosystems that can trigger abrupt climate changes worldwide, and these are reviewed here.

Greenland and Arctic ice melting

The geopolitical implications of loss of Arctic summer sea ice are potentially profound, but are beyond the scope of this assessment. From an ecological perspective, loss of sea ice will affect global albedo measures and wildlife, but would not have any effect on either sea level or freshwater intrusion into deep ocean currents. The more problematic issue for Arctic climate systems is the stability of the Greenland ice sheet (GIS), a frozen reservoir of fresh water locked into 2.85 million km^3 of ice that has existed since the late Pliocene period over 100,000 years ago. Should the entire ice sheet melt, the resulting melt water would raise global sea levels by approximately 7 metres (23 feet), while significant melting may affect the salinity and stability of the thermohaline circulation (THC) that drives the warm waters of the Gulf Stream (Alley *et al.* 2003).

Original estimates of global climate change assumed the long-term stability of the Greenland ice sheet, indicating that any significant melting of the ice sheet would take hundreds of years, if not millennia (IPCC 2001). More recent observations of ice melt indicated that the rate of net runoff was far higher than expected, and that loss of GIS was accelerating rapidly. The melt rate exceeded possible scenarios from ice sheet models, and scientists soon discovered that water intrusion to the glacial bases through moulins (cavities worn into the glaciers) resulted in lower friction between the glacial and underlying rock, effectively lubricating the movement of ice over stone. Other non-linear dynamics were also at work, such as force imbalances between calving sections of the glacier and inland sheets. The resulting changes in conditions, combined with higher summer temperatures in Greenland of some 2 degrees Celsius, have resulted in significant reductions in glacial cover and thickness in some regions (especially coastal areas, where summer temperatures are often above freezing). More recent estimates of GIS stability indicate that the GIS itself may largely disappear by the end of the century, an abrupt shift that is matched by past climatological records (Rignot and Kanagaratnam 2006; Solomon *et al.* 2007).

The potential impacts of GIS melting are twofold. The first and more obvious concern is rising sea level, a risk that has been downplayed in standard IPCC projections at least until the end of the twenty-first century (IPCC 2001, 2007). In the potential abrupt climate change scenarios of sudden warming, accelerated ice melt from Greenland could raise sea levels by anywhere from 1–4 metres in the short term or decadal measure, and up to 7 metres in a worst case scenario (Alley *et al.* 2003). Historical climate records indicate that abrupt rises in sea level are possible, and would cause obvious and severe dislocation in coastal areas globally. The areas at most risk of coastal flooding or inundation are easily mapped, although the cascading effects of large-scale dislocation must be assessed and measured according to regional social, economic and political systems.

The second impact of GIS melt is the possible interaction between a sudden influx of freshwater into the north Atlantic, and the stability of the THC of ocean currents. The THC consists of global undersea currents, moving vast quantities of energy from warmer to cooler regions, and past climate shifts are attributed to cessation of such energy movements. Climate researchers have hypothesized that the sudden cooling recorded at the end of the Younger Dryas Period 11,400 years ago, a drop of 10 degrees Celsius over a decade, was caused by the sudden release of melt water from the Saint Lawrence in North America (Alley *et al.* 2009; Rabe *et al.* 2011). The fresh water released into the north Atlantic decreased the density and salinity of the waters at the northern edge of the warm Gulf Stream, preventing the water from sinking. This action effectively shut down the global THC, and may have been the cause of the sudden, global cooling of the period. Recent scientific reports have indicated that the northern extent of the Gulf Stream has been shortening, with regional impacts of more severe winters in the British Isles and Scandinavia, although it is difficult to discern long-term trends from natural variation (Bryden *et al.* 2005). The risk of global warming turning to sudden global cooling is plausible, considering past and geologically recent climatological records, and the global impacts would be severe in terms of food security and shocks to fragile ecosystems (Schneider 2003).

Terrestrial and oceanic methane releases

Much of the focus on global GHG emissions remains on carbon dioxide, but increasing attention is being paid to methane (CH_4) concentrations in the atmosphere. Methane is released in greater quantities in the northern latitudes than elsewhere, and the rate of release has been increasing over the past decade. Explanations for the increase in methane releases initially rested on terrestrial release from melting lakes and permafrost. As warming trends increase more rapidly in the Arctic than elsewhere, ground-cover and lakes begin to melt, thereby releasing methane as a by-product of biotic decomposition. As methane cannot be released while these sources remain frozen, any warming can create a positive feedback effect of increasing methane release and increasing warming effects, which then cycle back into increasing warming. This so-called Clathrate Gun Hypothesis is thought to have contributed to or triggered the warming periods of the late Quaternary period (Kennett *et al.* 2002).

The Arctic areas of Siberia and North America contain large carbon deposits in the form of peat bogs and decomposed organic matter, and scientists estimate that thawing of the region may significantly increase Arctic methane contributions to the global carbon balance (Lawrence *et al.* 2008; Shakhova *et al.* 2010). As methane is over 20 times more effective than carbon dioxide as a greenhouse gas, as measured over a 100-year period, releases have a potentially disproportionate impact on climate change risks. Estimates of natural contributions from only a few years ago have already been revised upwards by approximately 50 per cent, and researchers warn that new 'surprises' in methane sources and feedbacks are likely in the years to come (Walter *et al.* 2006, 2007).

One such surprise has come from the Arctic oceans, particularly the shallow seas over the continental shelves (e.g. the Barents Sea). The Arctic seas contain large amounts of methane hydrate deposits, also known as methane clathrate or methane ice, a form of frozen water containing large amounts of methane within a crystal structure. This form of methane deposit is created by the decomposition of organic matter in low-oxygen marine environments, where a combination of low temperatures and high pressure create the hydrates. The Arctic deposits alone contain up to 400 gigatons of carbon, compared with the 700 gigatons of carbon present in the atmosphere. These deposits were formerly believed to be highly stable, and the IPCC (2007) report only made passing mention of their existence. Recent research in 2008, however, has indicated that continental marine deposits may be highly sensitive to changes in water temperature, and that high levels of methane are now being released from areas such as the Barents Sea and Arctic Ocean north of Russia (Rigby *et al.* 2008; Schiermeier 2008; Shakhova *et al.* 2010).

Should even a relatively small proportion of the Arctic methane hydrate deposits be disturbed, this process may create a powerful, positive feedback effect that would have substantial effects on climate change globally. Moreover, release of marine hydrate deposits would accelerate the release of CH_4 from northern permafrost, amplifying a process that, by itself, would be unlikely to trigger abrupt climate change. One outstanding problem is that the research into carbon releases in the Arctic is fairly new, and it will likely take years for conclusive research to establish the thresholds for methane hydrates. Whether such thresholds may have already been crossed by the time that the research becomes conclusive is another question, and one that must be posed to policymakers. These potential events in the Arctic (Greenland melting, Arctic sea ice, methane releases) were not included in the 2007 IPCC report as potentially significant influences of feedbacks in the global climate (IPCC 2007:372). Consequently, they are also missing from more recent security reports on climate change, which rely almost entirely on IPCC data to substantiate potential risks and to estimate future conditions.

Tipping points, cascading effects and impacts

Once the initial tipping points in certain systems are better understood, and certain abrupt change scenarios can be played out in terms of possible events, we can assess the potential impacts on various related systems globally. The cascading effects enter where change in one system (e.g. the Greenland ice sheet) results in conditions (rapid rise in sea levels) that have varying impacts elsewhere depending upon the vulnerability of the geographic region (e.g. sensitivity to flooding from low-lying coastal land), social systems (resilience to forced migration) and economic systems (fragility of infrastructure), all applied at scalable levels and according to analytical needs. Disruptions in social and economic systems then have their own cascading effects on related, complex systems.

Ecological systems

Abrupt changes in climate conditions have immediate impacts on environmental systems. Such abrupt shifts have occurred in the geologic past, forcing changes in habitable zones for flora and migratory routes for fauna, although the extent of human habitation and terraforming (e.g. land changes from agricultural or urban development) may severely restrict the ability of ecological systems to adapt to changing conditions. Ecologists predict that abrupt climate changes will result in accelerated extinction rates around the globe, potentially destabilizing ecosystems if keystone species (i.e. those essential to ecosystem stability) are lost. As ecosystems are already under severe pressure from habitat loss, shifts in temperature, rainfall and habitable zones will only worsen already vulnerable ecosystems.

The loss of ecosystems is not merely a matter of species loss, but represents an instability that will impact human-related systems. Human vulnerability to climate change relies in large part upon dependence to environmental systems, and the redundant human systems that can adapt to changing conditions. Despite what are perhaps common perceptions in the literature that regions in Africa and Asia are most vulnerable to climate changes, neither North America nor Europe have very adaptive systems that can account for sudden changes (Kaplan 1994; Paskal 2009). To cite a tangential example, the sudden loss of bee colonies in many regions has left agricultural crops without proper pollination, and without such an environmental resource crops are therefore even more vulnerable to other environmental changes (Ratnieks and Carreck 2010). Although we often consider single-factor impacts to underlying environmental systems (e.g. lower rainfall, warmer average air temperatures, etc.), the more likely consequences of abrupt climate change will be multi-factorial impacts, whose attendant risks will therefore be multiplied. It is only through increasing resilience in systems that adaptation can be possible.

Social, political and economic systems

It is at this point that the cascading effects of climate changes in the Arctic are likely to meet security definitions of Arctic states. Abrupt environmental changes will not only force a re-examination of geopolitics as transit and resource access shifts, but such changes may also overwhelm critically important systems underlying the security of states globally, such that shifts in one area may spark cascading effects that impact distant regions. For example, significant melt of ice sheets may impact energy infrastructure on the coasts, while disruptions to global ocean currents or massive releases of methane would spark situations of food insecurity, itself contributing to loss of livelihoods and government instability as far away as India. Unforeseen events, such as the 2010 fires in Russia, are examples of where the inability to identify potential risks in advance leaves both populations and critical systems vulnerable to new risks. The fires destroyed one-fifth of Russian grain crops, threatened a nuclear facility in Sarov and destroyed a naval air station and 200 aircraft in Kolomna (BBC 2010). The Arctic illustrates where significant environmental changes are already taking place, yet most analyses remain fixed on a regional focus when it comes to potential impacts. Instead, we should become more sensitive to future scenarios that fall outside currently recognized risk parameters (National Intelligence Council 2012).

Projects and initiatives are under way to provide strategic environmental assessments to policymakers, and to link the research community more effectively with interested parties in government, non-profit work and business. For example, the US Department of Energy sponsored a prototype early warning system from 2007 to 2010 for potential energy and environmental impacts at a global scale, the Global Energy and Environment Strategic Ecosystem (GlobalEESE). US military services are similarly attempting to address emerging risks from environmental changes in the Arctic, such as through the Navy's Task Force Climate (National Research Council 2011) and the Air Force's Minerva Program. The larger need for environmental security is of an open-source, international effort to provide research space for the growing environmental security community. Certain research centres already exist and provide international forums, but resource constraints and organization prevent such centres from drawing upon the larger academic, business and government communities of analysts in a systemic way. It may be argued that the best way to approach a complex set of systems with emergent order is to create a strategic foresight system with similar attributes. Interdisciplinary 'black swan' research is more likely to be successful if it can bypass organizational barriers and national parochialism, and instead perhaps create the sort of epistemic community that was successful in solving other transnational environmental issues (Litfin 1999).

Conclusion

Definitions for security (both energy and environmental) and climate change need to be constructed in such a way that policymakers have incentives to pursue mitigation and adaptation, not merely to focus on GHG emissions as the only suitable goal and outcome of the current United Nations climate negotiation (UNFCC) process. They must also allow integration of energy and environment as key concepts, both as contributing to forcing of environmental systems and as the possible solutions for cooperation. Such concepts will be inherently complex, involving large amounts of uncertainty and illustrating scenarios that contain multiple feedback effects and cascading perturbations ('ripple effects'). Simple models and a continued view that environmental systems lay outside of human activity impose artificial barriers on both understanding and solutions.

The point of this exercise is not to provide immediate and concrete answers. It is unlikely that this is even possible, owing to the large amount of uncertainty inherent in the discussion. Rather, we need to provide workable definitions and frameworks for approaching policy, ones that break from past Cold War-era definitions that assume state security, deliberate action and violent conflict. Should these definitions be used, we will be looking at the wrong places and at the wrong times.

Bibliography

Adger, W. N. (2000) 'Social and ecological resilience: Are they related?', *Progress in Human Geography*, 24(3):347–64.

Alley, R. B., Marotzke, J., Nordhaus, W. D., Overpeck, J. T., Peteet, D. M., Pielke, R. A., Pierrehumbert, R. T., Rhines, P. B., Stocker, T. F., Talley, L. D. and Wallace, J. M. (2003) 'Abrupt climate change', *Science*, 299:2005–10.

Alley, R. B., Brigham-Grette, J., Miller, G. H., Polyak, L. and White, J. W. C. (2009) *Past Climate Variability and Change in the Arctic and at High Latitude: A Report by the U.S. Climate Change Program and Subcommittee on Global Change Research*, Reston, VA: US Geological Survey.

Barnett, J. (2003) 'Security and climate change', *Global Environmental Change*, 13:7–17.

BBC (2010) 'Medvedev sacks officers over Russia fire failures'. Online. Available at: http://www.bbc.co.uk/news/world-europe-10871997 (accessed 4 August 2010).

Berkman, P. and Oran, Y. (2009) 'Governance and environmental change in the Arctic Ocean', *Science*, 324(5925):339–40.

Borgerson, S. G. (2008) 'Arctic meltdown: The economic and security implications of global warming', *Foreign Affairs*, 87:63–77.

Briggs, C. and Bath, J. (2009) 'Environmental security and neglected tropical diseases in Africa', in H. E. Purkitt (ed.) *African Environmental and Human Security*, New York: Cambria Press.

Briggs, C., Anderson, L. and Walji, M. (2009) 'Environmental health, security, and the long-term consequences of conflict', *Conflict, Medicine & Survival*, 25(2):122–33.

Brodie, B. (2007) *Strategy in the Missile Age*, Washington, DC: RAND Corp.

Brooks, N. (2003) *Vulnerability, Risk and Adaptation: A Conceptual Framework*, London: Tyndall Centre for Climate Change Research.

Brown, O., Hammill, A. and McLeman, R. (2007) 'Climate change as the "new" security threat: Implications for Africa', *International Affairs*, 83(6):1141–54.

Brown, V. (2004) 'Battle scars: Global conflicts and environmental health', *Environmental Health Perspectives*, 112(17):A994–A1003.

Bryden, H. L., Longworth, H. R. and Cunningham, S. A. (2005) 'Slowing of the Atlantic meridional overturning circulation at 25° N', *Nature*, 438:655–7.

Buhaug, H., Gleditsch, N. P. and Theisen, O. M. (2008) *Implications of Climate Change for Armed Conflict*, Washington, DC: World Bank Group.

CCSP (2008) *Abrupt Climate Change: A Report by the U.S. Climate Change Science Program and the Subcommittee on Global Change Research*, Reston, VA: US Geological Survey.

CNA Corp. (2007) *National Security and the Threat of Climate Change*, Arlington, VA: CNA Corporation. Online. Available at: http://www.cna.org/reports/climate (accessed 10 June 2013).

deFur, P. L., Evans, G. W., Cohen Hubal, E. A., Kyle, A. D., Morello-Frosch, R. A. and Williams, D. R. (2007) 'Vulnerability as a function of individual and group resources in cumulative risk assessment', *Environmental Health Perspectives*, 115(5):817–24.

Deudney, D. (1991) 'Environment and security: Muddled thinking', *Bulletin of the Atomic Scientists*, 47(3):22–8.

Dougherty, B. and Spanger-Siegfried, E. (2005) *User's Guidebook for the Adaptation Policy Framework*, Geneva: United Nations Development Programme.

Eakin, H. and Luers, A. (2006) 'Assessing the vulnerability of social-ecological systems', *Annual Review of Environment and Resources*, 31:365–94.

European Commission (2008) 'Climate change and international security: Paper from the High Representative and the European Commission to the European Council'. Online. Available at: http://www.consilium.europa.eu/ueDocs/cms_Data/docs/pressData/en/reports/99387.pdf (accessed 8 February 2009).

Floyd, R. (2008) 'The environmental security debate and its significance for climate change', *The International Spectator: Italian Journal of International Affairs*, 43(3):51–65.

Freudenberg, W. R. (1988) 'Perceived risk, real risk: Social science and the art of probabilistic risk assessment', *Science*, 242(4875):44–9.

Füssel, H. M. (2007) 'Vulnerability: A generally applicable conceptual framework for climate change research', *Global Environmental Change*, 17:155–67.

Gallopín, G. (2007) 'Linkages between vulnerability, resilience, and adaptive capacity', *Global Environmental Change*, 16:293–303.

Gleditsch, N. P. (1998) 'Armed conflict and the environment: A critique of the literature', *Journal of Peace Research*, 35(3):381–400.

Haines, A., Kovats, R. S., Campbell-Lendrum, D. and Corvalan, C. (2006) 'Climate change and human health: Impacts, vulnerability, and mitigation', *The Lancet*, 367:2101–9.

Halden, P. (2007) *The Geopolitics of Climate Change*, Stockholm: FOI Swedish Defence Research Agency.

Hauge, W. and Ellingsen, T. (1998) 'Beyond environmental scarcity: Causal pathways to conflict', *Journal of Peace Research*, 35(3):299–317.

Hey, C. (1998) 'EU environmental policies: A short history of the policy strategies in European environmental policy', in M. S. Andersen and D. Liefferink (eds) *The Pioneers*, Manchester: Manchester University Press.

Homer-Dixon, T. (1991) 'On the threshold: Environmental changes as causes of acute conflict', *International Security*, 16(4):76–116.

—— (1994) 'Environmental scarcities and violent conflict: Evidence from cases', *International Security*, 19(1):5–40.

IPCC (2001) *Climate Change 2001: The Scientific Basis. Contribution of Working Group I to the Third Assessment Report of the Intergovernmental Panel on Climate Change* (eds J. T. Houghton, Y. Ding, D. J. Griggs, M. Noguer, P. J. van der Linden, X. Dai, K. Maskell and C. A. Johnson), Cambridge: Cambridge University Press.

—— (2007) *Climate Change 2007: Contribution of Working Group I to the Fourth Assessment Report of the Intergovernmental Panel on Climate Change* (eds S. Solomon, D. Qin, M. Manning, Z. Chen, M. Marquis, K. B. Averyt, M. Tignor and H. L. Miller), Cambridge: Cambridge University Press.

Kam, E. (1988) *Surprise Attack: The Victim's Perspective*, Cambridge, MA: Harvard University Press.

Kaplan, R. (1994) 'The coming anarchy', *The Atlantic Monthly*, February:44–76.

Kennett, J. P., Cannariato, K. G., Hendy, I. L. and Behl, R. J. (2002) *Methane Hydrates in Quaternary Climate Change: The Clathrate Gun Hypothesis*, Washington, DC: American Geophysical Union.

Lantze, S. and Raven-Roberts, A. (2006) 'Violence and complex humanitarian emergencies: Implications for livelihoods models', *Disasters*, 30(4):383–401.

Lawrence, D. M., Slater, A. G., Tomas, R. A., Holland, M. M. and Deser, C. (2008) 'Accelerated Arctic land warming and permafrost degradation during rapid sea ice loss', *Geophysical Research Letters*, 35:L11506.

Lenton, T. (2011) 'Early warning of climate tipping points', *Nature Climate Change*, 1:201–9.

Litfin, K. (1999) 'Constructing environmental security and ecological interdependence', *Global Governance*, 15(3):359–78.

Marshall, M. (2013) 'Arctic ice low kicks off a cascade of tipping points', *New Scientist*, 217(2906):6–7.

National Intelligence Council (2012) 'Global trends 2030: Alternative worlds'. Online. Available at: http://www.dni.gov/files/documents/GlobalTrends_2030.pdf (accessed 27 March 2013).

National Research Council (2002) *Abrupt Climate Change: Inevitable Surprises*, Washington, DC: National Academies Press.

—— (2011) *National Security Implications of Climate Change for U.S. Naval Forces*, Washington, DC: National Academies Press.

Parker, C. F. and Stern, E. K. (2002) 'Blindsided? September 11 and the origins of strategic surprise', *Political Psychology*, 23(3):601–30.

Parry, M. L., Canziani, O. F., Palutikof, J. P., van der Linden, P. J. and Hanson, C. E. (eds) (2007) *Climate Change 2007: Impacts, Adaptation and Vulnerability. Contribution of Working Group II to the Fourth Assessment Report of the Intergovernmental Panel on Climate Change*, Cambridge: Cambridge University Press.

Paskal, C. (2009) *The Vulnerability of Energy Infrastructure to Environmental Change*, London: Chatham House.

Rabe, B., Karcher, M., Schauer, U., Toole, J., Krishfield, R., Pisarev, S., Kauker, F., Gerdes, R. and Kikuchi, T. (2011) 'An assessment of Arctic Ocean freshwater content changes from the 1990s to 2006–2008', *Deep-Sea Res. I*, 58:173–85.

Ratnieks, F. L. W. and Carreck, N. L. (2010) 'Clarity on honey bee collapse?', *Science*, 327(5962):152–3.

Rigby, M., Prinn, R. G., Fraser, P. J., Simmonds, P. G., Langenfelds, R. L., Huang, J., Cunnold, D. M., Steele, L. P., Krummel, P. B., Weiss, R. F., O'Doherty, S., Salameh, P. K., Wang, H. J., Harth, C. M., Mühle, J. and Porter, L. W. (2008) 'Renewed growth of atmospheric methane', *Geophysical Research Letters*, 35: L22805.

Rignot, E. and Kanagaratnam, P. (2006) 'Changes in the velocity structure of the Greenland Ice Sheet', *Science*, 311:986–90.

Schiermeier, Q. (2008) 'Fears surface over methane leaks', *Nature*, 455:572–3.

Schneider, S. H. (2003) *Abrupt Non-Linear Climate Change, Irreversibility and Surprise*, Washington, DC: OECD.

Schwartz, P. and Randall, D. (2003) *An Abrupt Climate Change Scenario and its Implications for United States National Security*, Global Business Network report. Online. Available at: http://www.gbn.com/articles/pdfs/Abrupt%20Climate%20Change%20February%20 2004.pdf (accessed 12 December 2008).

Shakhova, N., Semiletov, I., Salyuk, A., Yusupov, V., Kosmach, D. and Gustafsson, Ö. (2010) 'Extensive methane venting to the atmosphere from sediments of the east Siberian Arctic shelf', *Science*, 327(5970):1246–50.

SIDA (2008) *A Climate of Conflict*, Stockholm: Swedish International Development Agency.

Solomon, S., Qin, D., Manning, M., Chen, Z., Marquis, M., Averyt, K. B., Tignor, M. and Miller, H. L. (2007) *Climate Change 2007: The Physical Science Basis*. Cambridge: Cambridge University Press.

Walker, B. and Salt, D. (2006) *Resilience Thinking*, Washington, DC: Island Press.

Walter, K. M., Zimov, S. A., Chanton, J. P., Verbyla, D. and Chapin, F. S. III (2006) 'Methane bubbling from Siberian thaw lakes as a positive feedback to climate warming', *Nature*, 443:71–5.

Walter, K. M., Smith, L. C. and Chapin, F. S. III (2007) 'Methane bubbling from northern lakes: Present and future contributions to the global methane budget', *Philosophical Transactions of the Royal Society*, 365:1657–76.

Warner, M. (2009) 'Wanted: A definition of "intelligence"', in C. Andrew, R. J. Aldrich and W. K. Wark (eds) *Secret Intelligence*, New York: Routledge.

Wisner, B., Blaikie, P., Cannon, T. and Davis, I. (2005) *At Risk: Natural Hazards, People's Vulnerability and Disasters*, London: Routledge.

Zhang, J., Lindsay, R., Schweiger, A. and Steele, M. (2013) 'The impact of an intense summer cyclone on 2012 Arctic sea ice retreat', *Geophysical Research Letters*, 40(4):720–6.

Chapter 7

Climate change impacts, adaptation, and the technology interface

David Malcolm

Introduction

Environmental security represents one of the dimensions in a discussion of overall human security in the Arctic, according to the soon 20-year-old 1994 UNDP Human Development Report. The notion of environmental security plays an important role in understanding broader human security issues in an Arctic context. Key to understanding Arctic environmental security is the study of linkages between ecosystems and social systems in the circumpolar world. The purpose of this chapter is to explore these linkages in terms of both Climate Change Impacts and Adaptation (CCIA) and the importance of sustainable engineering and construction practices in Canada's northern regions. Engineering and construction practices in the Arctic must become appropriate adaptation responses to climate variability. This variability is expressed by large ranges of temperature fluctuation and the increased frequency of extreme events in the natural environment. CCIA and the technological responses to CCIA represent significant impacts on human security in the Arctic.

There is considerable argument as to what should, and what should not, be part of the human security dialogue beyond the conventional militaristic view of security. As outlined by Huebert (2004), indigenous views of Arctic security have been primarily concerned with survival, so that human security takes on the context of any influences that threaten that survival. He refers to human security in the Arctic in terms of three important issues: health security, cultural security and food security. Heininen (2004) refers to a prevalent view of civil security including human health and social and economic conditions as a replacement for the militaristic view of security in the region, whereby southern actors 'interpret the region as a potential military arena'. Both Huebert (2004) and Heininen (2004) refer to the natural and human environment, sometimes linked to pollution from military and industrial sources, as being an integral part of human security in the Arctic. Climate warming affects virtually all aspects of sustainable living in the North (Malcolm 2010) including infrastructure, transportation and health as well as environment and ecology. So climate warming affects all decision-making in the areas of governance and policy for sustainable living.

Climate impacts and adaptation in Canada's North

The importance and urgency of adapting to climate change impacts in Canada's North has been the focus of a considerable body of research. For example, the Arctic Climate Impact Assessment (ACIA) states that

> the Arctic, together with the Antarctic Peninsula, experienced the greatest regional warming on earth in recent decades, due largely to various feedback processes. Average annual temperatures have risen by about 2 to 3°C since the 1950s and in winter up to 4°C. The warming has been largest over the land areas.
>
> (Arctic Climate Impact Assessment 2005:3)

Other Arctic climate trends include rising river flows, declining snow cover, increasing precipitation (roughly 8 per cent across the Arctic over the past 100 years), thawing permafrost, diminishing lake and river ice, melting glaciers, retreating summer sea ice and changes in ocean salinity (Hassol 2004). The climate trends affect many natural ecosystems, such as by shifting distributions of plant and animal species northward, and impact both atmospheric and biogeochemical processes, e.g. penetration of UV radiation and carbon cycling (Häder *et al.* 2007; Zepp *et al.* 2007). This unprecedented warming brings extreme events and has caused the international insurance industry to take note of climate change and the ensuing increase in natural catastrophes (Vellinga and van Verseveld 2000; Diffenbaugh *et al.* 2005; Katz and Brown 2007; National Round Table on the Environment and the Economy 2009). In response to the impacts of these changes to human systems, we need to adapt our engineering, construction and maintenance practices to the changes in northern climates. The impacts of these changes are evident now, and will be even more acute during the coming decades. The Arctic Climate Impact Assessment provides a comprehensive look at the impacts of climate change in the Arctic, which helps us determine the need and time frames required for adaptation (Arctic Climate Impact Assessment 2005).

In Canada's North, we are faced with permafrost thawing and degradation, slumping of ice-laden landforms and changes in timing of freeze-up and melting of river ice and sea ice (Furgal and Prowse 2008). For example, in the central Arctic Ocean, submarine sonar measurements have revealed a 40 per cent reduction in ice thickness (Hassol 2004). Gas hydrates, sometimes considered to be an energy source that is twice as abundant as all the known oil, gas and coal on our planet combined, are buried in permafrost on land and in off-shore sea beds. The northward movement of the permafrost boundary due to Arctic warming trends releases the methane that is held in these hydrates. The methane adds to the total greenhouse gas emissions, bringing yet more global warming.

When thinking about research or practice in CCIA, including the impacts of climate change on the communities of northern Canada and the circumpolar North, it is important to consider five questions that were presented in the Preface of Volume II of the *Canada Country Study* (Maxwell 1997), as follows:

- What are the impacts of a changing climate and how will they affect me and my family throughout our lives?
- Are decisions being made today which will increase our vulnerability in the future because they are not taking such impacts into account?
- Will the approaches we use to adapt to the current climate still be workable in the future, or will new approaches be necessary to adapt to changes beyond our historical experience?
- Will the rate of changing climate allow enough time to adapt?
- Should society become more adaptable or flexible to changes in climate than it is now, and if so, how?

International efforts to alter the severity of climate change through greenhouse gas emissions have been guided by the 1997 Kyoto Protocol, which set targets to reduce greenhouse emissions and came into force in 2005. Post-Kyoto Protocol negotiations are taking place, and the most recent meeting of the United Nations Framework Convention on Climate Change, known as the Conference of the Parties (COP), was held at Cancun, Mexico, in late 2010 (known as COP16). Although COP16 did not produce an agreement of ambitious targets to reduce greenhouse gas emissions, it did reach agreement to set up a 'Green Climate Fund' and a 'Climate Technology Centre' and network in preparation for a second commitment period for the Kyoto Protocol.

Many studies of climate history, based on lake sediment and ice core research, for example, indicate that there have been several times over the past 100,000 years when global temperatures have been up to 2°C higher than today. However, greenhouse gas concentrations (e.g. CO_2 and methane) were never as high during this period as they are today. In any case, as we continue to see increases in greenhouse gas concentrations and temperatures, it is essential that northern societies learn to adapt to the impacts of climate change. This poses a great challenge and obligation to professional engineers in the circumpolar world. They must be the leaders in technology development and technology transfer to ensure optimal levels of overall human security during this time of global change.

Northern climate trends and extreme events

The Northern Climate ExChange at the Northern Research Institute, Yukon College, is in the process of developing an extensive Infosources Database that will be useful as society seeks to come to grips with how to adapt through CCIA research and practice. Northerners are already recording climate change trends like 'thinning of sea ice'. In fact, 'over the past 30 years, the annual average sea-ice extent has decreased by about 8 per cent, or nearly one million square kilometres, an area larger than all of Norway, Sweden, and Denmark combined, and the melting trend is accelerating' (Hassol 2004:25). Observations of climate change by the Inuit of Canada's North have been recorded and are discussed by Fenge (2001). Fenge (2001:80) reported: 'Already Inuit are altering their hunting patterns to accommodate changes to the ice regime and distribution of harvested species.' A detailed table is presented in Fenge (2001), which tabulates observations of environmental change by Inuit and Cree in different regions of the eastern Canadian Arctic on various aspects of weather, sea and river conditions and wildlife. These observations can be used to provide a qualitative picture of various aspects of climate variability as it affects building design and construction in the North.

Climate variability and weather extremes are increasing in the northern and southern regions of Canada. Although the residents of southern Canada seem not to be concerned about climate change in the North, nor about the importance of increasing the emphasis on climate change research in the North, they should realize that the weather in the South is largely influenced by climate conditions in the North. The North of Canada and the northern regions of Russia in Siberia are showing the most dramatic short-term effects of climate variability and climate warming (Global 2012). Permafrost slumping along riverbanks in the North is common, and forest fires are increasing in frequency and intensity. Forest fires can also increase the susceptibility of habitats to invasions by non-native species (Hassol 2004).

The impact of extreme weather events and climate change on the insurance industry on a worldwide basis has recently been published (Zaneti *et al.* 2004). In a special chapter on the impacts of climate change, the report admits that the higher losses since 1970 are mostly explained by economic, demographic and geographical factors, and that the impact on losses due to climate change is difficult to quantify. However, the report suggests that climate variability and extreme events will have greater effects in the future.

One characteristic of climate variability is an increase in extreme events; for example, forest fire activity is on the increase in the North (Hassol 2004). Thunderstorms have also been reported in the northern regions of the Northwest Territories (NWT) and Nunavut (Hassol 2004). As pointed out by Lahey (2002:81): 'It is not a few degrees' rise in average temperatures that will affect people the most; it is the corresponding increase in extreme weather.' This article in *Canadian Geographic* provides a good introduction to climate change science for the average person, including a discussion of the importance of mathematical computer models to describe present and future trends in climate change (Lahey 2002).

Sea level rise is another possible result of climate change, especially if climate-warming trends continue (Hassol 2004). Changes in sea level are especially important to vulnerable areas of coastline. Although most of the

Arctic coast of Canada is emerging, or rising up, due to geological forces, the area around Tuktoyaktuk is submerging, due both to geological forces and sea level rise, and the coastal erosion is threatening infrastructure and buildings near the coast. Sea level rise is caused by both seawater temperature changes (usually considered to be the major contributor to sea level rise), and by ocean water mass increase due to the melting of glaciers (Meier and Dyurgerov 2002) and the disintegration and melting of the West Antarctic Ice Sheet, as 'more open water allows stronger wave generation by winds, thus increasing wave-induced erosion along arctic shores' (Hassol 2004:78).

> Arctic vegetation zones are very likely to shift, causing wide-ranging impacts. [For example, the] treeline is expected to move northward and to higher elevations, with forests replacing a significant fraction of existing tundra, and tundra vegetation moving into polar deserts.
>
> (Hassol 2004:10)

The treeline may be expected to move northward from 5 to 7.5km per year and the southern boundary of the permafrost is also moving north (Maxwell 1997), causing soil instability and difficulties in engineering permanent structures and transportation infrastructure (National Round Table on the Environment and the Economy 2009). Climate warming in the North has dramatically increased the importance of permafrost engineering as a professional discipline (Andersland and Ladanyi 2004). Design in potentially unstable permafrost conditions is of special importance to all infrastructure development in the Arctic.

It has been reported by many northern residents that winters are getting warmer in the North. The season for ice roads is getting shorter, on the average between two weeks and eight weeks per year. This tends to cause thinner ice and more accidents. The overall effect seems to be negative as far as economic activity is concerned. Shorter seasons for ice road use, combined with longer periods between spring break up and autumn freeze up of the rivers, means longer and therefore more expensive air transportation for food and other perishable goods to mines and oil and gas facilities. There may be a positive impact on economic activity in the tourism area, since a longer ice-free period lengthens the season for tourist activity on the Alaska, Klondike and Dempster highways. On the other hand, 'reduced sea ice is very likely to increase marine transport and access to resources' (Hassol 2004:11) by lengthening the navigation season and increasing marine access to the Arctic's natural resources. For example, 'seasonal opening of the Northern Sea Route is likely to make trans-arctic shipping during summer feasible within several decades' (Hassol 2004:11). However, 'increasing ice movement in some channels of the Northwest Passage could initially make shipping more difficult' (Hassol 2004:11). Furthermore, warming also reduced the number of days per year in which travel on the tundra is allowed; for example, in Alaska it has dropped from over 200 to about 100 in the past 30 years, resulting in a 50 per cent reduction in days that oil and gas exploration and extraction equipment can be used.

Other model-based projections from the *Canada Country Study* (Maxwell 1997) for the Canadian Arctic indicated a possible warming from 5 to 7°C, including most of the Arctic Islands, and a warming of up to 10°C over central Hudson Bay. Such a warming would have a devastating effect on ice conditions in Hudson Bay to support polar bear habitats. Precipitation in the North is also likely to increase by up to 25 per cent.

Despite the age of the report, there is considerable information in the *Canada Country Study* (Maxwell 1997) that is still very pertinent and relevant to the design and construction of northern buildings and infrastructure, especially in the sections given to describing the physical environment and the socio-economic sectors. The discussion on the physical environment of Canada's North includes:

- Hydrology;
- Permafrost;
- Sea ice and icebergs (including glaciology);
- Sea level and coastal processes;
- Freshwater ice (including freshwater lakes and riverine processes).

One challenge of the coming decades will be to increase university- and community-based monitoring efforts of environmental change in northern regions to provide the measurements for comparison with the projections of computer-based climate models. Climate models are necessary because the climate statistics available from recent decades is insufficient to assess the probability of future climate averages or extremes (Lahey 2002).

Climate models contain mathematically simplified approximations of the key processes and relationships in the environment, among atmospheric, ocean, fresh water systems, ice and land surface variables, and include how these processes impact the climate over time. The models are driven by changes in CO_2, and other gases in the atmosphere, which may cause heating or cooling. The models are known by the GCM acronym, which stands for either 'General Circulation Models' or 'Global Climate Models'. The operation of a computer-based GCM is calibrated by putting past conditions into the model, and then observing whether the model will 'spin up' over time to give the climate conditions observed today. The model components are connected together so that atmosphere, ocean, land and ice interact on a daily basis. As the model runs forward, the modelled climate evolves to provide climate snapshots at future times.

Climate change impacts on water resources and the boreal forest

Climate change is expected to further intensify the pressure brought to bear on worldwide water resources by agriculture through the depletion of underground aquifers by intensive and wasteful irrigation practices (Brown 2003). Although one impact of climate change in Canada's North is expected to be increased precipitation, the increase in evaporative effect due to rising temperatures is expected to bring drier conditions. Water is emerging as one of the most valuable commodities of the twenty-first century.

The boreal forest regions of circumpolar countries are home to indigenous peoples in large numbers, and sustainability of northern indigenous communities in boreal regions is threatened by climate change (Malcolm 2010). Therefore, it is vitally important to examine the impacts of climate change on the boreal regions. Widespread changes in lake environments due to climate change have been documented in the boreal region of northern Canada (Schindler *et al.* 1997; Smol and Cumming 2000).

The northern boreal forests comprise approximately one-third of the area of all forest systems on the planet (Hassol 2004:50). In the Hassol 2004 report we read:

> Many impacts of climate change are already becoming apparent in the boreal forest: reduced rates of tree growth in some species and at some sites; increased growth rates in others; larger and more extensive fires and insect outbreaks; and a range of effects due to thawing permafrost, including new wetland development and collapsing of the ground surface and the associated loss of trees.
>
> (Hassol 2004:50)

Climate change impacts of methane hydrate dissociation

Gas hydrates have come to the forefront of the climate change debate. Methane hydrates form under certain ranges of temperature and pressure when a methane molecule is surrounded by a cage of water molecules. One volume of methane hydrate is equivalent to approximately 164 volumes of methane gas. Methane is many times as potent as a greenhouse gas as carbon dioxide. This becomes of considerable importance, since it is generally believed that the energy content of gas hydrates in permafrost and ocean floor sediments is very large, perhaps twice the total energy content of all known conventional hydrocarbon reserves on the planet today (Kerr 2004).

Canada's permafrost has significant reserves of methane hydrates, especially in the Mackenzie Delta region in the Arctic (Collett and Dallimore 2000;

Kerr 2004). Canada's gas hydrate research efforts are led by Scott Dallimore and a group of scientists at the Geological Survey of Canada, as well as by a small number of university researchers. Considerable exploration effort will be necessary to document the presence of methane hydrates. It is generally believed that there are considerably greater resources of natural gas hydrates in Siberia than in Alaska and northern Canada.

It is generally accepted that stable gas hydrate concentrations are to be found at 130m or deeper in permafrost, which would take hundreds and thousands of years to dissociate (Haq 2000). However, as the hydrate dissociates, gas is released into the permafrost material above it and may be trapped there for later diffusion up through the soil into the atmosphere. It is believed that most of the methane released through this process would be oxidized through biochemical activity as it goes up through the soil layers.

Release of large amounts of methane into the atmosphere due to permafrost melting and the warming of the seas could cause a runaway global warming effect. In his paper, Haq (2000:137) states: 'It is also feared that escalating methane emissions from the hydrates and the permafrost may soon lead to further strengthening of the on-going global warming trend and may cause unpredictable and abrupt changes in future climatic patterns.' Further discussion in this chapter is along the same line, i.e. 'a clearer understanding of what happens to the significant quantities of gas that might be catastrophically released from the hydrates becomes imperative. How much of this methane makes it to the atmosphere to force additional greenhouse warming?' (Haq 2000:139). In addition, fractures in geological strata known as gas chimneys fill a definite role in conducting methane from dissociating hydrate to the surface (Pecher 2002).

The science and technology of natural gas hydrates has recently been compiled in a volume that will be useful to engineers and scientists seeking to understand the details of gas hydrate dissociation (Carroll 2003).

Engineering for sustainable development in Canada's North

Canada's northern industries and communities are entering a new phase of development, as non-resource development accelerates in the North, and as climate warming changes many aspects of natural resource management and construction. For example, projected increases in permafrost temperatures and in the depth of the active layer are very likely to cause settling, and to present significant engineering challenges to infrastructure such as roads, buildings, and industrial facilities (Hassol 2004:94). Remedial measures are likely to be required in many cases to avoid structural failure and its consequences.

'The projected rate of warming and its effects will need to be taken into account in the design of all new construction, requiring deeper pilings, thicker insulation, and other measures that will increase costs' (Hassol 2004:88). Climate variability due to warming trends will continue to challenge the sustainability of engineered projects in the North. The engineering design and construction industries must become more oriented towards sustainable development goals and objectives. Workable designs must often be replaced by optimal designs, including value engineering, life cycle costing and appropriate risk management and assessment. Social, environmental and economic sustainability are the medium-term goals.

In the boreal North, Aboriginal communities are assisted in strategic planning and research into sustainability by such organizations as the Northern Research Institute of Yukon College, the Aurora Research Institute of Aurora College in the Northwest Territories and the Nunavut Research Institute, although only the northeastern fringes of the boreal region lie in Nunavut.

The sustainability goals are supported by many federal initiatives, including the Northern Development Strategy of the Government of Canada introduced in 2009. All of the federal Science-Based Departments and Agencies (SBDAs) have programmes that are active in Canada's North.

Many universities are active in northern research, most of which are members of the Association of Canadian Universities for Northern Studies (ACUNS). Their research is usually independent of community priorities, but partnerships with communities are becoming more common. University research in partnership with communities is funded by several federal programmes, including the Canadian Institutes for Health Research (CIHR) and the Community University Research Alliance (CURA) programme of the Social Sciences and Humanities Research Council of Canada (SSHRC).

In the North, communities and industrial developers alike are faced with permafrost degradation, slumping of ice-laden landforms, changes in timing of freeze-up, melting of river ice and sea ice, and many other aspects of global change. Climate warming brings greater variability and more extreme events, such as forest fires, landslides and storms. These impacts are occurring in the circumpolar North at an alarming rate, and it is necessary to begin immediately to understand the impacts and adapt to them (Malcolm 2002, 2010).

> Climate change is occurring faster than indigenous knowledge can adapt and is strongly affecting people in many communities. Unpredictable weather, snow, and ice conditions make travel hazardous, endangering lives. Impacts of climate change on wildlife, from caribou on land, to fish in the rivers, to seals and polar bears on the sea ice, are having enormous effects, not only for the diets of Indigenous Peoples, but also for their cultures.
>
> (Hassol 2004:96)

Climate change impacts have been documented by Inuit people (Fenge 2001) as well as by decades of research and observation based on Western science and engineering. These effects cause foundation instability, shortening of ice-road transportation seasons, unpredictable ice-road performance and maintenance, as well as extreme weather events never before experienced. Building construction, pipeline construction and other infrastructure construction practices need to be adapted to account for these effects, keeping the goals of sustainable development in mind.

The engineering design and construction industries must become more oriented towards sustainable development goals and objectives that will meet the needs of these Aboriginal communities. In this new phase of development, technology transfer offers a potential pathway to sustainable development in northern communities where workable designs must often be replaced by optimal designs, including value engineering, life cycle costing and appropriate risk management and assessment.

The importance of value engineering in the development of large, sustainable engineering projects has been in focus (Stephenson 2003), whereby Stephenson demonstrates that the concepts of value, quality and the principles of sustainable development are synonymous. Value engineering is important for optimizing across all essential qualities, including technical, environmental and social ones. Stephenson (2003) recommends that further research be carried out to show how value engineering can be used as a tool for fostering the innovation that is required to achieve sustainable development.

Technology transfer: imposed and invited

Technology transfer falls into the category of *breakthrough solutions* where engineering innovation is matched with the economic needs found in the developing world (Bone 2003). For the Canadian North, technology transfer offers hope for economic, political and social advancement of Aboriginal peoples that will serve to advance their economies, to enhance their culture and to create a healthy and sustainable economic system within their communities and across the North. Moreover, technology transfer is a leading change agent between Western and other societies.

In terms of Canada's Aboriginal population, technological transfer forms a vital link between Aboriginal/non-Aboriginal peoples, and, if done correctly as a partnership, technological transfer can have an extremely powerful and positive impact on this relationship. On the other hand, imposed technology often results in heavy social costs. Thus the term 'technology' in this context refers essentially to the *products* and *concepts* of 'Western' industrialization. For the North in the twenty-first century, the idea of technology can be best elucidated via a series of comparisons. Several salient examples follow:

- Instantaneous electronic wireless communication resulting in 'impersonal' contacts with 'outside society' versus interpersonal contacts with fellow community members.
- Knowledge shared beyond the group versus knowledge kept within the community.
- Communication by electronic media versus a primarily oral transmission of culture.
- Continual exposure to North American natural resource investment strategies, commercial cultural values and interpretative schemes versus Elder-based cultural sources.
- Continual exposure of 'country food' sources to northern food contaminants that are concentrated in northern marine and terrestrial mammals.
- Use of imported commercially processed foods that are alien to the region versus indigenous natural products of the land.
- Products purchased from remote sources versus hand fabrication in accord with cultural norms.
- Western medicines and 'health concepts' versus the natural, indigenous products employed in accord with traditional knowledge.
- Increased physical access to urban centres with a predominantly Euro-Canadian population.

All of the above have impacts affecting the very root nature of all aspects of a community's culture, the particular influences being dependent on a wide range of socio-cultural and demographic factors that may be regional or even community specific. Judgements as to the beneficial or deleterious nature of the impacts are multifaceted, and no definitive statement can be offered. In many cases, the results are ambiguous or beyond resolution into a single category.

To illustrate by a few examples, consider the following:

- Food processing technology – quickly prepared, readily available, microwave cooked, pre-packaged foods have raised the basic caloric intake, where it was sometimes marginal. However, the narrowed nutrient spectrum and rise in diabetes clearly mark this technology as a detriment to most remote communities.
- Industrial investment processes and non-renewable resource development technologies and practices in northern regions, which impact communities in a myriad of ways and collide with traditional culture and traditional knowledge practices. Traditional economies are impacted through the resulting wage economies. Wildlife migration patterns may be affected as well, depending upon the environmental impact of the industry in question.

- Mechanical high-efficiency snowmobiles – this industrial innovation has been integrated into remote Aboriginal communities with many positive consequences. In fact, it has been adapted to serve many aspects of traditional culture. The downside is its cost, thus making the hunting/trapping economy more expensive.
- Access to global electronic media – here the effects are equivocal. Loss of native language and traditional knowledge, intrusion of the highly commercial youth 'culture' and the stimulation to participate in the 'trend economy' are all often classed as deleterious. However, access to information about other isolated communities can lead to the formation of economic development partnerships with beneficial consequences to all parties.

Inuktitut and other regional languages may in fact be assisted by some forms of Western technology. For example, Microsoft Windows was available in Inuktitut in 2006 and now 96 languages are represented in Microsoft's language portal. Inuktitut Windows is part of Microsoft's new Local Language Program. It allows users to localize both Windows XP and Office 2010 to a specific language interface through a Language Interface Pack (LIP), that may be downloaded free of charge.

Applied technology: three case examples

We should tackle the issue of assessing technological impact on a case-by-case base, interpreting the results with regard to the community's own value-response schemas as well as to external non-northern standards. Such an approach does call for an expenditure of human and financial resources, but the results could be the basis of a strategy for enabling remote communities to minimize the negative consequences and amplify the positive ones. There are many case study areas of interest to engineering in the push for sustainability. Three such areas of applied technology are discussed below.

Distributed energy in the North

One suggestion is to address the problem of high cost and lack of reliability of electricity in northern Aboriginal communities, which often receive their power from small diesel-powered generators or find themselves at the end of a long transmission line having unacceptable losses in both voltage and reliability. One step towards this goal is exemplified by the Aboriginal Forum that took place in Vancouver, BC, 31 May–1 June 2004, entitled 'Beyond Oil and Gas: Developing Alternative Resources on Aboriginal Traditional Lands'. Small community-based co-generation industries for the co-production of thermal energy and electrical power generation can be designed and operated based upon forest biomass or the residues of other forest industries and fires. In some communities, run-of-the-river hydro developments can be developed economically without negative impacts on water levels or fish habitat. Electrical generation using solar energy through photovoltaic cells is becoming more and more common in remote locations, especially for operating low-power communications devices.

Technology transfer joint ventures

A second suggestion, a crucial one, is to establish best practices for consultation processes between industries and communities to enable beneficial technology transfer partnerships and joint ventures, and beneficial long-run sustainability in the communities involved. Such practices could be developed as a matter of government policy, not to hinder industrial investment, but to maximize its effectiveness, and to minimize costly time delays due to public intervention and environmental impact enquiries, while at the same time optimizing the particular cost/benefit scenarios for the communities involved.

Regional economic development networks

A third area for study is to work with communities and regions in the North to form regional economic development networks. The Mid Canada Research Institute has been active in this area. For example, the many small communities spread across the boreal north face similar problems: isolation from centres of economic generation, remoteness from resources and a lack of access to knowledge and expertise in strategic economic planning and coordination. Yet these communities may differ widely in culture, history, demography, corporate knowledge and sophistication of business infrastructure.

By establishing regional economic development networks among communities, value-added linkages can be created in which the 'products' of one community's business or industry become the 'feedstock' for the industry of another community (Bastedo 2009; Malcolm 2010). Such value-adding 'productivity chains' help overcome barriers to economic growth and employment facing smaller centres and isolated communities. These linkages are especially appropriate for the boreal region of Canada because of their proximity to the productive lakes and mineral resources in that region.

It should be possible to identify a 'cluster' of communities – one adjacent to a forestry resource, but lacking a technically skilled workforce; another with a better-educated, but unemployed, labour pool; and a third, remote from natural resources but close to business and transportation links. Separately, these communities are unlikely to have the capacity to overcome the barriers to joint action.

Conclusion

The twenty-first century is marked by a new phase of development where government, industry and Aboriginal communities must pull together to foster sustainable development in the North. But how should this formidable task be accomplished?

Technology transfer offers a potential solution to promoting sustainable development in northern Aboriginal communities. As custodians of technology, engineering companies have the expertise to apply technology towards sustainable development projects badly needed by northern communities. Supported by the federal government, these companies could design a series of pilot projects in partnership with individual northern Aboriginal communities to promote local economic development and increase adaptability of northern communities to the effects of climate change. Such projects could then contribute to the ongoing development strategy for and with the entire North with the goal being to raise the quality of life in these communities (and support local communities themselves to raise the quality of life on their own terms) and to narrow the gap between the Aboriginal and non-Aboriginal communities in northern Canada.

References

Arctic Climate Impact Assessment (2005) *Arctic Climate Impact Assessment*, New York: Cambridge University Press.

Andersland, O. B. and Ladanyi, B. (2004) *Frozen Ground Engineering*, 2nd edition, Hoboken, NJ: John Wiley and Sons.

Bastedo, J. (2009) *Greening the NWT Economy: Local Pathways to Economic Prosperity*, Yellowknife, NT: Ecology North. Online. Available at: http://www.ecologynorth.ca/wp-content/uploads/2013/02/Greening-the-Economy.pdf (accessed 6 June 2013).

Bone, R. M. (2003) *The Geography of the Canadian North: Issues and Challenges*, 2nd edition, Toronto: Oxford University Press.

Brown, L. B. (2003) *Plan B: Rescuing a Planet under Stress and a Civilization in Trouble*, New York: W. W. Norton & Company Inc.

Carroll, J. (2003) *Natural Gas Hydrates: A Guide for Engineers*, Burlington, VT: Elsevier Science.

Collett, T. S. and Dallimore, S. R. (2000) 'Permafrost-associated gas hydrate', in M. D. Max (ed.) *Natural Gas Hydrate in Oceanic and Permafrost Environments*, Dordrecht: Kluwer Academic Publishers.

Diffenbaugh, N. S., Pal, J. S., Trapp, R. J. and Giorgi, F. (2005) 'Fine-scale processes regulate the response of extreme events to global climate change', *Proceedings of the National Academy of Sciences*, 102(44):15774–8.

Fenge, T. (2001) 'The Inuit and climate change', *Canadian Journal of Policy Research*, 2(4):79–85.

Furgal, C. and Prowse, T. D. (2008) 'Northern Canada', in D. S. Lemmen, F. J. Warren, J. Lacroix and E. Bush (eds) *From Impacts to Adaptation: Canada in a Changing Climate 2007*, Ottawa: Government of Canada.

Global (2012) *Global Temperature Trends 2012*. Online. Available at: http://www.arctic.noaa.gov/detect/global-temps.shtml (accessed 29 March 2013).

Häder, D. P, Kumar, H. D., Smith, R. C. and Worrest, R. C. (2007) 'Effects of solar UV radiation on aquatic ecosystems and interactions with climate change', *Photochemical and Photobiolical Sciences*, 6(3):267–85.

Haq, B. U. (2000) 'Climate impact of natural gas hydrate', in M. D. Max (ed.) *Natural Gas Hydrate in Oceanic and Permafrost Environments*, Dordrecht: Kluwer Academic Publishers.

Hassol, S. J. (2004) *Impacts of a Warming Arctic*, Cambridge: Cambridge University Press.

Heininen, L. (2004) 'Security in the North', unpublished position paper presented at the Third Northern Research Forum, Yellowknife, 15–18 September 2004.

Huebert, R. (2004) 'Arctic security: Different threats and different responses', unpublished discussion paper presented at the Third Northern Research Forum, Yellowknife, 15–18 September 2004.

Katz, R. W. and Brown, B. G. (2007) 'Extreme events in a changing climate: Variability is more important than averages', *Climate Change*, 21(3):289–302.

Kerr, R. A. (2004) 'Gas hydrate resource: Smaller but sooner', *Science*, 303(5660):946–7.

Lahey, A. (2002) 'The model scientist', *Canadian Geographic*, 122:80–8.

Malcolm, D. G. (2002) 'Climate variability in Canada's North: Walking on thin ice', unpublished paper presented at Proc. International Winter Construction Symposium & Expo, Environment Section, Edmonton, 3–5 November 2002.

—— (2010) 'Speaking plainly about research, governance, and policy for sustainable living', *Pimatisiwin: A Journal of Aboriginal and Indigenous Community Health*, 8(1):171–83.

Maxwell, B. (1997) *Responding to Global Climate Change in Canada's Arctic, Volume II of Canada Country Study: Climate Impacts and Adaptation*, Downsview: Environment Canada.

Meier, M. F. and Dyurgerov, M. B. (2002) 'How Alaska affects the world', *Science*, 297(5580):350–1.

National Round Table on the Environment and the Economy (2009) *True North: Adapting Infrastructure to Climate Change in Northern Canada*, Ottawa: National Round Table on the Environment and the Economy.

Pecher, I. A. (2002) 'Gas hydrates on the brink', *Nature*, 420:622–3.

Schindler, D. W., Curtis, J. P., Bayley, S. E., Parker, B. R., Beaty, K. G. and Stainton, M. P. (1997) 'Climate-induced changes in the dissolved organic carbon budgets of boreal lakes', *Biogeochemistry*, 36:9–28.

Smol, J. P. and Cumming, B. F. (2000) 'Tracking long-term changes in climate using algal indicators in lake sediments', *Journal of Phycology*, 36:986–1011.

Stephenson, M. (2003) 'How can value engineering be used to enhance the sustainability of major engineering developments?' Master thesis, Graduate School of Environmental Studies, University of Strathclyde.

Vellinga, P. and van Verseveld, W. J. (2000) *Climate Change and Extreme Weather Events*, Gland: World Wildlife Fund for Nature.

Zaneti, A., Enz, R., Heck, P., Green, J. and Suter, S. (2004) *Natural Catastrophes and Man-Made Disasters in 2003*, Zurich: Swiss Reinsurance Company.

Zepp, R. G., Erickson, D. J. III, Paul, N. D. and Sulzberger, B. (2007) 'Interactive effects of solar radiation and climate change on biogeochemical cycling', *Photochemical and Photobiological Sciences*, 3(3):286–300.

Chapter 8

Bridging the GAPS between ecology and human security

Dawn R. Bazely, Julia Christensen, Andrew J. Tanentzap and Gunhild Hoogensen Gjørv

Introduction

When we began thinking about the relevance of the human security concept for ecology and ecologists, two books were riding high on the non-fiction best-seller lists: *Collapse* by Jared Diamond (2005) and *A Short History of Progress* by Ronald Wright (2004). Both books are informed by the most basic, widely accepted ecological theory: namely that there are limits to both the availability of resources and to population growth (Krebs 1988; Begon *et al.* 2006). Both books describe what happens when human societies come up against these limits.

Despite the broad appeal of these books, and their stories of how some human societies have experienced, or will likely experience, catastrophes arising directly from resource depletion, overall global trends in resource consumption by humans continue to escalate (Begon *et al.* 2006; Harte 2007). Since 1998 the World Wildlife Fund's *Living Planet Reports* have described this consumption and its impacts, using a variety of metrics for a wide range of resources, including cement, wood, fertilizer, water, natural habitat cover, and freshwater, marine and terrestrial species (WWF International 1998, 2012). The simplest assessment is that we need at least two and a half more planet Earths to sustain current rates of resource consumption by humans (McGlade 2004).

Many ecologists based in academic institutions continue to be flummoxed and frustrated by the apparent ongoing lack of awareness in our broader society of both the fundamental lessons about resource limitations that have been learned in the field of ecology and their implications for humanity and the planet (e.g. Dovers *et al.* 1996; Hollings 2000; Mason 2003; AFP 2006). Frustration about how little these lessons appear to inform both wider social and economic policy, as well as the individual consumption choices made by members of the public, has prompted ecologists to undertake critical analyses of the field of ecology that explore its (perceived) weakness as a science (e.g. Peters 1991; Weiner 1995) and provide prescriptions for how both the field and its broader impact could be improved (e.g. Belovsky *et al.*

2004). One eminent ecologist even penned a book of advice for citizens about ecology (Slobodkin 2003).

While Slobodkin's book was favourably reviewed in some quarters (e.g. Bosselman 2004), this was not the case for Denise Lach, a social sciences professor based, at the time, in a sustainability research centre with an interdisciplinary focus (Lach 2003). Lach concluded her critique by remarking: 'the author says that "ecology is in danger of becoming an uncomfortable blend of science and a passé but still trendy mass movement" (p.13). I'm sorry to report that this book does little to remedy this description of ecology for the citizen reader' (Lach 2003:890).

In light of Lach's critique, one of three main questions that we explore in this chapter is: why, despite all of the soul-searching, do the research findings of ecologists appear to have such little impact in our broader society?

This chapter took root during the 2004 workshop 'Human Security and the Arctic' held at the University of Tromsø. Workshop participants were challenged to assess the relevance of the human security framework for Arctic regions. The organizer, Gunhild Hoogensen Gjørv, pointed out that while most countries with Arctic regions are amongst the wealthiest in the world, many of their northernmost communities are notoriously under-resourced (Einarsson *et al.* 2004). Northern residents have reduced access to many social and health services, influencing health disparities, resulting in higher mortality rates, for example, compared with those of fellow citizens living in more southern regions of the same country. Additionally, Arctic peoples have been amongst the first to observe environmental change associated with a warming climate (ACIA 2004) and to assert that their traditional ways of life were being directly affected by the lifestyle choices of people living thousands of kilometres away (AMAP 2002, 2009; Watt-Cloutier 2005). During the Tromsø workshop, two further questions emerged: what is the connection between ecology and human security? And, if there is a direct link to be made, given what we know about the pace of change in Arctic ecosystems: how relevant is the human security framework for ecologists studying ecological change in Arctic ecosystems and mobilizing their knowledge of this change to our broader society?

In this chapter, we discuss these three overarching questions in the context of ongoing debates about how ecologists might best address the science–policy gap (e.g. Haeuber and Ringold 1998; Vaughan *et al.* 2007; Lowe *et al.* 2009; Brown *et al.* 2010). We propose that an interdisciplinary approach offers significantly increased potential for effectively mobilizing ecological theory, while recognizing that achieving an authentic interdisciplinary culture remains a challenge within prevailing academic institutional structures (Pfirman *et al.* 2007). Furthermore, we suggest that the human security framework provides an effective means for connecting ecologists with other stakeholders and actors (Tanentzap *et al.* 2009).

'Lonely ecologists': why do ecologists' research findings have such little impact among the broader public and in policy formation?

Ecology is 'the scientific study of the distribution and abundance of organisms' (Andrewartha 1961). Some variant of this definition, which was first articulated in 1866 and 1869 by the theoretical morphologist Ernest Haeckel, an admirer of Charles Darwin (Stauffer 1957), is to be found in all ecology textbooks. Haeckel wrote, as translated from the German,

> by ecology we mean the body of knowledge concerning the economy of nature – the investigation of the total relations of the animal to both its organic and inorganic environment; including, above all, its friendly and inimical relations with those animals and plants with which it comes directly or indirectly into contact – in a word, ecology is the study of all those complex interrelations referred to by Darwin as the conditions of the struggle for existence.
>
> (Stauffer 1957:141)

Andrewartha's (1961) definition implies 'relationships' and 'environment', which are terms made explicit in other major ecology texts, such as Krebs (1978:141, original emphasis), who defined ecology as 'the scientific study of the *interactions* that determine the distribution and abundance of organisms'.

In other words, all ecological research activity has, at its centre, an attempt to understand whether population numbers of any particular species are going up, going down or staying the same over time, and what is causing these changes, or their absence, to occur. Interestingly, in the entry for 'Ecology' in the *Encyclopedia of the Philosophy of Science* (Odenbaugh 2005), there is no mention at all of the most fundamental activity undertaken by ecologists: namely, *counting* living things. Without this basic counting, an ecologist could not fulfil the activities encompassed by the definitions of ecology. In their comprehensive discussion of ecology's activities, and how it must operate at multiple scales, Begon *et al.* (2006) make this central activity of 'counting' explicit.

Much observational data have repeatedly confirmed that all populations are eventually limited by the availability of a necessary resource, and these populations subsequently transition from an initial growth phase to a 'regulated' phase in which their numbers level-off (Krebs 1988; Box 8.1). This *regulated* phase can vary in its appearance, but often includes high mortality. A striking example of this regulation is provided by the feral Soay sheep population on the World Heritage Site of St Kilda, Scotland, in which up to 50 per cent of the population dies over the winter every three to five years, despite the absence of predators (Clutton-Brock and Pemberton 2004).

Box 8.1: Ecology's three uncomfortable messages: backed up by decades of research ...

Limits to growth: all populations are eventually limited by something (Figure 8.1). There is no such thing as limitless growth, and many ecologists understand that, at some point, the human population (Figure 8.2) will be regulated in some way so that growth declines, which can happen under a variety of scenarios.

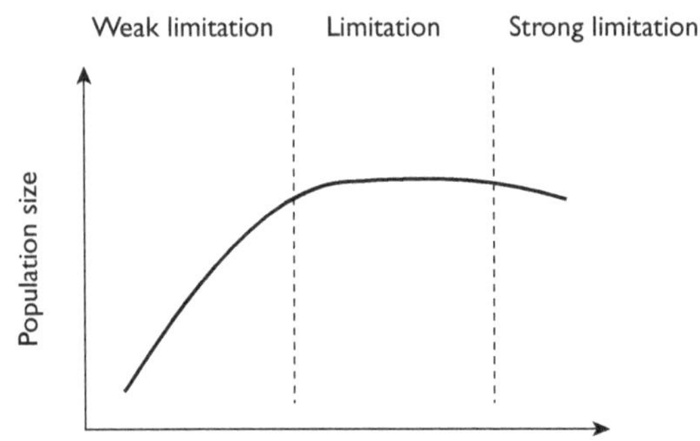

Figure 8.1 Population change over time, with initially no limits to growth, resulting in increased population size, followed by limitation, so that growth slows, and finally entering a phase of strong limitation, where population size starts to decline

Source: Andrew Tanentzap

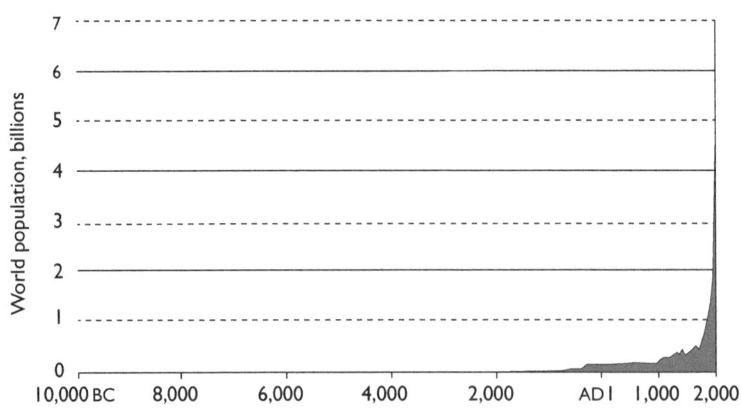

Figure 8.2 Past human population size over time

Source: Wikimedia Commons, EIT public domain

Complex networks and interactions, such as those found in food webs (Figure 8.3), mean that causal relationships are often difficult to determine precisely, and that future pathways are difficult to predict, due to the complexity of positive and negative feedbacks. While we project future scenarios precisely, we can often generate a range of options, although there is uncertainty about exactly which scenario will play out. This uncertainty is inherent in ecological systems. But many ecologists hold the view that our broader society has few means of dealing with this uncertainty when it comes to managing the environment in a realistic way.

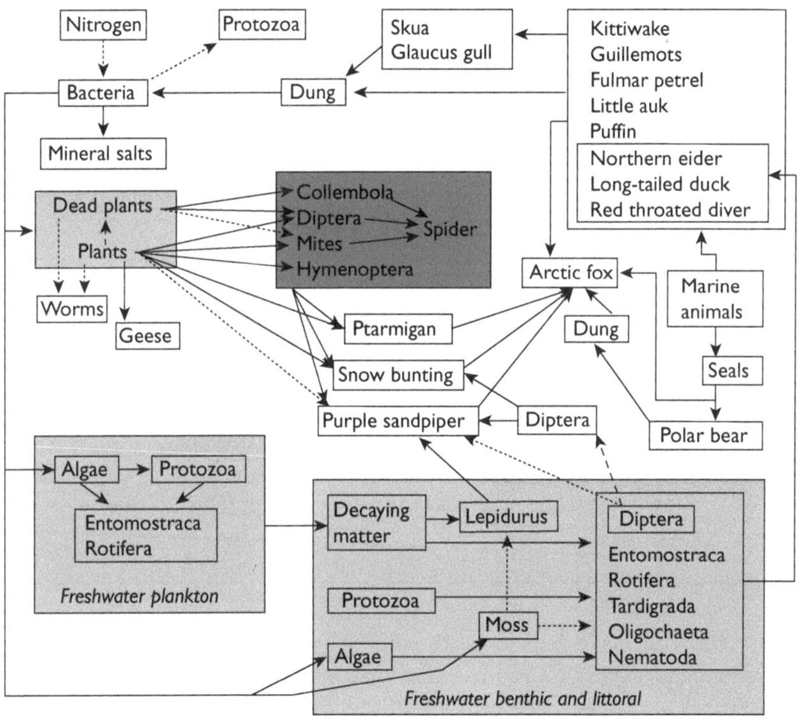

Figure 8.3 The food web described by Summerhayes and Elton for Bear Island, Norway, in 1923

Source: adapted from Summerhayes and Elton (1923); redrawn by Justinleif, Wikimedia Commons

The non-linearity of interactions observed in many ecosystems means, simply, that there is not a simple straight-line relationship between two factors. When it comes to buying clothes, we all know paying more for something does not necessarily mean that quality increases in step with price – one t-shirt can be made in the same factory as another, from the same fabric, but cost more if one is paying for a name-brand label. In ecosystems, these curvilinear relationships can result in increasing rapidity of change, which can go from happening very slowly, to very fast, especially when there is an unexpected disturbance, such as a hurricane or invasive species. Because of this, ecosystems can flip their state very rapidly and unexpectedly (Figure 8.4). It is these rapid flips that are of great concern to climate change researchers.

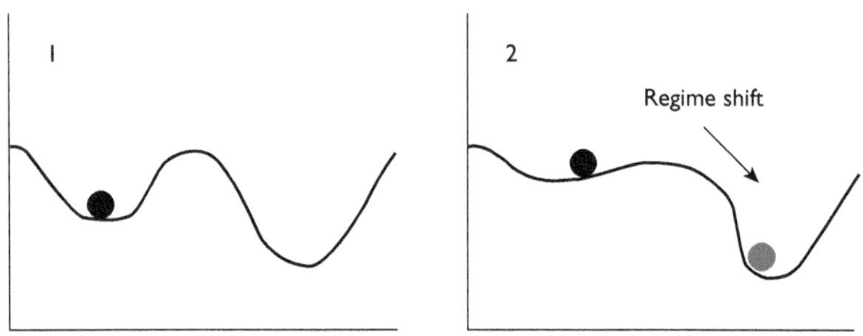

Figure 8.4 Diagrams illustrating an ecosystem shifting from one state (i.e. valley) to another. The ball represents the ecosystem and the line is the path or trajectory that it follows as conditions change. The slope of the line illustrates the concept that change can occur rapidly and be difficult to reverse

Source: Dawn Bazely

Clearly, humans have experienced population limits in the past (Wright 2004; Diamond 2005), and continue to do so in the present, in many different geographical locations. Furthermore, in the view of many ecologists, humankind will eventually be limited, at a global level, by the reduced availability of a whole host of resources, ranging from water to energy to food (Krebs 1988; Harte 2007). Many ecology and biology texts include sections about how the prevailing pattern of global human population growth exceeds exponential rates, how population trends vary geographically and the enormous challenges in predicting global human carrying capacity (e.g. Townsend *et al.* 2003; Begon *et al.* 2006; Evert and Eichhorn 2013). The literature that discusses what the actual human carrying capacity of the world may be generally acknowledges the limitations of the bare-bones ecological concept of carrying capacity for factoring in notions of human choice, values and politics into the equation (Cohen 1995; Lutz *et al.* 2001; Stokstad 2005; Harte 2007; Ehrlich and Ehrlich 2009).

There has clearly been mobilization of the main ecological concepts (Box 8.1) outside of the immediate field by non-ecologists (e.g. Barlow and Clarke 2002). The seminal document that has driven the notion of ecological limits in the policy arena is *Our Common Future*, the Report of the Brundtland Commission of the World Commission on Environment and Development (WCED 1987). It introduced the concepts of sustainability and sustainable development, which form the cornerstones of many United Nations programmes, as outlined in various documents to be found at the United Nations Environment Programme Sustainability webpage (UNEP 2013). The Ecological Footprint Analysis (Wackernagel and Rees 1996) further mobilized ecology theory and allowed for the notion of sustainability to be defined in concrete terms: in short, it estimates the land and water area required to produce the resources consumed, and to absorb the wastes produced, by a human population using available technology. The footprint approach has been implemented by WWF International since 2000 in its *Living Planet Reports* (WWF International 2000).

Nevertheless, given that rates of resource consumption continue to outstrip supply rates, it is worth briefly examining three possibilities that may shed light on the ongoing communication gap about basic ecological theory between the ecological research community and other communities:

Possibility 1: Ecologists are generally confined to science ghettos of their own making, speaking mainly to each other, and, consequently, their lessons are not being broadly communicated rapidly enough.

Possibility 2: No matter how clearly the key, uncomfortable messages of ecology are communicated (Krebs 1988; Box 8.1), the majority of people are predisposed to ignore them.

Possibility 3: Even if the messages of ecology are communicated and understood, for whatever reason most members of the global human population, like the members of the civilizations described in Diamond (2005), are unable to change either their individual behaviours or their society's collective behaviour, and are simply doomed to reach environmental limits that will result in 'crashing' populations. The recent Franny Armstrong (2010) film *The Age of Stupid* explored this idea and posed the question: 'If we could see so many signs of climate change, for so long, why didn't we act?'

Are ecologists mainly talking to themselves?

The notion that most academic ecologists are guilty of preaching to the converted is a charge that has been levelled at environmentalists (Shellenberger and Nordhaus 2004). In his speech *The Death of Environmentalism and the Birth of the Commons Movement*, made to the Commonwealth Club of California on 8 December 2004, Adam Werbach, executive director of the Common Assets Defense Fund, stated that environmentalists must 'talk to people outside of their movement … Environmentalists find that their greatest audiences are themselves' (*Chicago Tribune* 2005).

Across academia in general, and locally in the ecological research community, there is a long-term, ongoing conversation about the need to break out of ivory towers and silos, often recommended via engagement in 'interdisciplinary' collaborations (e.g. Haeuber and Ringold 1998; Vaughan *et al.* 2007; Lowe *et al.* 2009). Groups such as the Resilience Alliance (Hollings 2000) represent efforts by ecologists to form interdisciplinary collaborations in order to promote sustainable development principles and policies informed by ecological research findings. Hollings (2000:7), himself an ecologist, stated that 'ecologists have been largely ignorant of human behavior, organizational structures, and institutional arrangements that mediate the relationships between people and nature'.

Can academics break out from their silos through increased interdisciplinary collaborations? There is a large, diverse, yet disparate literature that explores the nature of the science–policy gap (e.g. Dovers *et al.* 1996; Brown *et al.* 2010). Ecologists fall on the science side of this gap. Certainly, there are many within-university examples of efforts to engage academics in interdisciplinary conversations on campuses globally. For example, in 2004, the Liu Institute, University of British Columbia, held a monthly seminar series entitled *Science and its Policy Impacts*, aimed at examining 'how scientific knowledge shapes policy agendas and how communication between the two could be improved at a variety of levels' (Liu Institute 2004).

Another way to bridge this gap is by participating in interdisciplinary research teams comprised of diverse groups of academics and other stakeholders. In this environment, the challenging issue of language, in which different groups ascribe different meanings to the same terms, and may be talking at cross-purposes, must be addressed. Norton (1998) pinpointed breakdowns in communication arising from the failure of ecologists to make the linkages between ecological functions of ecosystems and social values clear, and also to the reluctance of ecologists to mix values issues with scientific study. He called for an integrated language of management that is: '1 adaptive, 2 perspectival, 3 multiscaled, 4 operationalizable, 5 normative in content and 6 communication enhancing' (Norton 1998:350).

Interdisciplinary, team-based research represents a very different approach to research from that which prevails across most academic disciplines (Leshner 2004). Unfortunately, the time required for learning to communicate with academics outside of one's own field represents a significant amount of research time for which there may be little incentive. The challenges associated with assessing interdisciplinary activities compared with in-discipline research for the purposes of career advancement within academia are so significant that the USA-based Council of Environmental Deans and Directors has produced a set of guidelines advising on interdisciplinary hiring, tenure and promotion (Pfirman *et al.* 2007). Nevertheless, there is ongoing interest in and support for interdisciplinary collaboration, and the National Research Council of the USA has recently commissioned a project overseen by the Board on Behavioral, Cognitive and Sensory Sciences, of the US National Academies, on *The Science of Team Science* (BBCSS 2012).

Are ecological messages so unpalatable that they cause most people to cover their ears?

Ecological research is generally based in biology or biology-related departments and faculties in the natural and physical sciences. Consequently, a first step in examining just how well the messages of ecology are being communicated and mobilized is to assess how well the public comprehends and understands science in general. This is a central question in the research field of science literacy as well as the public understanding of science. Various researchers have pointed out that the level of ecological literacy and, more broadly, science literacy among the populations of the USA and elsewhere is largely unknown (Hutton 1996; Popli 1999; Jordan *et al.* 2009; Ladle and Gillson 2009). There is renewed interest in scientific literacy because of interest in the public perception of the science of climate change and its ecological implications (Leiserowitz 2009; Kahan *et al.* 2012).

If improved ecological literacy among the broader public is desirable (Otto 2011), then how could ecology be made more accessible, not only to the broader public, but also more widely across academia? The language of ecology is peppered with technical terms such as *feedbacks*, *non-linear dynamics* and *disturbance*. All of these terms have highly specific, often mathematically defined meanings. In addition to the ecological concept that limits will eventually be reached by every population (Krebs 1988; Box 8.1), complexity is accepted within the field, even if it is not understood by all ecologists. Most practising ecologists would probably prefer to study a nice simple food chain rather than a messy, complex web in an ecosystem with multiple trajectories and pathways full of uncertainty (Box 8.1). While few ecologists directly observe ecosystems exhibiting 'non-linear' dynamics, where they flip from one stable-state to another, rapidly and unexpectedly, there are well-documented examples of this occurring. Meanwhile, ecologists are concerned that this very thing is occurring with global climate change: namely that ecosystems will change too rapidly for human response, posing a very real concern amongst not just ecologists, but the broader scientific community. Harte (2007) explains how this process is likely to play out, given that human population growth is dynamic and non-linear, and not passive and linear, and reinforces destabilizing feedbacks (Figures 1–4 in Harte 2007). When read by a broader audience, this paper is likely to elicit one of two kinds of reactions: (1) readers may be flummoxed by the complexity of the issue, and simply dismiss it as incomprehensible and lacking relevance to their daily life, or (2) they may grasp that even though it is a complex topic, that the human race is headed towards a doom-and-gloom scenario, and therefore dismiss it because it is too depressing to think about, they feel powerless to act to prevent the outcome. Box 8.1 is our attempt to explain these concepts in some very plain language – up to a point.

If we believe that ecologists are improving the communication of their messages through vehicles such as increased interdisciplinary collaborations (Leshner 2004), then what other reasons may exist to explain why members of broader society might fail to act on the information (see Armstrong 2010)? Wright (2004) wrote: 'The notion that growth can be boundless is clearly the Big Lie of our times.' In an *op ed*, 'Sustainable planning is cure for growth delusion' (19 July 2010, *Irish Times*), Gerry Crilly wrote: 'Misguided ideologies – "more is good" and "bigger is better" – have led to an unsustainable existence, globally, nationally and locally.' The 'growth delusion' is defined by Lloyd (2009:516) as: 'The twin problems of peak oil and climate change [being] … underpinned by our belief that economic growth can be supported indefinitely by a finite Earth.' In his discussion of rationality, Lloyd (2009) considers how the human brain can simultaneously accept two directly contradictory thoughts. Kahan *et al.* (2012) recently explored this *cognitive dissonance* in their research into the relationship between science literacy and concern about the impacts of climate change. They found that the most scientifically literate individuals tended to be the least concerned about climate change (Kahan *et al.* 2012), most likely because of the implications for their political views, personal lifestyle and ecological footprint.

In *Managing Without Growth*, environmental economist Peter Victor (2008) presents an economic argument rooted in the messages of ecology, which runs counter to the dominant global economic framework of unlimited growth for all, also known as 'the growth delusion'. Clearly, broad-scale political and other actions informed by ecology's three uncomfortable messages (Box 8.1) must happen outside of the narrow field of ecology and be multi-scale and multi-stakeholder. One possibility is that these actions require the kind of political will that is simply unattainable at the national or global level. Diamond (2005:496) articulated this view as:

> No one at the UN or in the First World governments is willing to acknowledge the dream's impossibility … What will happen when it finally dawns on all those people in the Third World that current First World standards are unreachable for them, and that the First World refuses to abandon those standards for itself?

What is the connection between ecology and human security?

Human security is a concept developed and championed by the United Nations (Commission on Human Security 2003). Environmental security is one of its many aspects (Dalby 2002). Some of the other commonly listed dimensions of human security include: economic, health, food, personal, community and political security (Tadjbakhsh and Chenoy 2007). These areas can be further broken down to even more specific areas, such as housing security (Seifert 2005; Christensen 2011).

A fundamental tenet of human security is that its very *meaning* is defined by local peoples and communities (Commission on Human Security 2003). In other words, the *framework* of human security is indelibly rooted in what it means, or what is required, for the individual members of any given community to be secure. As a concept and framework, the components that make up human security are *context-dependent*. Notions of human rights are also explicit in the genesis of human security as a concept. Human security provides a framework in which local peoples can identify issues and solutions that will increase their security, and many policies, pathways and options become available. Local knowledge is, consequently, a fundamental aspect of generating human security, and in this way is analogous to the ecological concept that each and every ecological community and ecosystem has the potential to follow multiple trajectories and outcomes within the same set of broad constraints and drivers.

There is a diversity of views among security researchers about the appropriate scope and application of the human security concept (Dalby 2002; Krause 2004; Griffiths 2008; Hoogensen Gjørv, this volume). On the one hand, some view human security as a concept that is applicable within a narrow range of short-term, crisis-type circumstances (e.g. Krause 2004). While others (e.g. Tadjbakhsh and Chenoy 2007) take a longer-term, broader view, and acknowledge that short-term crises are nearly always years in the making, so that human security must be a complex, multi-faceted project with a long timeline.

Upon introduction to the human security concept (e.g. Tanentzap *et al.* 2009), most ecologists are more likely to appreciate its longer-term, broader dimensions. This may be attributed to one of the messages of ecology being the tendency of ecological systems to make rapid state-change switches, that may appear to come out of nowhere, but that, in fact, are the products of a slow accumulation of small changes (Krebs 1988; Box 8.1). Furthermore, ecosystem functioning and dynamics are always multi-factorial in terms of their drivers of change or stability (Krebs 1988).

Once explained, the link between ecology and human security is clear to ecologists. It is self-evident that even if all of the land mines in the world were to magically disappear, and all armed conflicts were to be instantly resolved, so that the 'freedom from [immediate] fear' aspect of human security was addressed, the prevailing ecological conditions in many of these regions would result in their inhabitants being challenged by high levels of ecological unpredictability and instability, such as increased drought or crop failures. This lack of 'environmental security', or, as it is sometimes termed by ecologists, 'ecosystem health' and 'ecological integrity', will likely be a contributing factor in determining future political stability, as suggested in ecological disaster scenarios where future wars will be fought over access to water (De Villiers 2000).

This connection was well understood by the environmentalists Shellenberger and Nordhaus (2004:12):

> Why, for instance, is a human-made phenomenon like global warming – which may kill hundreds of millions of human beings over the next century – considered 'environmental'? Why are poverty and war not considered environmental problems while global warming is? What are the implications of framing global warming as an environmental problem – and handing off the responsibility for dealing with it to 'environmentalists'?

Political and sociological analyses connecting ecosystem health, resource scarcity and violence have actually gained much currency in the last 20 years:

> El Salvador would be a hopeless place even without war. It is a crowded, deforested, eroding land that has lost most of its fertility and wildlife. Salvadoran peasants only twenty years ago ate white-tailed deer, armadillo, spider monkey, and cottontail rabbit as inexpensive daily fare; today, most wild animals except crop-eating rodents are extinct, and corn tortillas and beans make up the average daily diet.
>
> (Krauss 1991:57)

While these kinds of ecology–security connections are being made with increasing frequency, until recently political commentators did not usually implicate ecosystem stability directly in human security. For example, in Krauss' (1991) broad, comparative survey of Central America, the chapter on Costa Rica, 'Central America's One Democracy', makes no mention of the country's high degree of forest cover and intact biological diversity compared with other countries in the region, nor how this may affect its political outcomes. When asked about this omission, Cliff Krauss (Pers. comm., 3 March 2005) agreed that Costa Rica's rainforests and the wealth resulting from ecotourism was likely closely linked to the country's stable, democratic government. But, he pointed out that Nicaragua also has a large amount of intact forest cover, yet nonetheless an appalling record on human rights and governance. In the latest WWF *Living Planet Report* (2012), El Salvador was estimated to have the third lowest biocapacity in Latin America, ahead of Jamaica and Haiti, while Nicaragua's biocapacity was higher, but still well below the average for Latin America: it's complicated!

Apparent similarities in the level of ecosystem integrity in a country where, at the same time, there may be broad differences in the quality of human life is not a reason for dismissing the notion that functional ecosystems, which provide various services and benefits, such as clean water (Costanza *et al.* 1997), are a contributing factor to human security. In cases where similar conditions exist in countries with vastly different political stability and quality of life, as rated in various United Nations Human Development Index (HDI) tables and reports (UNDP 1990), there will always be multiple driving factors, and every system has a different history and trajectory. Sagar and Najam (1998) criticized the HDI for failing to include sustainability and ecological indicators related to depletion of natural resources. Were measures of ecosystem integrity to be taken into account in comparing any two countries with differing political systems (having good vs. poor governance), and similar ecosystem integrity, the ecological view would likely be that any future improvement in human security and governance in the country with a poor rating may be more likely to be maintained due to greater ecosystem resilience (Elmqvist *et al.* 2003).

A comparison of maps and satellite imagery illustrates the connection between ecology, human security and state security, as represented by the latter's political borders. The world seen on Landsat images is more a reflection of the major biomes or habitat types, such as boreal forest and tundra (Evert and Eichhorn 2013), than countries. However, at a larger scale, the political borders of neighbouring countries located in the same major biome type are easy to detect (e.g. Kuemmerle *et al.* 2006). This difference in land cover often reflects the country's political history. Hispaniola, the island shared by Haiti and the Dominican Republic, provides an example in which vegetation cover and other ecosystem attributes are very different between the two countries and the border can be clearly distinguished (Wilson *et al.* 2001). These altered biome states on either side of the border will have very different hydrological (water-cycling) and nutrient regimes.

The links between ecology and the environmental security component of human security can be found in the debates occurring in the resource scarcity and conflict literature, where both sides attempt to argue if and how resources and conflict might be related (if at all), and how these perceptions affect our understanding of security (e.g. Matthew *et al.* 2002). Additionally, links are increasingly being made between environment, justice and ethics by those in the legal (Keiter 1998; Clarkson and Wood 2010) and health (Donohoe 2003) fields. Many universities have created programmes and institutes with the mandate of bringing an interdisciplinary, broad-based approach to dealing with the 'wicked problems' of climate change and poverty (Jentoft and Chuenpagdee 2009). Similarly, at international agencies as diverse as the

World Bank, which supports some principle of sustainability (although many might dispute the commitment to this), and the World Conservation Union (IUCN), the value of linking poverty and the environment has also been endorsed (Matthew *et al.* 2002; World Bank 2005).

Should ecologists engage with the human security framework?

> If you want advice on how to save the planet don't walk into an ecology department.
>
> Hugh Possingham, March 2009, Fenner Conference on Environment (Brown *et al.* 2010:48)

We view the human security framework as providing a more effective means of addressing the fundamental mismatch between a) what ecology tells us about limits to growth and b) the goal of sustainable development, compared with other approaches that have been proposed. The human security framework allows a number of the shortcomings identified by researchers examining the science–policy gap with respect to ecology to be addressed explicitly. The framework specifies a process by which many different groups, individuals, viewpoints and human values can be included in the act of policy formation (Hoogensen *et al.* 2009). By making the link between all actors clear, as well as the web-like, iterative nature of the policy formation process, the human security framework could enable ecologists who may be fearful of mixing their Popperian 'objective' scientific method approach (Odenbaugh 2005; Lowe *et al.* 2009) with human values (Norton 1998; Brown *et al.* 2010) to participate in policy discussions without fear of their research being somehow misused and abused in a subjective manner.

Furthermore, by clarifying how different groups participate in policy formation, the human security approach provides ecologists and scientists with a generic, adaptable mechanism for addressing the lack of clear policy guidelines that is common in these fields (Bocking 2002). Many ecologists tend to confuse the policy process with the political process. As Norton (1998) has pointed out, they are often suspicious of bringing any kind of human values into the equation, and believe that, since the lines between science and policy advocacy can become blurred, they should be cautious about participating in policy advocacy (Lackey 2007). In the latter case, the perception that science is a 'neutral, objective' activity needs to be placed in the context of a values-based choice being made to pursue this particular approach to scholarship (see Lélé and Norgaard 2005)!

Additionally, because the human security framework links the individual to the community and beyond, we view it as being more accessible than other approaches to mobilizing the messages of ecology. A case in point is the technical language of the resilience alliance (Hollings 2000) – at a recent policy conference about the Canadian Arctic, an eminent northern politician asked 'What does that mean for me?' after a presentation that sought to explain the resilience–stability–adaptation framework (Bazely, *in litt.*, 2 June 2009). Additionally, by making the political context of the issues clear, the human security approach clarifies how the framing of environmental and ecosystem issues is a political, and not just a conceptual and cultural, issue.

> 'When we use the term "environment" it makes it seem as if the problem is "out there" and we need to "fix it,"' said Susan Clark, Executive Director of the Columbia Foundation, who believes the Environmental Grantmakers Association should change its name. 'The problem is not external to us; it's us. It's a human problem having to do with how we organize our society. This old way of thinking isn't anyone's fault, but it is all of our responsibility to change.'
>
> (Shellenberger and Nordhaus 2004:12)

The human security approach provides a way for individuals to locate themselves in both local and global networks, and has the potential for mapping out transformational pathways for personal and community limit-setting. As social beings, humans are constantly setting limits and boundaries around our behaviour. Can those individuals in Global North countries having large individual or family ecological footprints implement behavioural limits that will be beneficial to people living in distant locations, that they will likely never meet? In fact, sociological data, from the *Cultural Creatives Project* in the USA (Ray and Anderson 2000; Ray 2008), are encouraging in this respect, because they indicate that a large proportion of the population of the USA does, in fact, care about these issues, with 83.2 per cent of those interviewed agreeing with the statement 'Each generation's duty is to make the world a better place for future generations' (Ray 2008).

Is human security as a framework relevant for understanding the impacts of ecological change on Arctic peoples?

Arctic communities face major environmental challenges, resulting not only from the already-stressful environment to which they have adapted over millennia, but also increasingly from socially induced changes to the relationship between culture, survival and freedom (Watt-Cloutier 2005; Krupnik *et al.* 2011). Additionally, the effects of environmental change on Arctic ecology will have enormous consequences for Arctic peoples, in the same way that a range of human-induced impacts, on both the natural and agricultural ecosystems of Africa, will continue to have huge repercussions for communities seeking to recover from the effects of war, famine and epidemics such as HIV-AIDS (Bilsborrow and DeLargy 1990; Dudley *et al.* 2002; Lewis 2005; Berhe

2007). Given the situation, an obvious question to ask is: 'Does the human security framework have any relevance for examining the impacts of ecological changes in the Arctic?'

Franklyn Griffiths (2008) asked a related, but slightly different, question in his 2008 conference paper: 'Is human security a good fit for the Arctic?' Griffiths, a political scientist who comes to human security from its home discipline, concluded, for various reasons, that human security is actually not that good a fit with current Arctic issues. Two of these reasons were that human security takes a tremendously individualistic perspective, and that at the time of writing no one at the Inuit Circumpolar Council (ICC) was explicitly using the terminology. We agree with the latter observation, and the GAPS (Gas, Arctic Peoples and Security, International Polar Year 2006–2011) project that we developed after the 2004 Tromsø Workshop had an explicit goal of using securities frameworks, including that of human security, to inform research in an Arctic context around oil and gas development (Hoogensen *et al.* 2009).

As previously stated, the human security framework parallels much of ecological thinking. Therefore, to say that human security is first and foremost about the individual, and therefore lacking in other dimensions, an argument that we have heard from other social scientists, in our view fails to take account of its broader implications and processes, which allow for linkage arrows pointing in all directions from the individual to the community and beyond.

The present realities and future potential impacts of climate change, which are occurring globally, will, if we are not explicit, obfuscate all local initiatives to improve individual and community wellbeing in the North. Without saying as such, political scientists such as Abele *et al.* (2009) have clearly set the stage for incorporating the obvious intersections between human security and ecological change in the Arctic:

> Global warming has certainly created economic opportunity in the North but it also presents difficulties, even dangers, for all human endeavours. Melting sea ice is opening northern waters to international traffic on a scale never seen before. This development, in concert with intensifying global interest in the vast oil, gas and mineral resources of the circumpolar basin, puts the regulation of Arctic shipping, the establishment and policing of boundaries and the management of potentially substantial offshore drilling on the policy agenda. At the same time, the negative impacts on local ecosystems and traditional activities such as hunting require greater attention from policy makers, and there is a need for measures to help northerners adapt to the changes already being felt.
>
> (Abele *et al.* 2009:4)

Not only does the human security concept provide a means for engaging with the impacts of ecological change on northern communities, but the interdisciplinarity and emphasis on policy translation necessitated by the approach are, in our view, critical tools for ensuring the efficacy of northern research.

Conclusions

In 2004, this chapter set out to evaluate the relevance of the human security framework for ecologists. Along the way, we undertook an experiment that brought ecologists into an interdisciplinary research team setting in the GAPS IPY project (Hoogensen *et al.* 2009), and used the human security framework to guide our research in local communities in the Arctic. The ecologists amongst us (Bazely and Tanentzap) were forced out of our comfort zone, and of being guilty of mainly speaking with fellow ecologists (*sensu* Shellenberger and Nordhaus 2004). The interdisciplinary research team setting required communication amongst academic disciplines that only rarely interact, and forced everyone to speak in much plainer language, not only when doing field work in local communities in the Arctic, but also with our academic colleagues.

We learned valuable lessons about planning research from a broader, community-based standpoint and found that it was difficult, time-consuming and challenging, but, in the long run, perhaps more effective in translating and mobilizing academically generated knowledge. Very importantly, the ecologists also learned to navigate the often-blurry lines between science and policy and politics. A key aspect of the latter was our experience and engagement with the field of political ecology (Forsyth 2003), which has often been a source of legitimate criticism of environmental science and ecology, with respect to its perceptions of neutrality and objectivity. It was particularly instructive for us to learn that political ecology research, like ecological research, often lacks explicit policy outcomes (Walker 2006; Bazely *in litt.*, 2 June 2009). Additionally, the diverse views across academia, including amongst political ecologists, about how interdisciplinary collaborations aimed at solving 'wicked problems' might proceed, have all underscored for us the value of the human security approach in highlighting and emphasizing a diversity of outcomes and opinions. In this regard, it is analogous to views in ecology about ecosystem resilience being increased by species diversity (Elmqvist *et al.* 2003).

References

Abele, F., Courchene, T. J., Seidle, F. L. and St-Hilaire, F. (2009) 'Introduction and overview', in F. Abele, T. J. Courchene, F. L. Seidle and F. St-Hilaire (eds) *The Art of the State, Volume IV: Northern Exposure: Peoples, Powers and Prospects in Canada's North*, Montreal Institute for Research on Public Policy (IRPP).

ACIA (2004) *Impacts of a Warming Arctic: Synthesis Report of the Arctic Climate Impact Assessment*, Cambridge: Cambridge University Press.

AFP (Agence France-Presse) (2006) *Russian Ecologists Despair over Lack of Govt Vision*, 9 May 2006. Online. Available at: http://www.terradaily.com/reports/Russian_Ecologists_Despair_Over_Lack_Of_Govt_Vision.html (accessed 23 July 2010).

AMAP (2002) *AMAP Assessment 2002: Persistent Organic Pollutants in the Arctic*, Oslo: Arctic Monitoring and Assessment Programme (AMAP).

—— (2009) *AMAP Assessment 2009: Human Health in the Arctic*, Oslo: Arctic Monitoring and Assessment Programme (AMAP).

Andrewartha, H. G. (1961) *Introduction to the Study of Animal Populations*, London: Methuen.

Armstrong, F. (2010) *The Age of Stupid* (videorecording), New York: Docurama Films.

Barlow, M. and Clarke, T. (2002) *Blue Gold: The Battle Against Corporate Theft of the World's Water*, New York: New Press.

BBCSS (2012) *The Science of Team Science Project*. Board on Behavioral, Cognitive, and Sensory Sciences, National Research Council, Washington, USA. Online. Available at: http://sites.nationalacademies.org/DBASSE/BBCSS/CurrentProjects/DBASSE_080231#.UTuzvaVQEyE (accessed 30 January 2013).

Begon, M., Harper, J. L. and Townsend, C. R. (2006) *Ecology*, 4th edition, Oxford: Blackwell Publishing.

Belovsky, G. E., Botkin, D. B., Crowl, T. A., Cummins, K. W., Franklin, J. F., Hunter Jr., M. L., Joern, A., Lindenmayer, D. B., MacMahon, J. A., Margules, C. R. and Scott, J. M. (2004) 'Ten suggestions to strengthen the science of ecology', *Bioscience*, 54:345–51.

Berhe, A. A. (2007) 'The contribution of landmines to land degradation', *Land Degradation & Development*, 18:1–15.

Bilsborrow, R. E. and DeLargy, P. F. (1990) 'Land use, migration, and natural resource deterioration: The experience of Guatemala and the Sudan', *Population and Development Review*, 16:125–47.

Bocking, S. (2002) *Linking Science and Policy for Urban Nonpoint Source Pollution in the Great Lakes Region*, report for the International Association for Great Lakes Research, unpublished. Online. Available at: http://www.iaglr.org/scipolicy/nps/ (accessed 6 December 2002).

Bosselman, F. P. (2004) 'Ecological science for lawyers: A review', *Southeastern Environmental Law Journal*, 12:143–58.

Brown, V., Gutknecht, J., Harden, L., Harrison, C., Hively, D., Jørgensen, C., Levi, T., Pflugeisen, J., Rovengno, P., Wang, Y., Wiedenmann, J. and Mangel, M. (2010) 'Understanding and engaging values in policy relevant science', *Bulletin of the British Ecological Society*, 41:48–56.

Chicago Tribune (2005) 'Editorial: A new environmentalism', 21 February 2005. Online. Available at: http://articles.chicagotribune.com/2005-02-21/news/0502210053_1_coal-burning-environmentalists-ted-nordhaus (accessed 22 February 2005).

Christensen, J. B. (2011) *Homeless in a Homeland: Housing (In)security and Homelessness in Inuvik and Yellowknife, Northwest Territories, Canada*. Doctoral Dissertation. Montreal: McGill University.

Clarkson, S. and Wood, S. (2010) *A Perilous Imbalance: The Globalization of Canadian Law and Governance*, Vancouver: University of British Columbia Press.

Clutton-Brock, T. H. and Pemberton, J. (eds) (2004) *Soay Sheep: Dynamics and Selection in an Island Population*, Cambridge: Cambridge University Press.

Cohen, J. E. (1995) 'Population growth and Earth's human carrying capacity', *Science*, 269:341–6.

Commission on Human Security (2003) *Human Security Now*, p. 168. Online. Available at: http://reliefweb.int/sites/reliefweb.int/files/resources/91BAEEDBA50C6907C1256D19006A9353-chs-security-may03.pdf (accessed 23 November 2003).

Costanza, R., d'Arge, R., de Groot, R., Farber, S., Grasso, M., Hannon, B., Limburg, K., Naeem, S., O'Neill, R. V., Paruelo, J., Raskin, R. G., Sutton, P. and van den Belt, M. (1997) 'The value of the world's ecosystem services and natural capital', *Nature*, 387:253–60.

Crilly, G. (2010) 'Sustainable planning is cure for growth delusion', *Irish Times op ed*, 19 July 2010. Online. Available at: http://www.irishtimes.com/newspaper/opinion/2010/0719/1224275016809.html (accessed 21 July 2010).

Dalby, S. (2002) *Environmental Security*, Minneapolis: University of Minnesota Press.

De Villiers, M. (2000) *Water: The Fate of Our Most Precious Resource*, Toronto: Stoddart.

Diamond, J. (2005) *Collapse: How Societies Choose to Fail or Succeed*, New York: Viking Penguin.

Donohoe, M. (2003) 'Causes and health consequences of environmental degradation and social injustice', *Social Science & Medicine*, 56:573–87.

Dovers, S. R., Norton, T. W. and Handmer, J. W. (1996) 'Uncertainty, ecology, sustainability and policy', *Biodiversity and Conservation*, 5:1143–67.

Dudley, J. P., Ginsberg, J. R., Plumptre, A. J., Hart, J. A. and Campos, L. C. (2002) 'Effects of war and civil strife on wildlife and wildlife habitats', *Conservation Biology*, 16(2):319–29.

Ehrlich, P. and Ehrlich, A. (2009) 'Enough already', *New Scientist*, 191(2571):47–50.

Einarsson, N., Larsen, J. N., Nilsson, A. and Young, O. R. (2004) *Arctic Human Development Report*, Akureyi: Stefansson Arctic Institute.

Elmqvist, T., Folke, C., Nyström, M., Peterson, G., Bengtsson, J., Walker, B. and Norberg, J. (2003) 'Response diversity, ecosystem change, and resilience', *Frontiers in Ecology and the Environment*, 1(9):488–94.

Evert, R. F. and Eichhorn, S. E. (2013) *Raven Biology of Plants*, 8th edition, New York: W. H. Freeman and Company.

Forsyth, T. (2003) *Critical Political Ecology*, London: Routledge.

Griffiths, F. (2008) 'Not that good a fit? "Human security" and the Arctic', *Arctic Security in the 21st Century Conference, April 2008: Report*, Vancouver: Simon Fraser University School for International Studies, pp. 56–60.

Haeuber, R. and Ringold, P. (1998) 'Ecology, the social sciences, and environmental policy', *Ecological Applications*, 8(2):330–1.

Harte, J. (2007) 'Human population as a dynamic factor in environmental degradation', *Population & Environment*, 28:223–36.

Hollings, C. S. (2000) 'Theories for sustainable futures', *Conservation Ecology*, 4:2. Online. Available at: http://www.consecol.org/vol4/iss2/art7/ (accessed 3 February 2010).

Hoogensen, G., Bazely, D., Christensen, J., Tanentzap, A. and Bojko, E. (2009) 'Human security in the Arctic: Yes, it is relevant!', *Journal of Human Security*, 5:1–10.

Hutton, N. (1996) 'Interactions between the formal UK school science curriculum and the public understanding of science', *Public Understanding of Science*, 5:41–53.

Jentoft, S. and Chuenpagdee, R. (2009) 'Fisheries and coastal governance as a wicked problem', *Marine Policy*, 33:553–60.

Jordan, R., Singer, F., Vaughan, J. and Berkowitz, A. (2009) 'What should every citizen know about ecology?', *Frontiers in Ecology and the Environment*, 7(9):495–500.

Kahan, D. M., Peters, E., Wittlin, M., Slovic, P., Ouellette, L. L., Braman, D. and Mandel, G. (2012) 'The polarizing impact of science literacy and numeracy on perceived climate change risks', *Nature Climate Change*, 2:732–5.

Keiter, R. B. (1998) 'Ecosystems and the law: Towards an integrated approach', *Ecological Applications*, 8(2):332–41.

Krause, K. (2004) 'Human security: More than just a good idea?', in M. Brzoska, M. and P. J. Croll (eds) *Promoting Security: But How and For Whom?* Brief 30, Bonn: Bonn International Center for Conversion, pp. 43–6. Online. Available at: http://www.bicc.de/publications/publicationpage/publication/promoting-security-but-how-and-for-whomcontributions-to-biccs-ten-year-anniversary-conference/ (accessed 21 February 2013).

Krauss, C. (1991) *Inside Central America: Its People, Politics and History*, New York: Simon and Schuster.

Krebs, C. J. (1978) *Ecology: The Experimental Analysis of Distribution and Abundance*, 2nd edition, New York: Harper and Row.

—— (1988) *The Message of Ecology*, New York: Harper and Row.

Krupnik, I., Allison, I., Bell, R., Cutler, P., Hik, D., Lopez-Martinez, J., Rachold, V., Sarukhanian, E. and Summerhayes, C. (eds) (2011) *Understanding Earth's Polar Challenges: International Polar Year 2007–2008*, University of the Arctic, Rovaniemi, Finland/CCI Press (Printed Version), Edmonton, Alberta, Canada, and ICSU/WMO Joint Committee for International Polar Year 2007–2008.

Kuemmerle, T., Radeloff, V. C., Perzanowski, K. and Hostert, P. (2006) 'Cross-border comparison of land cover and landscape pattern in Eastern Europe using a hybrid classification technique', *Remote Sensing of Environment*, 103:449–64.

Lach, D. (2003) 'Nonsolutions for nonecologists', *Bioscience*, 53:888–90.

Lackey, R. T. (2007) 'Science, scientists, and policy advocacy', *Conservation Biology*, 21(1):12–17.

Ladle, R. J. and Gillson, L. (2009) 'The (im)balance of nature: A public perception time-lag?', *Public Understanding of Science*, 18:229–42.

Leiserowitz, A. (2009) *International Public Opinion, Perception, and Understanding of Global Climate Change*. Report of the Yale Project on Climate Change Communication, Yale University, USA. Online. Available at: http://environment.yale.edu/climate/publications/international-public-opinion-perception-and-understanding-of-global-cl/ (accessed 9 March 2013).

Lélé, S. and Norgaard, R. B. (2005) 'Practicing interdisciplinarity', *BioScience*, 55:967–75.

Leshner, A. I. (2004) 'Science at the leading edge', *Science*, 303:729.

Lewis, S. (2005) *Race Against Time: Searching for Hope in AIDS-Ravaged Africa*, Toronto: Anansi Press.

Liu Institute (2004) 'Science and its policy impacts', Monthly Seminar Series poster.

Lloyd, B. (2009) 'The growth delusion', *Sustainability*, 1:516–36.

Lowe, P., Whitman, G. and Phillipson, J. (2009) 'Ecology and the social sciences', *Journal of Applied Ecology*, 46:297–305.

Lutz, W., Sanderson, W. and Scherbov, S. (2001) 'The end of world population growth', *Nature*, 412:543–5.

Mason, B. (2003) 'Ecologists attack endangered-species logjam', *Nature*, 426:592.

Matthew, R., Halle, M. and Switzer, J. (2002) *Conserving the Peace: Resources, Livelihoods and Security*, International Institute for Sustainable Development and IUCN – The World Conservation Union. Online. Available at: http://www.iisd.org/publications/publication.asp?pno=477 (accessed 27 August 2010).

McGlade, J. (2004) 'How many Earths?', *Our Planet (UNEP)*, 15(4):19–20.

Norton, B. G. (1998) 'Improving ecological communication: The role of ecologists in environmental policy formation', *Ecological Applications*, 8(2):350–64.

Odenbaugh, J. (2005) 'Ecology', in S. Sarkar and J. Pfeifer (eds) *The Philosophy of Science: An Encyclopedia, Volume 1*, London: Routledge.

Otto, S. (2011) *Fool Me Twice: Fighting the Assault on Science in America*, Pennsylvania: Rodale.

Peters, R. H. (1991) *A Critique for Ecology*, Cambridge: Cambridge University Press.

Pfirman, S., Martin, P., Berry, L., Fletcher, M., Hempel, M., Southard, R., Hornbach, D. and Morehouse, B. (2007) *Interdisciplinary Hiring, Tenure and Promotion: Guidance for Individuals and Institutions*. Council of Environmental Deans and Directors. Online. Available at: http://www.ncseonline.org/programs/education-careers/cedd/projects/faculty-development# (accessed 22 April 2010).

Popli, R. (1999) 'Scientific literacy for all citizens: Different concepts and contents', *Public Understanding of Science*, 8:123–37.

Ray, P. H. (2008) *The Potential for a New, Emerging Culture in the U.S.* Online. Available at: http://www.wisdomuniversity.org/CCsReport2008SurveyV3.pdf (accessed 8 March 2013).

Ray, P. H. and Anderson, S. R. (2000) *The Cultural Creatives: How 50 Million People are Changing the World*, New York: Three Rivers Press.

Sagar, A. D. and Najam, A. (1998) 'The human development index: A critical review', *Ecological Economics*, 25:249–64.

Seifert, R. (2005) *Home Sick: How Medical Debt Undermines Housing Security*, Boston: The Access Project.

Shellenberger, M. and Nordhaus, T. (2004) *The Death of Environmentalism: Global Warming Politics in a Post-Environmental World*. Online. Available at: http://grist.org/article/doe-reprint (accessed 13 January 2005).

Slobodkin, L. B. (2003) *A Citizen's Guide to Ecology*, Oxford: Oxford University Press.

Stauffer, R. C. (1957) 'Haeckel, Darwin, and ecology', *The Quarterly Review of Biology*, 32(2):138–44.

Stokstad, E. (2005) 'Will Malthus continue to be wrong?', *Science*, 309:102.

Summerhayes, V. S. and Elton, C. S. (1923) 'Bear Island', *Journal of Ecology*, 11:216–33.

Tadjbakhsh, S. and Chenoy, A. M. (2007) *Human Security: Concepts and Implications*, New York: Routledge.

Tanentzap, A. J., Bazely, D. R., Williams, P. A. and Hoogensen, G. (2009) 'A human security framework for the management of invasive nonindigenous plants', *Invasive Plant Sciences and Management*, 2:99–109.

Townsend, C. R., Begon, M. and Harper, J. L. (2003) *Essentials of Ecology*, 2nd edition, Oxford: Blackwell Publishing.

UNDP (1990) *Human Development Report 1990*, New York: Oxford University Press.

UNEP (2013) *United Nations Environment Programme – Environment for Development*. Online. Available at: http://www.unep.org/sustainability/ (accessed 8 February 2013).

Vaughan, H. H., Waide, R. B., Maass, J. M. and Ezcurra, E. (2007) 'Developing and delivering scientific information in response to emerging needs', *Frontiers in Ecology and the Environment*, 5(4):w8–w11.

Victor, P. (2008) *Managing Without Growth: Slower by Design, Not Disaster*, Cheltenham: Edward Elgar.

Wackernagel, M. and Rees, W. E. (1996) *Our Ecological Footprint: Reducing Human Impact on the Earth*, Philadelphia: New Society Publishers.

Walker, P. A. (2006) 'Political ecology: Where is the policy?', *Progress in Human Geography*, 30(3):382–95.

Watt-Cloutier, S. (2005) *Petition to the Inter American Commission on Human Rights Seeking Relief from Violations Resulting from Global Warming caused by Acts and Omission of the United States, 7 December 2005*. Online. Available at: http://www.inuitcircumpolar.com/files/uploads/icc-files/FINALPetitionICC.pdf (accessed 25 July 2010).

WCED (World Commission on Environment and Development) (1987) *Our Common Future*, Oxford: Oxford University Press.

Weiner, J. (1995) 'On the practice of ecology', *Journal of Ecology*, 83:153–8.

Wilson, J. S., Brothers, T. S. and Marcano, E. J. (2001) 'Remote sensing of spatial and temporal vegetation dynamics in hispaniola: A comparison of Haiti and the Dominican Republic', *Geocarton International*, 16(2):7–18.

World Bank (2005) *World Bank Group: Environmental Themes Webpage*. Online. Available at: http://go.worldbank.org/X0NHVF3TI0 (accessed 2 March 2005).

Wright, R. (2004) *A Short History of Progress*, Toronto: Anansi Press.

WWF International (1998) *Living Planet Report: Overconsumption is Driving the Rapid Decline of the World's Natural Resources*, Gland: WWF International.

—— (2000) *Living Planet Report 2000*, Gland: WWF International.

—— (2012) *Living Planet Report 2012: Biodiversity, Biocapacity and Better Choices*, Gland: WWF International.

Part III

Health security

Chapter 9

Telemedicine as a tool for improving human security

Line Linstad

> The delivery of health care services, where distance is a critical factor, by all health care professionals using information and communication technologies for the exchange of valid information for diagnosis, treatment and prevention of disease and injuries, research and evaluation, and for the continuing education of health care providers, all in the interests of advancing the health of individuals and their communities.
>
> (WHO 1998:10)

Introduction

Information and communications technology (ICT) has changed and will continue to change our communication patterns. It is a tool that makes the world 'smaller' and increases our access to information and services on a truly global scale. The potential exists for increasing and broadening people's access to healthcare services and information. From an Arctic human security perspective, telemedicine can be seen as a tool that could increase the personal well-being of Arctic peoples, both within the individual states composing the Arctic, but also across the region. This chapter aims to define and explore the assumptions that, if met, would place telemedicine in the human security framework as a tool that can serve to operationalize the human security concept. The main assumptions that I have made are that:

1. Health security today is closely linked to the state;
2. Counter-intuitively, it is probably easier to develop international cooperation on health matters that are not directly linked to the process of diagnosing and treating patients;
3. Health security is one of several sectors that comprise a 'web of security' in a region;
4. The national health legislation and the national reimbursement systems may often constrain the development of regional healthcare services;
5. Telemedicine can be used as a tool to enhance health security in the Arctic region.

My goal is to explore and evaluate these assumptions en route to reaching a conclusion about the relationship between telemedicine and human security.

Traditional and alternative views of security

Health, medicine and access to healthcare have not been a focus of the security studies community. The development of security studies as an academic field of inquiry, separate from broader international relations theories and political philosophies, took a decidedly narrow focus regarding what could be considered relevant to security discourses. Despite centuries of diverse theoretical traditions that recognized security from the position of individuals as well as states (Rothschild 1995; Hoogensen 2005a), security was reduced to one actor in one context, looking at the survival of the state in the international system.

This focus on one actor as an independent actor within a system fighting for its survival has been difficult to sustain as the only way to understand security. If the focus is only states, even these abstract actors develop complex relationships with other states, observable as 'regions'. The Arctic presents an interesting case in this regard. The Arctic is a region composed of multiple states. But it is unique as a region, as most of the states that are considered Arctic have only a small part of their territory within the Arctic circle (Hoogensen 2005b). Theorists such as Barry Buzan developed the concept of the 'security complex' (Buzan 1990:13) to provide a way of understanding that security is a phenomenon that must be studied at a regional level. Buzan characterizes a regional security complex as 'a set of states whose major security perceptions and concerns are so interlinked that their national security cannot reasonably be analysed apart from one another' (Buzan 1990:13). In their book *Security: A New Framework for Analysis*, Barry Buzan, Ole Wæver and Jaap de Wilde (Buzan *et al.* 1998:11) emphasize the notion that

> all states in the system are enmeshed in a global web of security interdependence. But because most political and military threats travel more easily over short distances than over long ones, insecurity is often associated with proximity. *Most states fear their neighbours* more than distant powers; security interdependence as a whole is far from uniform. The normal pattern of security interdependence in a geographically diverse, anarchic international system is one of regionally based clusters, which we label security complexes [emphasis added].

The above understanding of security is largely rooted in the realist tradition of international relations and reflects some of the dominant thinking in security analysis. It focuses on the state-based and military aspects of security. As already discussed in the introductory chapters of this book, after the end of the Cold War many scholars argued for the need to widen the focus of security studies so as to integrate more sectors. The likelihood of military aggression has, today, greatly declined in several global regions, including the Arctic (Human Security Report 2005). It is highly improbable that any Arctic state will turn to war as a means of resolving conflict. Arctic regional security is less based on fear than it is on cooperation.

The 'realist' school and its approach to international politics and security has been frequently criticized. This traditional approach, critics say, failed to predict and explain the end of the Cold War. It also is an approach that over time (and particularly with the advent of neorealism) has not been interested in, let alone adequately accounted for, what security means at the individual level. Furthermore, it does not take into account the behaviour of the states in the international system as a synthesis of collective human behaviour. The reality, that state interests are usually a reflection of the views of the elite, is often overlooked. These views are products of the elite's construct or perceptions of what constitutes threats to the survival of the state, and what will be required to maintain their elite position within that state. The behaviour of the state may therefore be viewed as a means of protecting the elite's position in society, while the elite may not care for the rest of the society. By acting this way, the elite may be failing to guard the human rights of other state citizens. In examining matters of security in the Arctic, there is a case to be made for applying a broader approach than that of the traditional realist approach to security studies. By broader, I am referring to the need to extend the analysis into sectors such as the economy, the environment, and issues of culture and national identity, and include the perspectives of a broader range of citizens. Buzan *et al.* concluded:

> the liberal project does seem to have succeeded in marginalizing military security and along with it the approach of traditional security studies. But in so doing it has raised new security problems that can only be handled in a multisectoral framework.
>
> (1998:212)

It was a bold step for the Copenhagen school, represented by Buzan *et al.*, to push the debate to include more sectors (beyond the military and state) to better understand how security operates and is practised. Their revised definition of the security complex stems from the need that when studying security matters in one sector, it is important to keep in mind that all sectors are interdependent. For example, globalization of the world economy has impacts on the environment that leads to concerns about 'sustainable growth'.

Buzan *et al.*'s development of the notion of security complexes harmonizes it with 'the extended definition of security' as:

> a set of units whose major processes of securitization, desecuritization, or both are so interlinked that their security problems cannot reasonably

> be analyzed or resolved apart from one another. The formative dynamics and structure of a security complex are normally generated by the units within it – their security perceptions of, and interaction with, each other. But they may also arise from collective securitizations of outside pressures arising from the operation of complex metasystems, such as the planetary environment or the global economy.
>
> (1998:201)

The low likelihood of war occurring in the Arctic does not prevent the consideration of the Arctic region as a security complex. Various Arctic states share common security problems, from ensuring sovereignty, to environmental politics stemming from issues such as pollution and climate change, the rights of indigenous people to live their lives according to cultural traditions and issues of access to employment and healthcare. The Arctic can be characterized by extremes in population, where it is a vast area with 'crowded' cities and rural areas that are scarcely populated. This pattern of population distribution can result in centralizing pressures that force people to migrate to the cities. By centralizing pressures I mean that access to social goods such as schooling, healthcare and employment is better (and becomes increasingly centralized) in these urban rather than rural areas.

The Arctic is unique as a security complex. It is a region itself, but it is also a region of many regions, each of which is a part of a different security complex. The Scandinavian countries are in the European security complex, while the US and Canada can be seen as part of the North American security complex. Russia is part of the European security complex and is a security complex itself, while at the same time being composed of many different nations and regions. Additionally, some Arctic regions of Russia are in Asia. Different Arctic states may therefore have common security challenges in several sectors and they may also have security challenges (threats) coming from other parts than the Arctic. How one applies either of Buzan *et al.*'s (1998) definitions depends on what we are analysing. It can also lead one to lose focus, and forget what a core objective should be – the security of citizens within a region.

Human security and the need for a wider approach to the study of security

The outline of the Report of the Commission on Human Security (CHS) states that 'as security challenges become more complex and various new actors attempt to play a role, we need a shift in paradigm. The focus must broaden from the state to the security of people – to human security' (CHS Outline 2003:1). The CHS report requests that all suggestions made in support of a human security approach be accepted by the member states. These states are the core actors, able to suggest measures, and accept and implement change on a national and international scale. Other relevant actors, such as non-governmental organizations (NGOs), are not recognized for their capacity to mobilize national and international forces (never mind what they might be doing at sub-state/society and individual levels). Human security is thus framed within a state-centric logic, almost aligning itself with a traditionalist's view on security or, at best, the widened approach of Buzan *et al.* (1998), despite the rhetorical acceptance of other actors playing a significant role. Although, as the CHS report (CHS Outline 2003) states, 'Human security complements state security, furthers human development and enhances human rights', their remains a need to recognize how non-state actors (who are acknowledged as having a role) indeed contribute to human security.

Linking human security and health security: telemedicine as a tool for facilitating cooperation

In exploring whether telemedicine could or should be placed within a human security framework, I will first explore the assumption that 'health security today is closely linked to the state'. When I use the term 'the state' I use it synonymously with 'government'. I view state action to be a result of government will. To do so, I will begin where traditional security leaves off – with the defence of the state.

One could argue that if the state succeeds in making its inhabitants secure from health threats, then this may increase its legitimacy. If it fails, this may have a direct impact on the ruler's legitimacy. This is in line with what the UN report states as a historical fact: 'Health security and military security are directly related. Indeed from a historical perspective, the legitimacy of rulers has depended on their capacity to protect the health of the public, through military and other means' (CHS 2003:97).

In the military sector, states are 'required' to be able to protect their territory from external aggression. This has led to the existence of armies requiring personnel, and without access to sufficient manpower, the state cannot protect itself. Likewise in the health sector, if the state wants to protect its inhabitants from health security threats, there is also a need for manpower. Without doctors and nurses there would not be any healthcare. Thus, both in the military and the health sectors there is a need to secure national self-sufficiency using manpower. In the military sector, states can increase security through alliances. In healthcare this may be possible too. This raises the question of whether states should organize their hospital services across borders. This is possible and may even be cost effective. The main constraints may be legal and financial.

If the government wants to achieve legitimacy through health politics and health security (ensuring the provision of healthcare and mitigating against health threats), this has to be done through political strategies and measures. I will divide health politics into three parts:

1 The concern for reducing threats from international epidemics – from this follows a need for a political strategy on how to counter this kind of threat.
2 The concern for developing and managing a healthcare system of high quality – from this follows a need for political strategies that aim to attain a healthcare service of good quality that is accessible for patients and cost effective.
3 The concern for developing and maintaining an educational standard for staff – from this follows a need for strategies that will educate enough personnel for the service, leading to self-sufficiency.

What are the possible threats in health security?

1 Epidemics;
2 Denial of access to healthcare services;
3 Insufficient access to healthcare personnel.

While there may be other threats, I focus on these three, because they are likely to be among the most important, as acknowledged by the World Health Organization (WHO). Telemedicine is an obvious tool for use in both preventing the spread of epidemics and increasing access to healthcare. There are two main parts to telemedicine. The first focuses on the use of technology to diagnose and treat patients at a distance, and the second focuses on technology that may be used as a tool for educating staff and patients. This second part will be addressed in my discussion of threats arising from denial of access to healthcare services.

The threat of epidemics

The 2003 CHS report states that the awareness of:

> global interdependencies in health is growing among the public and political leaders. Public financing for global health has begun to increase ... Non governmental organizations are proliferating, and media coverage of health and security has increased markedly. New institutional arrangements are being established, such as the Global Fund to Fight AIDS, Tuberculosis, and Malaria.
>
> (CHS 2003:102)

In order to counter these epidemic-based threats to human security, international cooperation is essential. When states are aware that an epidemic is moving rapidly, they are probably more willing to cooperate in order to increase health security. The CHS report (2003:102) points out that 'Reducing health threats to human security will require unprecedented cooperation among the public and political leaders'. The Commission therefore proposes that:

> The world urgently needs primary health services and national disease surveillance systems – formally networked into a global system. Information, data and analysis can help to identify disease outbreaks and strengthen efforts to control their spread. So every country should have primary health services and disease surveillance capacities ... These national systems should be linked through networks in a truly global system, to allow for rapid knowledge sharing and rapid responses.
>
> (CHS 2003:139)

In meeting and combating the threat that is constituted by the spread of epidemics, telemedicine can facilitate immediate access to important information through international early warning systems. For example, the threats arising from the spread of tuberculosis from Russia to Europe have led to several projects in the Arctic, including those that incorporate telemedicine. The spread of strains of tuberculosis that are resistant to normal medication poses a genuine threat to both the European security complex and to the Arctic security complex. AIDS is spreading in Russia too. People are travelling more widely, and the sex industry in Russia is the source of a potential threat (Dobriansky 2003). This case illustrates how threats to health security do not stop at borders. Further evidence of this can be found in more recent discussions about the uses of telemedicine for potential increases in vector-borne diseases resulting from climate and other environmental changes, such as meningitis (UN-SPIDER 2008). An increased awareness about how to manage disease effectively is needed, and telemedicine plays therefore a crucial role in health security. These technologies and processes are relevant for Arctic populations, as well as populations in other regions of the world, not least in the global South where diseases such as malaria and cholera are in dire need of management and control.

Infrastructure is needed in order to implement telemedicine in the Arctic, where a combination of satellites and cables are required. The infrastructure serves as a 'communication road' through which patients and health personnel can connect with each other. They can communicate through the Internet, by videoconferencing and by e-mail. E-mails may be accompanied by attachments, which can be sound and/or picture files. A sound file could be an ultrasound file or sound of the heart. Today the Arctic has a good coverage of satellites. However, this infrastructure has maintenance costs, the satellites are private and users must pay for access to this infrastructure. These costs can serve as barriers to the use of telemedicine. The possibility that increased cooperation among Arctic states can reduce costs and increase the stability of the infrastructure is an area to be explored. Additionally, if users in other sectors, such as education, cooperate, this can increase the volume of traffic in the network and thereby reduce costs. However, these benefits can only be realized through cooperation among states. This is a question that was debated in the Arctic Council in October 2003. The implications of this technology are relevant for states, regions and the broader international community.

Health threats arising from denial of access to healthcare services

The development of national healthcare systems is closely linked to the nation-building process and to the ability of the state to act as a resource allocator. This is certainly the case in Norway. In Norway, the process has been driven by the conflict between centralizing and decentralizing forces, in which local authorities claim more independence while the central government asserts its need for control. The recent exercise in hospital reform is a result of this on-going conflict. This reform process has centralized the control of the Norwegian hospital services. Because Norway is by virtually all standards considered to be a wealthy country, the government has many resources to allocate, but medical needs can have no limits, which has resulted in the need for cost-control. The health reforms that have taken place is an experiment aimed at improving the government's ability to control escalating costs in the hospital sector, and is not limited to the Norwegian healthcare system. Politicians across the Arctic are interested in creating a healthcare service that produces services of good quality whilst remaining cost effective (Helse- og Omsorgsdepartement 2009).

Patient consultations remain the core activity of national healthcare systems. This system is a product of both political values and ideas in the nation-building process and of the medical view on what a healthcare system is supposed to deliver. The healthcare system has clearly defined borders and, within the national system, hospitals may have a responsibility for certain catchment areas. Additionally, doctors who are general practitioners may be responsible for particular districts within a region. What would motivate a move to international cooperation in this field? Perhaps a need for greater capacity for treatment, and pressure for better access to healthcare. Increased Internet use has led to patients having better access to medical information as well as to treatment possibilities in different hospitals. This increased patient awareness may affect the nature of their questions to healthcare personnel and lead to increased demands for treatment.

Several years ago the Norwegian government opened a 'heartbridge' between Norway and Britain with the goal of reducing the number of patients waiting for heart surgery. There are also other examples of patient advocacy groups in Norway that have managed to lobby the Norwegian parliament to increase healthcare spending for certain groups. However, this kind of increased service tends to have an *ad hoc* character, and the changes are usually never integrated into a 'national service' or regional framework that would better assist access for people, particularly in more remote areas. This may be related to the nature of the state's responsibility for the actual service being delivered, and the way this responsibility must be transformed into a national and regional system of laws and financial systems. The questions are: how effectively can telemedicine address the gap in access to information for patients, and how can it support the effectiveness of practitioners in more isolated areas?

Conclusion

Health security is crucial to human, state and regional security. Populations in the Arctic are subject to widely different health threats and concerns (rapidly increasing diabetes, tuberculosis and suicide rates among Inuit populations, significantly reduced life expectancy for men as well as excessive alcohol and tobacco consumption in Russia, and not least aging populations across the Arctic (World Bank 2008; Health Canada 2012)). Healthcare access and healthcare personnel are crucially important for quality healthcare delivery, which in turn affects both human and state security. Russia's potential as a global power is weakened by poor health outcomes in its population (World Bank 2008). The crucial role healthcare plays in security has been overlooked, but fits well into the dialogue on security on societal, national and regional levels. As Buzan *et al.* (1998) claimed in their seminal work *Security: A New Framework for Analysis*, security is about existential threats. If we do not exist, there is not much else to worry about. A state cannot exist without a population; a population will cease to exist without adequate access to healthcare. This is true for all states, regions and parts of the world.

If governments across the Arctic can better integrate telemedicine into the way that healthcare is organized, telemedicine could increase access to and delivery of healthcare. Telemedicine can also cross territorial borders since its technology does not stop at borders. Access to healthcare depends on how governments prefer to organize their health system. So it is the organizational model that has to change if access is to be increased by international cooperation. Telemedicine is just one of many tools that can be used in an organizational change to increase access to healthcare and therefore health security in isolated Arctic communities.

References

Buzan, B. (1990) *The European Security Order Recast: Scenarios for the Post-Cold War Era*, London: Pinter Publishers.

Buzan, B., Wæver, O. and de Wilde, J. (1998) *Security: A New Framework for Analysis*, London: Lynne Rienner Publishers.

CHS (Commission on Human Security) (2003) *Human Security Now: The Final Report*, New York: The Commission on Human Security.

CHS Outline (2003) *Outline of the Report of the Commission on Human Security*. Online. Available at: http://unocha.org/humansecurity/chs/finalreport/Outlines/outline.pdf (accessed 25 March 2013).

Dobriansky, P. J. (2003) *The Emerging Security Threat of HIV/AIDS: Russia*. Department of State for Global Affairs. Washington, DC: US Department of State.

Health Canada (2012) *First Nations and Inuit Health: Diseases and Health Conditions*. Online. Available at: http://www.hc-sc.gc.ca/fniah-spnia/diseases-maladies/index-eng.php (accessed 25 March 2013).

Helse- og Omsorgsdepartement (2009) *St.meld.nr 47 (2008-2009): Samhandlingsreformen: Rett behandling-på rett sted-til rett tid*. Oslo: Den kongelige helse- og omsorgsdepartement.

Hoogensen, G. (2005a) *International Relations, Security and Jeremy Bentham*, London and New York: Routledge.

—— (2005b) 'Bottoms up! A toast to regional security?', *International Studies Review*, 7:269–74.

Human Security Report (2005) *Human Security Report: War and Peace in the 21st Century*, Oxford: Oxford University Press.

Rothschild, E. (1995) 'What is security?', *Daedalus*, 124(3):53–98.

UN-SPIDER (2008) *Disaster Medicine, Telemedicine, and Integrated Vector Management*. Online. Available at: http://www.un-spider.org/health-support-guides/disaster-medicine-telemedicine-and-integrated-vector-management (accessed 25 March 2013).

WHO (World Health Organization) (1998) *A Health Telematics Policy: In Support of WHO's Health-for-All Strategy for Global Health Development*, Geneva: WHO.

World Bank (2008) *Better Outcomes Through Health Reforms in the Russian Federation: The Challenge in 2008 and Beyond*, World Bank: Europe and Central Asia Human Development Department/Russian Federation Country Management Unit.

Chapter 10

Health and human security

Communicable diseases in the post-Soviet Arctic

Lars Rowe, Elana Wilson Rowe and Geir Hønneland[1]

Introduction

The idea of human security was developed to encourage states and international organizations to direct similar amounts of energy and resources to so-called 'soft issues', such as health, women's issues and child safety, as they would towards the more familiar aspects of national security, such as border defence and the maintenance of militaries. The concept of human security has been applied primarily in the developing world, or the 'Global South'. As stated in Chapter 4, this tends to imply that human security issues do not exist in First World or Northern states. The present focus of human security concerns can, consequently, mask a host of shared concerns, and hinder those of us who are working actively on issues such as health, protection of women and children and migration from collaboration across the developed/developing world divide. Public health problems, such as the spread of HIV/AIDS and tuberculosis in the Former Soviet Union (FSU) and the efforts at international cooperation outlined in this chapter, form an issue of human security around which such collaboration could occur. The goals of the 2004 Tromsø conference, *Human Security in the Arctic*, and of this volume are to explore the extent to which the concept of human security may contribute to Arctic research and vice versa, and to outline present and future threats in the Arctic regions of Northern states.

In this chapter, we draw upon the earlier research of Hønneland and Rowe (2004) in which the Task Force on Communicable Disease Control in the Baltic Sea Region (referred to as the Task Force below) was evaluated. The Task Force was a joint venture among 11 states surrounding the Baltic Sea, and its mandate was to establish local-scale projects in the health systems and prison health services addressing the control of tuberculosis and HIV/AIDS.[2] In our research we conducted interviews with approximately 100 individuals within the health sector, the bulk of which were based in Russia and the Baltic countries, in order to assess their perceptions of the Task Force specifically, and the challenges facing the health sector generally. Our main objective in this chapter is to place the threat of communicable diseases in the FSU in a human security framework.

1. First, we outline the threats that communicable diseases, predominantly AIDS and tuberculosis, pose in Northwest Russia and describe the efforts of the Task Force to address these issues.
2. Second, we describe the general attitudes of Russian participants involved in the Task Force programmes to elucidate some of the dispositions that can facilitate or challenge international collaboration around health issues in Russia.
3. Third, in order to clarify the challenges faced by those aiming to develop international collaboration, we describe the reception by Russians of a tuberculosis prevention strategy that was created initially by the World Health Organization (WHO) to combat tuberculosis in developing countries. This example illustrates some of the challenges associated with effective sharing of health lessons between the developing and developed world.
4. We then describe how the Task Force's emphasis on and support for programmes focusing on the spread of communicable diseases amongst marginalized groups, such as drug users, prisoners and prostitutes, was received by Russian and Baltic programme participants. This is particularly critical to assessing the potential role of the human security discourse in Russia, because human security has often been used as a tool to direct attention towards society's most vulnerable members.
5. Finally, in order to augment the prior discussion of the role of attitudes and dispositions, we highlight structural peculiarities that can challenge the implementation of new approaches to public health in the FSU.

AIDS and tuberculosis in Russia and the Baltic states

At the end of the 1990s, the health situation in Russia and the Baltic states was causing serious concern among medical experts and officials in the West. Life expectancy had decreased dramatically since the break-up of the Soviet Union, mainly as a result of diseases caused by malnutrition, smoking and alcohol consumption (Task Force 2000:9–11; McKee 2001). Furthermore, HIV/AIDS had emerged as an increasingly widespread cause of suffering in the post-Soviet era. Tuberculosis, a disease that in Western societies had been more or less eliminated or at least controlled effectively, was also re-emerging. Faced with a looming tuberculosis epidemic, centuries-old fears of infectious diseases spreading like wildfire from person to person and country to country were being rekindled. Although Russia and the Baltic states were considered to be most at risk from tuberculosis and HIV/AIDS, some went so far as to suggest that the epidemics could destabilize the political climate in Northern Europe as a whole.

The most severe effects of both the HIV/AIDS and tuberculosis epidemics have been felt in Russia. In the case of tuberculosis, after levelling out at around 30 cases on average per 100,000 of the population at the beginning of the 1990s, the rate rose dramatically towards the end of the decade, reaching approximately 80 cases per 100,000 people. The HIV figures reflected even greater infection levels, rising steeply towards the end of the 1990s, and prompted dramatic statements from Western medical experts and news agencies.[3] While reported cases per million of the population were just below 25 in mid-1997, a disturbing 130 cases per million of the population were recorded by the end of 1999. It is now estimated that the rate of HIV prevalence in the region of the former Soviet Union (Russia, Eastern Europe and Central Asia) has increased by 250 per cent (UNAIDS 2011). While serious, the situation was less dramatic in the Baltic states. In Estonia, the number of tuberculosis cases was relatively low, but the country suffered an outbreak of HIV among injecting drug users in 2000. The situation in Latvia paralleled that of Russia for both tuberculosis and HIV. In Lithuania, although the tuberculosis rates reached those of Russia and Latvia, the HIV rate remained as low as Sweden's, on the Western shores of the Baltic Sea.

The entire situation was deemed unacceptable by Western experts, particularly the emergence of a multi-drug resistant tuberculosis strain that was caused by insufficient or interrupted treatment of ordinary tuberculosis. Terms like 'worst-case scenario' and 'unprecedented public health risk' were used,[4] as were terms like 'katastroika' and 'mortality crisis'. The worried Western community emphasized the correspondence between economic recession and high tuberculosis and HIV/AIDS incidence in the former Soviet areas. It was claimed that the social disruption could threaten the Eastern part of the Baltic Sea region and might destabilize the area not only socially, but politically, too. It was assumed that the health crisis would damage inter-state relations in the Eastern areas of the region. In the absence of basic security in terms of human health, it was said, basic security in the wider social and political sense could not be achieved. Arguments such as these both informed and motivated the international effort to combat the spread of communicable diseases in the Northwest of Russia. Thus, the rhetorical style of the Task Force initiative clearly demonstrates the emphasis on a 'soft security' approach, and illustrates that the language of human security has already been adopted in some situations in the Arctic.

What, then, was to be done to reduce the spread of communicable diseases in the region? The traditional method of preventing the cross-border spread of infectious diseases is to enforce a strict quarantine regime in combination with an equally strict medical screening at national borders. But as nation states become increasingly reliant on commercial and cultural interaction in the process of globalization, these methods have lost much of their appeal. Specifically, the post-Cold War ambition of encouraging rather than discouraging cross-border contact in the European North rendered border control measures both impractical and ideologically unacceptable.[5]

The Task Force was established as a joint venture among the 11 states surrounding the Baltic Sea for the period 2000–2002. We traced how this massive international effort was implemented and perceived in the recipient countries. This discussion is relevant to questions about human security explored elsewhere in this book in several ways. First, the Task Force addressed a problem closely linked to human security, namely the threat to public health posed by communicable diseases. Second, an analysis of the recipient countries' perception of the Task Force and its activities may shed light on their willingness both to acknowledge their own problems as well as their will to accept solutions from outside. This can inform us about impediments to the strengthening of health systems in post-Soviet areas, which will need to be overcome to improve the human security situation. It also illustrates the difficulties with the transferring of concepts and methods from the more established foreign aid system in the Global South to what some may call a 'new disaster zone' in post-Soviet areas.

General receptiveness to Western ideas

Two distinctly different perceptions of the Western ideas proposed under the Task Force emerged between our Baltic and Russian interviewees. On the one hand, a substantial majority of the Baltic and Russian project participants expressed more or less total agreement with their Western partners about basic principles of medical and health care. On the other hand, a number of the interviewees expressed serious doubts about the Task Force and similar Western endeavours, emphasizing instead the proud heritage of Soviet medicine. Representatives of the more extreme variant of this view tend to be of a very specific category: male Russian civil servants of a certain age. However, as our overview shows, there is a widespread tendency in the post-Soviet area – even among those generally positive to Western ideals and trans-national collaboration – to stress that the problems addressed by the Task Force are 'not really ours' and to find the Task Force's focus on certain marginalized groups in society hard to defend.

'I'm simply amazed...': a welcome assistance

We found many examples of Russian resistance to, and lack of familiarity with, the Western principles introduced by the Task Force. However, we believe that it is important to emphasize that such scepticism was not globally representative, at least judging from our interviews with a large number of Baltic and Russian project participants. Task Force collaboration was typically described as follows: 'I'm simply amazed at what we have achieved'; 'The partnership has been a unique experience'; 'This is an enormous thing (*ogromnoye delo*)'; 'The Task Force has been a gigantic financial support and a gigantic learning process'; and so on. On occasion we asked interviewees to elaborate on collaborative problems. After failing at first to understand the question, they burst out in laughter at the idea that there should be any problems at all.

We acknowledge that there are methodological challenges associated with quotes such as these. Are Russian and Baltic project affiliates really happy with the projects and in complete agreement with their Western partners, or are they simply saying what they think was expected of them in the interview setting? The Task Force projects are accompanied by substantial financial input, which might inhibit the project participants' likelihood of expressing dissatisfaction with them (for fear of jeopardizing funding). Also, the project participants might wish to appear 'polite' to two interviewers from the main Task Force donor country. On the other hand, we interpret the reported satisfaction to the effect that at least dissatisfaction is not substantial enough to make these people want to 'risk' further financing (to the extent that expressing dissatisfaction would incur a risk) or appear 'impolite' to the interviewers.

'Why are they so interested in us?': the 'Cold Peace' perspective

A large gap separated the general satisfaction with Task Force projects expressed by most of our interviewees and the fundamental distrust of the West declared by other project participants. In one of the interviews, the distrust in such collaboration projects was apparent as soon as the interviewee started to explain the background to his project (a standard opening question in our interviews). He explained how his first professional contacts with foreigners had been quite disappointing: the Finns had not shown due respect for Russian expertise, which he found 'rather humiliating', and little came out of the plans. He seemed to think that international cooperation was a nuisance and was pleased to continue alone,[6] but soon thereafter 'along came the Swedes'. Although they also had a 'degrading attitude' to the Russians, this time around he 'accepted' the project proposed by the foreigners.[7] However, he asked himself: 'Why are they so interested in us? What is their hidden agenda?' He indicated the direction of his suspicions by saying: 'Every country's Ministry of Foreign Affairs is a legal intelligence agency.' He refused to elaborate on this, but we interpret him here – together with the general spirit of the interview – to the effect that he regarded the Task Force in a 'Cold Peace' perspective: the Western states were attempting to invade yet another arena, in this case the health service, to exploit Russian (or Soviet) expertise and press reforms on Russia. He also indicated that figures on communicable diseases in Russia were boosted to attract money from the federal budget or foreign sponsors: '[the high figures are] just a little lie for the politicians (*malenkaya lozh dlya politikov*)'.[8]

Similar views were expressed by senior Task Force officials, although generally more discreetly. For instance, one high-ranking Russian official was dismayed at the way that Task Force members from the West had 'carved up' the Russian regions among themselves at a meeting:[9] 'it evoked a feeling of amusement (*chuvstvo yumora*) in me – it is, after all, our country (*eto zhe nasha strana*)!' The way he recounted this story, however, reflected exasperation rather than amusement. Despite the ambiguity in his tone, he clearly resented Westerners coming to Russia and 'doing as they like'. Representatives of different Western countries quarrelling over which Russian regions are 'theirs' is a good illustration of this. Finally, a topic raised by a few of our interviewees concerned the general ignorance of the foreigners who come to Russia with a view to reform its post-Soviet health sector: 'The Americans walked in – no, they trampled in – without the slightest knowledge about how things work here.' Despite the divergent attitudes outlined above, a marked majority of our interviewees expressed a desire to distance their regions and states from the origin of the communicable disease problem.

'Not really ours': distancing oneself from stigmatizing problems

One noteworthy feature of our interviews was the tendency of the Russian and Baltic project participants to make comments, without being asked, about the origin of the problems addressed by the Task Force. A Russian HIV/AIDS specialist began his project presentation by pointing out that HIV came to the Russian Republic of Kalmykia in the late 1980s, when a Soviet sailor returned home from a visit in Angola. Discussing the HIV/AIDS problems in Kaliningrad Oblast with us, another Russian civil servant was at pains to emphasize that the virus had not originated there: 'it's actually from Poland'. In Arkhangelsk Oblast, we were informed that the virus was 'imported' from the Volga and St Petersburg areas, so the region itself was originally 'clean'. Our interviewees in the Baltic countries generally identified Russia as the source of most of the problems addressed by the Task Force. They also said that some of the underlying issues contributing to the spread of communicable diseases had been 'imported' to the post-Soviet area from outside. Drug addiction is an obvious example, frequently referred to in our interviews, but prostitution was also placed in the same category of problems by one of our interviewees:

> Prostitution is not really a Russian phenomenon. In Russia, we treat our women [well and give them] good food and drinks and flowers, and what happens afterwards … who knows? But we don't pay them to have sex with us. That's not how things are done in Russia.

The interesting thing here is not so much whether it is 'true' that a problem is imported. It is widely known that HIV did not originate in the Soviet

Union, and that social problems like drug use and HIV have flowed into the Baltic states in recent years from the Kaliningrad and St Petersburg areas in Russia. The interesting thing is that many of the people in Russia and the Baltic states who are involved in Task Force projects want to distance themselves from the problems addressed by Western initiatives of this sort. Not only do Western governments and organizations occasionally 'ride roughshod over Russia and the Baltics' (often with simplistic 'magic formulas' and little knowledge of Russian and Baltic society, at that), they stigmatize the countries in their references to them as 'AIDS-ridden' and 'overflowing with drugs and prostitution'. The general reaction seems to be something like 'OK, we do have a problem at the moment, but it is not really something that is intrinsic to our society. We got them when the old borders were opened; they were never really ours.'

Both these and the 'Cold Peace' attitudes previously discussed have several points in common. When our interviewees asked 'Why are they so interested in us?' they were alluding first to the putative interest of Western governments to exploit Russia or damage its competitiveness, and second, to the perception in the West of Russia as a country that was particularly in need of aid. We encountered various degrees of such displeasure with Western aid in our interviews. Especially in the Baltic states, we occasionally met slight indifference combined with something akin to surprise that the Nordic countries had got it into their heads to address these problems.[10] Many project participants seem to be ambivalent about accepting this aid. On the one hand, they recognized that they have a problem with communicable diseases in their country and that collaborative projects might help. There is also the pecuniary side, as they often profit financially from such projects (either personally or as institutions). On the other hand, they did not like being the recipients of foreign aid and saw it as a reflection of Western arrogance to focus on problems in Russia, especially since similar or even greater problems are found in the West. A recurrent argument in our interviews was that HIV/AIDS is a huge problem also in the US, 'but nobody speaks about that'.[11] It was quite fascinating to observe how interviewees changed back and forth from expressing gratitude and satisfaction with the Task Force projects, to scolding the West for singling out Russia as a 'problem area' while being in denial of its own problems. We recognize that the idea of human security, with its emphasis on acknowledging the insecurities of vulnerable populations that exist within all states and spread across state boundaries, could lessen the barriers that result from a foreign aid recipient country feeling particularly singled out or stigmatized. However, it is also fair to say that such an effect from the human security concept can only be hoped for at an official level. Concepts like 'human security', 'soft security' and the like are not likely to outweigh the dominant 'Cold Peace' discourse in Russia, which is more or less deeply rooted in most post-Soviet societies.

Transferring from the Global South: DOTS in the FSU

In this section, we pay special attention to the introduction of the WHO's tuberculosis treatment programme, Directly Observed Treatment with Short-course chemotherapy (DOTS), as a particularly compelling illustration of both resistance to and receptiveness of Western medical ideals in the post-Soviet area and of the general attitudes outlined in the previous section. The Task Force has not been a major player in introducing DOTS to Russia and the Baltic states, but the tuberculosis projects under the Task Force are generally oriented towards the promotion of DOTS. However, many of its tuberculosis control projects articulate with pre-existing joint programmes run by Baltic and Russian regional authorities and various international organizations.

The DOTS strategy has been promoted by the WHO since the early 1990s. The current form of the DOTS strategy was formalized in 1994, although it was based on several decades of experience. DOTS is considered one of the most effective interventions in global public health ever from both a health and economic viewpoint.[12] Its average cure rate is nearly 80 per cent, ranging from 70 to 95 per cent, and DOTS is currently the primary strategy for stopping tuberculosis (WHO 2006). Its low cost per patient and effectiveness without hospitalization means that it is a feasible approach even for poor countries. Without DOTS, the annual tuberculosis rates were expected to rise by 41 per cent from 1998 to 2020 (from 7.8 million to 10.6 million cases per year); achievement of WHO targets would prevent 23 per cent or 48 million cases in the period from 1998 to 2020 (Dye *et al.* 1998:1886). Furthermore, DOTS provides the best-known defence against the development of multi-drug resistant strains of *Mycobacterium tuberculosis* (Kochi 1997:228). In other words, DOTS is a strategy in the fight against tuberculosis that has produced well-documented positive effects. According to Western experts, without improvements in case detection and cure, the incidence of tuberculosis in Russia is expected to rise by about 10 per cent per year, i.e. a doubling every seven years. The introduction and expansion of DOTS is expected to halt the rise and, by 2015, a 10 per cent annual decline in incidence is predicted if DOTS is widely implemented (Coker 2001:435). However, the reception of DOTS in the FSU has not been altogether positive.

The reception of DOTS in the FSU

A noticeable feature of the introduction of DOTS in the post-Soviet area is that the Baltic states and the Russian rim regions in the Northwest have embraced the WHO strategy wholeheartedly. In contrast, tuberculosis experts who are centrally located in the Russian 'capitals'[13] of Moscow and St Petersburg have displayed opposition, voicing reasons that are in line with those of Mikhail

Perelman (2000), outlined below. In practice, the Western anti-tuberculosis strategy clashes in many ways with the Russian tradition. It is important to note that the Russian approach to treatment and detection of tuberculosis has a long tradition and has long been perceived as being very effective.

One of the leading Russian experts on tuberculosis control, Mikhail Perelman, has articulated the basis of Russian opposition to DOTS in a two-fold argument (Perelman 2000). First, he claims that certain aspects of the DOTS regime are not applicable in the Russian setting. Second, he opposes the use of the acronym DOTS in general, because it is meant to signify something different from traditional Russian tuberculosis control in particular. Perelman objects to DOTS because he believes that it excludes other available tuberculosis treatments and important nuances, such as the combined use of several drugs and the varying intensities of different phases of treatment. According to Hans Kluge, WHO manager for tuberculosis in Russia, the WHO and the Russian institutions reached an agreement that, in Russia, the terms DOTS and DOTS-Plus will no longer be used (Perelman 2000:1102).

Perelman's opposition to DOTS reflects more than different approaches to medicine, and although there are marked differences between the WHO-sponsored approach and the traditional Russian position (primarily with regard to the use of hospitalization, radiology and surgery), the tone in which the article is couched expresses a sense of indignation on behalf of Russia and traditional Russian (or Soviet) medicine. More than anything, it seems that Perelman wants to show that the WHO has not invented the TB-wheel with its DOTS strategy. This sentiment is reflected, for example, in his reference to 'the "new" Western acronym for this well-known principle of therapy'. This indignation was expressed by one interviewee, a major Russian tuberculosis expert, located in one of the 'capitals'. She stated:

> Ah, this is where [my Western collaboration partners] call me an old Communist. You know, DOTS consists of five elements ... We're not in any way against any of them. No sensible person would be. And we're grateful to the West for bringing it here. As a result, tuberculosis has become an important issue in Russia. But we do have a tradition in Russia, and not everything about it is bad!

She went on to state that DOTS does not address the social problems associated with tuberculosis and points to other states that have solved their tuberculosis without the use of DOTS, such as Finland, Sweden and Norway.

In other passages in Mikhail Perelman's article there is also a sense of national pride and resentment towards DOTS, as explicitly 'Western' (not merely 'foreign'), and to the ignorance of 'Western' experts of traditional Russian tuberculosis control:

> Recently, seminars and conferences which include foreign experts or training in other countries have become additional elements of post-graduate education. Educational materials from the World Health Organization (WHO), the International Union Against Tuberculosis and Lung Disease (IUATLD), Western Europe and the USA, are popular among physicians and have resulted in fruitful discussions. *However, Russian phthisiologists have often been disappointed at the scientific level of presentations given by foreign lecturers.*
>
> [emphasis added] (Perelman 2000:1098)

From the point of view of Russian tuberculosis experts, Russia has a proud heritage of tuberculosis control. The system has experienced problems since the break-up of the Soviet Union, but Russia is no developing country starting from scratch in its battle with the disease.

Both the fact that DOTS is generally used in developing countries, and the simplicity with which it is often presented by the WHO, seem to exasperate Russian experts. A sense of resentment arose during DOTS' introductory phase, when Westerners were seen as lumping Russia together with developing countries (DOTS was originally a 'developing-country strategy'), though this particular grudge was rarely mentioned in other connections. One exception concerned a discussion that allegedly had taken place on a CIA report about HIV/AIDS at a Task Force meeting. The report – and Task Force members at the meeting – expressed concern about the development of HIV/AIDS in, among other countries, India, Ethiopia and Russia. 'They compared us with these countries!' was the Russian reaction, combined with exasperation that the HIV/AIDS threat from the USA 'never' seems to be an issue.

In the West, particularly among international organizations such as the WHO, it is often a point in itself to ensure that information is provided as simply and accessibly as possible. The risk here is that Russian experts may feel they are being treated like amateurs and look for inconsistencies in the Western approach to flag their credentials. Perelman's (2000) mention of the disappointment of Russian specialists with the scientific simplification of Western experts reflects the same sentiment: Russian scientists are not supposed to communicate in a simple and understandable manner; they are expected to ensure that their material is accurate and of the requisite complexity. In sum, Perelman's scepticism to DOTS is not only an expression of patriotism or anti-Western sentiment ('we oppose DOTS because it comes from the West'), but a combination of national (or possibly personal) pride and a reaction to the way in which the alternative to the traditional Russian approach has been presented, i.e. as a completely different approach that cannot be implemented only partly or adjusted to local requirements. From the point of view of the WHO, it is either all DOTS or no DOTS at all.[14] In some of our interviews, 'WHO arrogance' was offered as an explanation of Russian resistance to DOTS.[15]

As already mentioned, neither the Baltic states nor the Northwestern rim of Russia are as dismissive of DOTS. The general picture drawn by our interviewees in these regions is encapsulated in the following paraphrased statement: 'The Finns [or Norwegians, or Swedes, or whatever] arrived, convinced us, and everything is going so much better now.' Representatives of the Russian regions that have introduced DOTS complain not about the West, but about the Russian tuberculosis establishment: 'In Russia, [DOTS] came from the regions (*s regionov shlo*). The problem lies with our Russian experts. They have to rid themselves of certain prejudices. The stereotypes have to be changed.' Another regional tuberculosis expert expressed moral indignation at the national establishment: 'As doctors, we are supposed to help people where we can. We promised that when we took the [Hippocratic] oath.' What she seems to be saying is that those who dismiss a proven strategy such as DOTS as unscientific, e.g. for patriotic reasons, behave immorally. So was DOTS introduced in a number of Northwest Russian federal subjects simply because people from the West succeeded in convincing the Russians about the excellence of this strategy? There are certainly other elements to the story. The relevance of the centre–regions conflict is obvious: the DOTS issue is used by representatives of the outlying Russian regions to oppose the superiority of the centre, understood as the federal capital of Moscow and the 'Northwestern capital' of St Petersburg. Funding available through organizations supporting DOTS was also an important incentive as funds at the regional and federal levels have become scarcer.

Prostitutes and prisoners: promoting an emphasis on marginalized groups

The concept of human security promotes the acknowledgement of the insecurities of vulnerable groups, which can be obscured by national statistics or poorly addressed by inappropriately designed policies. The Task Force also promoted an emphasis on marginalized groups – an emphasis that brought about another set (second category) of 'problems'. While Russian and Baltic project participants agreed in principle that the especially rapid spread of tuberculosis in prisons and amongst drug users and prostitutes should be addressed, they either thought it was inappropriate or that they would have trouble 'selling' it at home. In the first case, the Russian and Baltic project participants found these projects worthwhile in principle, but problematic in this particular setting. Their concerns were more rationally founded and less emotionally based than in some of the cases reviewed above (in which Russian interviewees tended to question the intentions of Western governments). In the second case, the project participants were personally in favour of the projects, but noted that it is difficult for the Russian or Baltic publics and their superiors to understand the priorities.

The typical example of the first of these types of problems is the priority given to prisons in the Task Force. A separate programme for prisons was established under the Task Force in 2001, and many of the projects under the other programmes (concerning, for example, tuberculosis or HIV/AIDS) are also oriented towards the prison system. Specifically, physical reconstruction has been performed to prevent the spread of communicable diseases: isolates have been built, and cells, kitchens, laundries and other prison facilities have been renovated. Our interviewees tend to understand the rationale behind these projects, i.e. they do not think their Western partners have suspicious intentions or are insensitive to the situation. They probably also understand that the health gains might be considerable. However, some of them displayed exasperation at the fact that this part of society is singled out for financial assistance from the West. As expressed by one high-ranking Russian official:

> We do not understand why prisons are given priority. In Russia, we have always given priority to women, pregnant women, children and the elderly. The prisoners are bandits! In official meetings [with Western partners], the Russian [representatives in Task Force projects] do not object, but many of them really don't understand the priority given to prisons. It is also difficult to explain this to the public. For example, in Murmansk a lot of money is given to an isolate in a prison colony, while there is little to give to civilian tuberculosis institutions. People don't understand this. I can hear the voice of the people [this can't be right]! Another example: In Petrozavodsk, there were problems with the town's water pipes. The same problem was found in their prisons, but after the Task Force was there, conditions in the prison are better than in the town!

In this particular case, our interviewee claims to speak on behalf of the Russia people, but also seems to be expressing his personal opinion. When he says that 'we do not understand why prisons are given priority', and in the next sentence '[i]n Russia, we have always...', he seems to want to voice the opinion of Russians in general (implying that he might happen to be in favour of these projects personally). However, we feel that his opinions on others' behalf are not far from his own personal belief, either. It is also interesting to note his claim to the effect that 'nobody objects to these projects in conversations with Western partners, but they are actually against them'. Others aired similar doubts about the prison projects, though they emphasized that they themselves were favourably disposed towards them. For instance, some of our Baltic interviewees argued that opposition to prison projects could be explained by these countries' Soviet heritage (and thus was 'not really Baltic'): 'The mentality was such [in the Soviet Union] that society did not have to take care of prisoners. For many people, it is very difficult to understand why prisoners should get more financial support per capita than the unemployed.' Others, again, blame opposition to the closed nature of the prison sector during the Soviet era:

> The real problem is that people are not used to having prisons, not to say prison hospitals, in their neighbourhood. Estonia is a very 'new' society. We are used from Soviet times for prisons to be very closed. Nobody knew what was going on there. Now the prisons have become part of society, but it takes time for people to get used to this. People don't understand why prisoners should be given good conditions. This is a process of familiarization which will take time.

Other typical examples of the second category of problems, where project participants in the recipient countries agree on the particular action but find it difficult to 'sell' to the post-Soviet public, concern information campaigns for safe sex and harm reduction measures targeted at intravenous drug users. 'Parents have a tendency to dislike this kind of information, they find it unnecessary and feel it does more harm than good,' a Russian medical worker noted on efforts to provide safe sex information for schools.

Likewise, some of our interviewees mention problems associated with harm reduction projects aimed at prostitutes and intravenous drug users. There is a tendency among Russian officials, they say, to think that drug abuse, and in particular prostitution, are not really widespread in their country (cf. statements above to the effect that prostitution is 'not really a Russian thing', and that figures on communicable diseases are inflated to attract funding from federal authorities and foreign donors). Similarly, some of our interviewees observed that officials tended to 'close their eyes' to drug abuse, often because they simply do not know how to tackle it. As far as prostitution is concerned, it might be not so much that officials do not think it exists, but, as a male-dominated bureaucracy, they silently 'approve' of it and therefore resent foreign interference. Somewhat along the same lines, we were told that the police or other governmental officials even work as pimps for street prostitutes.

Finally, our interviewees also suggested that drug addicts in Russia and the Baltic states are generally regarded as criminals, more deserving of punishment than assistance. This has in some instances impeded the implementation of projects to distribute clean needles to intravenous drug users.[16] In one Northwest Russian city, permission was obtained from the police to run a 'needle bus', but when it started up (the police had been informed of the opening date in the application), the police came and arrested the drug users.[17] The result, in all these cases, is that project participants in the post-Soviet countries carry out their tasks in a partly hostile environment, without the understanding or support of the authorities or the public. Sometimes there is outspoken criticism of the priorities made by the Task Force. One of our Baltic interviewees said that the needle exchange project seemed more a goal in itself for the Task Force than a means to an end (improved health security). He meant, in other words, that the Task Force insistence on culturally sensitive projects such as needle exchange is somewhat 'contrived'. However, attributing all of the problems associated with a focus on marginalized groups to one of attitude or disposition ignores structural and institutional issues, which exert significant influence over individuals' abilities to take on new ideas and to act in ways to improve health care.

Structures in post-Soviet health care

The primary aim of the Task Force was to fight the spread of communicable diseases in the Baltic Sea region. Intuitively, a massive disease-focused campaign might seem the logical way to proceed. However, a purely disease-based approach would not have been sufficient. The structure of the health-care system needed to be addressed as well. There were many structural impediments to project implementation that became evident during the lifetime of the Task Force. Many of these countries are still transitioning from previous, Soviet practices. The challenges posed by the old Soviet tradition are many, both in terms of the prevailing culture and social structure. In this section, we explore the challenges to reforming prison health care as a continuation of some of the issues outlined above and discuss the challenges to public health care in terms of communicable diseases, particularly the relationship between general practitioners and specialists.

Civilian health care and the prison authorities

The strong, vertical segmentation of government was a characteristic trait of Soviet public life. The different branches of the state lived their own lives, in many ways isolated from the rest of society. The prison authorities were not an exception to this rule, and this has become a problem when dealing with issues associated with controlling the transmission of communicable diseases. Overcrowded prisons, their ineffective ventilation systems and usually poor hygiene are a breeding ground for tuberculosis (Stern 1999:13). The relatively high prevalence of HIV caused by intravenous drug use and other high-risk behaviour also increases the likelihood of further infections or the progression of inert to active and contagious tuberculosis (Farmer 1999:72). Obviously, the infected inmates are not only a danger to each other. Prison sentences last only so long, and sooner or later infected prisoners are released into the general population. It is vital that the civilian health services are kept informed of prison conditions, as infected inmates will soon become their patients. Effective communicable disease control requires well-developed surveillance systems, which enable resources to be put in place where they do most good. Reliable surveillance cannot, of course, be achieved if a major source of infection, namely the prison system, does not coordinate with the civilian health authorities. The need to integrate prison-based health care with civilian health care is also part of international standards on the treatment of prisoners:

> Dealing with prisoners with active TB cannot be done in isolation. The imperatives of treatment, the minimum six-month course, the permeability of the prison walls, all require the prison treatment to be enmeshed in a wider system. Rule 22(1) of the UN Standard Minimum Rules for the Treatment of Prisoners requires that 'the medical services (in prisons) should be organized in close relationship to the general health administration of the community or nation'.
>
> (Stern 1999:30)

Until recently, it has been difficult for the civilian health system and the prison authorities to work together. The Task Force has been credited for its substantial efforts to establish cooperation between civilian health and prison authorities. Once again, though, we need to remember that the Task Force is by no means the only initiative that has emphasized the necessity of cooperation among civilian health and prison authorities.[18]

In Russia, progress has been slower than in the Baltics, though there is movement. One aspect of public–prison cooperation, which Western partners need to be aware of, is that prison and civilian health services belong to two different hierarchies. The penal system in Russia is under the Ministry of Justice of the Russian Federation, and is thus a federal agency. It has a higher status than the civilian health service, which is a regional or a municipal agency, depending on the type of services provided. For example, the tuberculosis dispensaries are usually part of the oblast or provincial administration. In Russia, the psychological and legal barriers to cooperation between regional and federal structures are significant, and inter-agency and inter-governmental cooperation during Soviet times was unheard of.

We observed these legal barriers in action in Arkhangelsk Oblast, where the need to improve laboratory services has been urgent for some time both in the civilian sector and penal system. To a Westerner, the obvious, most efficient and cheapest solution would be to establish a dedicated laboratory to serve both systems, but according to a Russian interviewee, a combined solution would in fact be in conflict with Russian law. She said it was very difficult to get Western partners to comprehend the situation. Another interviewee said that 'in the prisons there are too many rules and regulations, which they have to follow. It is difficult for them to be flexible. They have their own difficulties, and live by their own laws.' As one Western interviewee put it, 'they lose synergies and competence by retaining the dual system. The two systems "steal" competent medical workers from each other, and, besides, neither of them are sufficiently busy to need a whole laboratory for themselves', articulating a view that rejects the resignation of the Russian explanation. An example from St Petersburg shows that innovative solutions can be found. There, the city authorities provide medical help for the federal prison, the Kresty. This assistance is not formalized, but it shows that cross-sector cooperation is possible.

Despite the problems, all of our Russian interviewees stressed that relations between the prison authorities and the public sector have been improving. In general, public health services and the prison authorities are happy to see old barriers gradually disappear, but it seems to be happening by dint of sheer necessity. Most often, the pressure for change has come from the civilian authorities and foreign donors, but there are also examples of the opposite. In Arkhangelsk, the executive head of the prison authority was described as a pioneer. He had called directly for closer collaboration with the civilian health authorities, which was something of a sensation at the time. A representative from the civilian health authorities in Arkhangelsk called this 'a progressive step!' and 'a good example for all [Russian] regions!' Our impression is that, in Arkhangelsk, both the public and, most importantly, the prisons themselves realize they need to work together. As an Arkhangelsk prison hospital official put it: 'Without the assistance of and cooperation with the civilian sector, we would not have solved any of our problems.' It would have been impossible to hear anybody say this just a few years back.

Primary health care, specialist rule and new strategies

Structural barriers for preventing the spread of tuberculosis and HIV/AIDS are also prominent in the civilian health-care system. The Soviet health-care system, or 'Soviet socialized medicine' as it was labelled by observers from Capitalist countries, was upheld by the massive, albeit weakening, Soviet economy. The system itself was demanding in terms of financial and human resources, and it faced disaster when funding collapsed together with the Union as a whole. The result was a health-care system poorly adapted to meet the needs of a new, open economy that emerged in the wake of the demise of the Soviet Union.[19] Not only did the new economic realities conflict with those of the old system, but the threat from new and re-emerging infectious diseases could only be met by a more flexible health-care system.

The new situation required a shift from health *control* to health *service delivery*. Earlier, the curative element was strong, while preventive health was not highly prioritized.[20] Whereas Soviet-style health care had its control functions and wide-reaching powers to isolate patients and treat them with force if necessary, post-Soviet society has neither the economic nor the legislative muscle to maintain this level of 'health inspection'.[21] It takes time and money to re-build a health sector, and while it is not within the scope of this chapter to discuss Soviet health care or its transition in the post-Soviet era, it can be said that the dissolution of the Soviet Union left behind a highly centralized system, in which experts were assembled within a few, sizeable institutions for the treatment of specific diseases. There was little concern for the local level, and the status of the general practitioner (GP) was, and remains, low. As wages are higher for specialized doctors, one effect is that many doctors tend to specialize. The subsequent outcome is too few GPs at the level of local health-care delivery. All these factors represent hurdles for any attempt to cope with new and re-emerging infectious diseases.[22]

Therefore, the Task Force approach to communicable disease control was not solely aimed at combating the diseases as such, but also on promoting structural change to facilitate greater empowerment of GPs at the local level and expanded cooperation between GPs and specialists. The effort to rectify the structural shortcomings of the post-Soviet health-care systems was directed at strengthening primary health-care delivery (Task Force 2000:33). For people with infectious diseases, early treatment is essential. It requires that the first medical worker met by a patient, a front-line doctor or nurse, be sufficiently trained to make the correct diagnosis, prescribe the proper medication and follow up the patient over time. The Soviet-style system was not designed for this. For instance, TB patients could not see a tuberculosis specialist without a referral from their local doctor, and a referral depended on the local medical worker's ability to identify tuberculosis as a possible cause for the patient's grievances. Often, local personnel had not received sufficient training and failed to comprehend the symptoms and provide proper treatment. Much time was therefore often lost before the patient commenced treatment, and many therefore remained unaware of the risk they posed to people around them. HIV/AIDS infection carries the same risks, of course.

Many projects were therefore designed for primary health-care worker training, for doctors as well as nurses, in the diagnosis and treatment of infectious diseases. A substantial effort was put into producing information material in Russian and Baltic languages on tuberculosis control, HIV and other sexually transmitted diseases, vaccination and vaccination coverage, and prudent use of antibiotics. Exchange visits of primary health-care workers from municipalities in Western and Eastern countries have taken place. Most interviewees give a favourable impression of these efforts. Again, the situation is more difficult in Russia than the Baltic countries, as Soviet traditions remain more firmly entrenched in Russia than in other former Soviet republics.

The clarity of the Task Force's strategy for primary health care was highlighted by many officials. A Latvian interviewee said that, although Latvia had a primary health-care service already, the Task Force had generated new ideas and helped bridge the divide between different specialities in the health-care system. But as one Russian interviewee pointed out, 'they [specialists] tend to mistrust GPs and are very protective of their particular field of work'. The Task Force therefore encouraged specialists to dampen their 'protectionism' and open up for wider, cross-discipline communication. To take an example of the difficulties that can arise when coordination is poor, it is obviously vital for doctors to know if a tuberculosis patient has tested positive for HIV, but this information has not been readily available. We found an improvement in Arkhangelsk, where they had established a database for epidemiological information in the region that was open to all types of specialists. It was considered an important step forward, as specialists earlier had merely archived information, as they had neither incentive nor possibility to share it.

The structural problems encountered in Russia may seem insurmountable in contrast to the relatively dynamic situation in the Baltic states. It is clear that no single international collaboration can have any realistic ambition of transforming Russia. There is little doubt that such a transformation must originate from inside, and it will surely only be a result of an internal acknowledgement of its rationale and necessity.

Conclusion: human security as a political tool

In this chapter, we have outlined what could be described as a human security threat – the spread of tuberculosis and HIV/AIDS in the FSU – and explored international collaborative efforts aimed at addressing this problem, specifically the introduction of DOTS and the work of the Task Force on Communicable Disease Control in the Baltic Sea Region. Our research highlights some of the practical difficulties associated with importing health methods from the so-called 'Third World', namely DOTS, into the Arctic in general, and the FSU in particular. These difficulties, to a large extent, stem from a general unwillingness on the part of Russian medical specialists to accept what they perceive as being the inherent devaluation of their skills and medical tradition that importing 'Global South methods' implies.

On a more general level, we would argue that Russian scepticism towards implementation of Western concepts designed to understand and ameliorate problems in the 'Global South' may have broad applicability. For our own part, we include our own scepticism to importing the human security concept, which also was developed with the Global South in mind, as an analytical tool in Arctic research. At this point we do not view the concept as being inherently useful in identifying, analysing or resolving issues in our research in the FSU through operational means. That being said, the discourse of human security may be useful in helping those involved in addressing health issues talk about their shared problems. The ability to discuss and communicate using a shared conceptual language should not be undervalued and may be a first step towards a more fruitful, less polarized set of positions around which to facilitate international cooperation. Likewise, the concept of 'security' may be a more familiar one to elected state officials, bureaucrats and civil servants, and human security could prove to be a useful rhetorical tool, should it gain further political currency, in maintaining a focus on health as a pressing and global issue. For human security to become an integral part of Russian thinking about health, though, the concept must be modified to be in step with the Russian reality, which can only happen with the direct involvement of those living there.

Notes

1 Lars Rowe and Geir Hønneland are researchers at the Fridtjof Nansen Institute in Norway. Elana Wilson Rowe is a senior research fellow at the Norwegian Institute of International Affairs.

2 Of the 230 projects, the overwhelming majority were implemented in Russia and the Baltic states (Estonia, Latvia and Lithuania). Hønneland and Rowe were engaged in the contextual evaluation of the Task Force from 2002 to 2004, studying the political, structural and historical factors constituting the environment in which Task Force projects were implemented. This work resulted in a book (Hønneland and Rowe 2004) and a report (Rowe 2004).

3 Not only was the sheer number of articles on the subject high, but the sense of an impending catastrophe characterized some of them. For instance, papers were proclaiming 'Russia in danger of social catastrophe', 'HIV-bomb strikes Russia' and 'The Nordic countries fear tuberculosis and diphtheria from Eastern Europe'. Examples from *Verdens Gang* (NTB), 24 June 1999, *Verdens Gang*, 21 May 2000, and *Aftenposten*, 25 October 1999.

4 To cite the foundation document for the Task Force on Communicable Disease Control in the Baltic Sea Region, under the heading 'An Unprecedented Public Health Threat':

> One of the fears of all actors in the field of public health in Russia and the Baltic countries is that multi-drug resistant TB moves into HIV positive populations. If, or when this happens, a likely scenario is that immuno-suppressed HIV positive people will become infected and rapidly infectious. The spread of multi-resistant TB could be very rapid indeed all over this region.
>
> (Task Force 2000:11)

5 In the Task Force foundation document, a brief discussion of border control as a means of combating communicable diseases concludes that 'border control and quarantine of individuals are certainly not measures to be taken in the interest of communicable disease control in the [Baltic Sea] region' (Task Force 2000:12).

6 This is our interpretation, not the interviewee's own words. In general, his views were reminiscent of the Soviet perception of two separate sciences, Soviet and Western (denouncing the concept of a 'universal' science). His reference to the Russian proverb that 'every sandpiper praises his own swamp' (*kazhdy kulik svoyo boloto khvalit*) signifies, as we see it, that the two scientific worlds will (and should) exist side by side since neither will (or should) dominate the other. At a more prosaic level, the statement signals that Finnish haughtiness in this case was a 'natural' thing, just as a basic assumption of the 'Cold Peace' discourse is that it is 'natural' for states constantly to be at loggerheads with each other. Other views of his expressed during the interview support our assumption that this is a person who is not too fond of the widening East–West contacts that have taken place in the post-Cold War period. His remark that 'I hate it when foreigners try to speak Russian' is illuminating in this respect. Likewise: 'If we had still lived in the Soviet Union, all the [Task Force] projects would have been implemented in a jiffy!'

7 Note here the contrast with those who happily receive Western assistance.

8 He was not referring to Russian statistics as such – 'we have the most accurate statistics in the world' – but to the tendency to argue that official statistics show just 'the tip of the iceberg'. However, it is unclear what exactly he meant by 'our statistics'. Russia has a long tradition of producing one set of 'official' statistics and one set of 'departmental' (*vedomstvennaya*) statistics for internal use. Russian 'hyper-diagnosis' of communicable diseases was also mentioned by other Russian interviewees, primarily in connection with the difficulty of making a correct diagnosis with outdated technology.

9 This is a well-known fact in East–West cooperation in Northern Europe after the Cold War. For instance, Finland has particularly close historical and cultural ties with the

Republic of Karelia in Russia and therefore focuses efforts on this particular federal subject. Norway, for its part, shares borders with Murmansk Oblast and is hence most interested in that region. Likewise, the Western countries around the Baltic Sea have tended to have 'their own' Baltic state, with which contacts have been particularly close.

10 This different level of interest may be because the Baltic states are generally more prosperous than Russia, they have far smaller administrative systems and their problems are more manageable. Also, they might dislike being grouped together with Russia in the category of countries needy of foreign support.

11 Interviewees indicated related problems in the US: 'Sexually transmitted diseases are a huge problem in the United States. Syphilis is rampant among women and children there.' Implicitly: 'Why are you so interested in us when problems are huge also in the West?'

12 In the words of the Director of the Global Tuberculosis Programme of the WHO: '[N]o other new health intervention of this decade has achieved such significant results or been of such thoroughly proven effectiveness in the field' (Kochi 1997:225).

13 In Russia, St Petersburg and Moscow are often referred to as the 'capitals' (*stolitsy*), although only the latter has status as the country's official capital. Peter the Great moved the Russian capital from Moscow to his newly built 'Window on Europe' in the early eighteenth century, and the Bolsheviks moved it back to Moscow after the 1917 Revolution. Furthermore, in today's Northwestern Russia, St Petersburg is a regional 'capital' inasmuch as the government offices of the Northwestern okrug have been there since its inception in 2000.

14 This is reflected in our interview with an employee of the Moscow office of the WHO: 'We cannot compromise on DOTS. For us, this is a political matter.'

15 It should be observed that this 'arrogance' refers back to the late 1990s when the WHO first started its efforts to introduce DOTS in Russia; we have no evidence that this is the general perception of the WHO today.

16 Again, it should be emphasized that these problems are not representative of the situation more generally. Our interviews left us with the impression that problems with needle exchange programmes were most intense in their early phases, but that some sort of settlement with the police was reached in most cases.

17 This story was told to us by an informant who had no direct involvement in this particular project. When we later interviewed the Russian manager of the project in question, we hoped to have him confirm this information. However, as he seemed extremely eager to emphasize the good things about his project and downplay any problems, we did not press him on this issue. All attempts to approach the episode in conversation met with an insistence that 'we have a very good cooperation with the police'. We assumed that he was afraid of talking about implementation problems, thinking it might reduce the chances of further project financing.

18 Several interviewees commented on this. On the other hand, many others tend to give credit to whatever joint undertaking is being discussed at the moment, exaggerating thereby its impact. In our interviews, the Task Force was discussed. That said, in this chapter we describe Western health interventions in more general terms, using the Task Force as our case. The reader should be aware that the general developments described here are due to many international efforts and Russian or Baltic initiatives, but also a result of the Task Force collaboration.

19 For an interesting history of the Soviet recession from the 1970s until the final collapse and ensuing reform policy in the last decade of the twenty-first century, see Kotkin (2001). For an introduction to the Soviet health-care system, see Field (1967).

20 The term 'preventive health' is meant to describe principles for good individual health in Western countries, such as healthy diets, non-smoking and low alcohol consumption. The Soviet health-care system was not designed to encourage these principles, as it relied to a large degree on inspection measures and control functions to maintain health standards.

The result is that Russians in general have a lower health consciousness on the individual level than what is perceived as necessary in Western countries. As one Russian interviewee put it: 'There is a big difference between Western and Russian attitudes to individual health. In Russia, after 70 years of Communist rule, we still perceive health issues, also individual health, to be a matter of the state.'

21 When describing this old health inspection or hygiene control, one Lithuanian interviewee commented that there were benefits to the system: 'We were able to control the population, and this is an important aspect of communicable disease control.' Another interviewee made a similar statement, expressing regret with the fact that, of late, human rights considerations prevent prison authorities from retaining tuberculosis-infected inmates for terms exceeding their sentences if they had not completed their treatment.

22 According to one interviewee, there are about 1,800 trained general practitioners in Russia, but only 350 of them have stayed on in primary health care. The same interviewee pointed out that the low wages for general practitioners was an obvious reason for the low number of general practitioners.

References

Coker, R. (2001) 'Control of tuberculosis in Russia', *Lancet*, 358:434–5.

Dye, C., Garnett, G. P., Sleeman, K. and Williams, B. G. (1998) 'Prospects for worldwide tuberculosis control under the WHO DOTS Strategy', *Lancet*, 352:1886–91.

Farmer, P. (1999) 'Cruel and unusual: Drug resistant tuberculosis as punishment', in V. Stern (ed.) *Sentenced to Die? The Problems of TB in Prisons in Eastern Europe and Central Asia*, London: International Centre for Prison Studies, King's College.

Field, M. G. (1967) *Soviet Socialized Medicine: An Introduction*, New York: Free Press.

Hønneland, G. and Rowe, L. (2004) *Health as International Politics: Combating Communicable Diseases in the Baltic Sea Region*, Aldershot: Ashgate Publishing.

Kochi, A. (1997) 'Tuberculosis control: Is DOTS the health breakthrough of the 1990s?', *World Health Forum: An International Journal of Health Development*, 18:225–32.

Kotkin, S. (2001) *Armageddon Averted: The Soviet Collapse, 1970–2000*, Oxford: Oxford University Press.

McKee, M. (2001) 'The health consequences of the collapse of the Soviet Union', in D. Leon and G. Walt (eds) *Poverty, Inequality and Health: An International Perspective*, Oxford: Oxford University Press.

Perelman, M. I. (2000) 'Tuberculosis in Russia', *International Journal of Tuberculosis and Lung Disease*, 4:1097–103.

Rowe, L. (2004) *Report from the Steering Committee for Evaluation of the Task Force on Communicable Disease Control in the Baltic Sea Region, FNI-report 9/2004*, Lysaker: Fridtjof Nansen Institute.

Stern, V. (1999) 'Introduction: An overview and some issues', in V. Stern (ed.) *Sentenced to Die? The Problems of TB in Prisons in Eastern Europe and Central Asia*, London: International Centre for Prison Studies, King's College.

Task Force (2000) *Healthy Neighbours*, Task Force on Communicable Disease Control in the Baltic Sea Region, Oslo: Ministry of Health.

UNAIDS (2011) *UNAIDS Worlds AIDS Day Report*. Online. Available at: http://www.unaids.org/en/media/unaids/contentassets/documents/unaidspublication/2011/JC2216_WorldAIDSday_report_2011_en.pdf (accessed 23 March 2013).

WHO (2006) *The Stop TB Strategy*. Online. Available at: http://www.who.int/tb/strategy/stop_tb_strategy/en/index.html (accessed 23 March 2013).

Part IV

Human security

Women and indigenous groups

Chapter 11

Aboriginal self-determination and resource development activity

Improving human security in the Canadian Arctic?[1]

Gabrielle Slowey

> The potential for abundant natural resources and a changing climate, with resulting opportunities and challenges, are transforming the Arctic. It has become a region of active economic and political cooperation and competition, and of scientific research.
>
> (Department of Foreign Affairs and International Trade 2009)

The Arctic is hot. The renewed focus on the region has occurred not only in Canada but also the other seven states that border the Arctic Circle, as well as other foreign powers interested in the region's potential, including Asia and Europe. Discussions about the security of the Arctic in Canada centre on matters of sovereignty and how best to protect Canada's national interests. In the 2007 speech from the throne, the Canadian government made it clear that the Arctic would be a priority, promising to increase Canada's presence on the international stage by modernizing its military training facilities, expanding surveillance and building a world-class scientific research station. But what was of particular surprise in that same speech was that it also outlined a 'northern strategy' that would essentially focus on increasing the *human security* of the people living in the region. Promising to strengthen Canada's north, the government pledged to protect the environment, promote economic development and devolve governance 'so that northerners have greater control over their destinies' (*The Canadian Press* 2007). Reflecting on the promises made in 2007, I pose the question: does self-government advance the human security of Aboriginal peoples living in the Canadian north? And does self-government represent a step forward towards a broader understanding of, and improving, Canadian security in the Arctic? Looking at experiences set across two Arctic communities, one in the Yukon and another in the Northwest Territories (NWT), this chapter argues that, yes, self-government is critical because it improves substantially the human security of Aboriginal peoples by providing Aboriginal peoples and their governments the opportunity to improve the socio-economic conditions of the community. Moreover, even though Aboriginal governments themselves are not sovereign (unlike the provincial or territorial governments that are constitutionally distinct under the Canadian system of federalism), because they function as governments they must be engaged with on a government level by other government bodies, which further strengthens our overall sovereignty in the region. Self-government therefore reinforces the security and sovereign position of the Canadian state by solidifying its political and physical presence in the region.

To begin, this chapter considers the importance of land claims and self-government and locates both in the human security literature. Although neither Aboriginal people nor the Canadian government use the language of human security in negotiating or promoting land claims or self-government, they share common characteristics. Land claims in Canada, like human security, aim to improve the material quality of Aboriginal life as they focus on issues of economics, energy, land, resources and traditional pursuits. Self-government contains further elements of human security including health care, housing, education, justice, food and environment, along with personal, community and political elements of life. In effect, when combined, land claims and self-government are about increasing Aboriginal human security. Hence this chapter argues that both are necessary to achieve Aboriginal self-determination, and their absence undermines the security of indigenous peoples.

Drawing on research conducted in two different Arctic communities, I demonstrate ways in which self-government, especially in areas where there is increased oil and gas activity, dramatically improves the ability of communities to address their own human security issues. That is, where self-government exists, Aboriginal communities are more secure. Finally, this chapter concludes that self-government, like human security, is a work in progress. That is, it is not an end goal but a process through which Aboriginal peoples must work to increase their overall quality of life and raise their standard of living, and to that end self-government represents an important and critical step forward.

Self-government and human security: making a connection

In Canada, land claims are intended to help remedy the imbalance in the state–Aboriginal relationship. They are mechanisms through which the state can attempt to resolve contentious issues concerning land and its ownership. Hence, land claims are primarily designed to resolve the question of title outstanding between Aboriginal peoples and the state. While the Crown owns all the land, it negotiates land claims as a way to settle claims to title over these lands made by Aboriginal groups. That is, where no treaties were previously signed or exist, the state embarks on a process of new or modern treaty negotiations (referred to as comprehensive claims), and where treaties existed previously (also known as historic treaties) they seek to remedy any terms that remain unfulfilled (referred to as specific claims). The overarching objective on the part of government is to ensure that any lingering dispute over title to land is clarified. In contrast, the motivating factor for many Aboriginal peoples is the desire to regain control over land and increase local decision-making power.

For the most part, modern-day treaties are the outcome of comprehensive land claims and, as such, form the basis for a new relationship between the state and Aboriginal peoples. In terms of what they entail, comprehensive claims focus on two main items: (1) land that is necessary to keep culture alive and dynamic; and (2) cash that is intended to make business (as opposed to government) the primary engine for economic growth. Comprehensive claims can also include co-management regimes, environmental assessment panels, resources and royalty-sharing arrangements. A clear purpose therefore is enhanced socio-economic amelioration of Aboriginal peoples. The notion that land claims will clarify matters of title is tied to the notion that capitalists require certainty for development projects to occur and corporations are attracted to areas where ownership of land is not an issue for development. Where outstanding claims persist, there is too much uncertainty, which represents an undesirable investment environment (Slowey 2008). Unresolved territorial disputes also pose a potential threat to Canada's national security, as uncertainty over who owns the land could, theoretically at least, undermine Canada's claim to Arctic lands. Hence, as Canada spreads its message to the world that a considerable portion of the Arctic is Canadian territory inhabited by Canadian citizens, land claims further bolster this claim by eliminating any uncertainty as to who owns the land, who lives in the Arctic and which state ultimately and legitimately oversees both.

Land claims are also, however, about improving the human security of Aboriginal peoples. Though the traditional notion of security is associated with the promotion of state sovereignty, it is also about protecting citizens. That is, the notion of security is tied not only to the state but extends also to its citizens. Hence security is not exclusively about the state in the same way that human security is not exclusively about citizens; rather, security, be it traditional or human, is about the relationship that exists between the state and its citizens. For instance, in matters of national security, the protection of land remains essential. Nowhere is this more evident than in the Arctic where, because of its vastness, the state relies heavily on the local Aboriginal people. In particular, the Canadian Rangers are a reserve sub-component of the Canadian Armed Forces that consist primarily of local, Aboriginal peoples who volunteer their time to perform surveillance and sovereignty patrols across the Canadian Arctic. Representing much of the military presence in the region, the Rangers serve in part as the eyes and ears of the nation, as well as support for local people as well as visitors, by doing integral work in search and rescue.

For Aboriginal peoples in the Arctic (and elsewhere in the world) the land remains critical, as it is the central element of many Arctic peoples' culture, representing both their past and their future. As Connie Deiter and Darlene

Rude (2005) argued in their policy paper *Human Security and Aboriginal Women in Canada*, Aboriginal peoples' understanding of human security goes beyond safety from international threats. Instead, 'their human security is intimately tied to the land, and the cultural traditions derived from fishing, hunting and gathering necessary to foster a safe and healthy life for the future generations of their people' (Deiter and Rude 2005:ix). For the Aboriginal peoples of the Canadian Arctic, the traditional, narrow perception of security as a purely military phenomenon is insufficient because security is not simply about Canada protecting its Arctic territory, but Canada protecting the security of the people who live there – who live on the land and who rely upon it. It is about protecting their physical safety as well as their economic and social well-being ('The Responsibility to Protect' 2001:15).

While the language of human security is very much rooted in the field of international relations and tied to foreign affairs, the extent to which it is relevant in domestic analysis is undeniable. Although human security is a people-centred approach to foreign policy, it can also be viewed as an approach to domestic policy. As Owen explains:

> The vast majority of human security threats are indeed most severe in the developing world. This, however, is an empirical reality, not an ideologically driven theory. There are certain regions within developed countries that also have human security problems – many northern aboriginal communities offer examples – and these should be highlighted.
>
> (2008:447)

In fact, northern Aboriginal peoples may be considered to be among the most 'insecure' in Canada. Statistically, while Canada consistently tops the United Nations list for the highest quality of life in the world, in 2013 Canada dropped below the top ten (CBCNews Canada 2013). In 2005 Canada ranked number five in the world; however, when the condition of Aboriginal peoples living 'on-reserve' was taken into account, the country ranked approximately 48 out of 174 countries on the UNDP scale (CBCNews Canada 2005). Poverty, infant mortality, unemployment, suicide, crime and child welfare are much higher for Aboriginal peoples than any other group in Canada. Additionally, quality of medical care, housing and educational training are much lower for Aboriginal peoples living on-reserve. Given their obvious 'insecurity', land claims can function as a means through which to improve the socio-economic well-being of Aboriginal peoples and their communities.

In terms of its political dimension, a key component of human security is decision-making or, in this context, self-government. Although the James Bay Northern Quebec Agreement (JBNQA) of 1975 included promises of self-government, Cree-Naskapi community governments are not constitutionally entrenched and operate on the basis of federally delegated powers. That is, they exercise powers delegated to them under the 1984 Cree-Naskapi Act. In fact, in that same year, the Inuvialuit signed a land claim agreement that also failed to address matters of self-government. Since the official proclamation of a federal self-government policy in 1995, self-government agreements are now negotiated concurrently with land claims. And twenty years after the JBNQA was finalized, a self-government agreement formally accompanied the land claim settlement. In 1995, the Yukon managed to negotiate the first modern self-government agreement as part of its comprehensive land claim. This is significant because it means that land claims are no longer only about land and cash compensation and title but also about restoring power to the Aboriginal peoples.

Self-government, like human security, is without a clear and precise definition. Instead, it is a composite of elements that vary according to the community or situation at hand. What self-government is about, in essence, is increasing governance in Aboriginal communities, because, ultimately, self-government reconfigures the location of power with respect to decision-making abilities from the state to the community. In their recommendations to the Government of Canada, Deiter and Rude (2005) suggest first and foremost that the focus of the government be to recognize and affirm First Nations' right to self-government by implementing First Nations' land and resource management agreements (Deiter and Rude 2005:5). Where land and resources are at issue, the implementation of land claims and the transfer of powers of self-government are imperative if communities are going to maximize opportunities associated with development. Indeed, land claims and self-government are central to the struggle against insecurity. And when it comes to oil and gas development, no group in Canada lives in greater insecurity than Aboriginal peoples in the Arctic.

To illustrate the significance of land claims and the difference made by self-government, it is useful to compare and contrast concerns expressed in communities where land claims have been settled but where self-government has or has not been achieved. That is, while land claims have been finalized in many communities, the topic of self-government remains outstanding in quite a few, thus challenging the security of Aboriginal peoples. Although there are significant differences existing among the northern territories, ranging from issues of devolution to the pre-eminence of resource development, it is useful to consider the experiences of peoples living in remote communities, in light of their experiences and anxieties surrounding self-government. That is, it is interesting to note the changes that self-government has brought to communities like the Vuntut Gwitchin[2] First Nation (VGFN) of Old Crow, Yukon, when juxtaposed against those perspectives espoused from a community that has yet to achieve self-government; namely, the resource-rich region of Tuktoyaktuk in Inuvialuit territory in the NWT.

Land claims and self-government are increasingly about bringing political and not just economic power back into the communities. Nowhere is this more evident or necessary than in the high Arctic where the oil rush is on again and Aboriginal peoples are engaging in self-government, either in practice or in negotiation. To wit, while the Inuvialuit have historically focused on economic development, their 'corporate' governance model renewed steps towards a form of self-government that suggest it is an important next step in the process of self-determination. The experience of the Inuvialuit, when highlighted against that of the Vuntut Gwitchin of Old Crow, suggests that self-government is moving in the right direction. To begin, this chapter explores the case of the Inuvialuit and then turns to a discussion of the VGFN.

Self-determination and resource development: increasing human security?

Inuvialuit

In June 1984, the Inuvialuit, a group of 3,500 Inuit people living at the top of the NWT, signed the first comprehensive land claim in the region. The deal awarded them 91,000sq. kilometres in land and CAD$152 million dollars in cash compensation (a tax-free payment paid over a period of thirteen years, accompanied by a one-time payment of CAD$10 million for an economic enhancement fund and CAD$7.5 million for a social development fund). This includes 13,000sq. kilometres of land to which they also own subsurface (mineral) rights. The agreement also provides the Inuvialuit with a say in wildlife and environmental management through participation on a variety of co-management boards and councils (e.g. Notzke 1994; White 2002). As Notzke explains, the introduction of the co-management regime under the Inuvialuit Final Agreement (IFA) extended their influence over development and resource management occurring in the region (1994:302). Notzke (1994) goes so far as to suggest that the co-management regime established under the IFA refers to the sharing of power and responsibility between government and local resource users, and to various levels of integration of local and state level management systems. While participation agreements certainly provide 'considerable latitude for conveying socio-economic benefits on the Inuvialuit' and even though the Inuvialuit exercise a lot of power and autonomy in terms of possessing the ability to unilaterally amend rules and procedures with respect to their land and its usage, the scope of power assigned by the land claim ultimately remains limited to land and tied to economic or environmental decisions made by the Inuvialuit Regional Corporation or one of its affiliates (e.g. Inuvialuit Development Corporation or Inuvialuit Land Corporation) (Notzke 1994:221).

At the time of negotiation, the Inuvialuit did not enter into self-government negotiations because the federal comprehensive claims policy did not allow for self-government negotiations to occur at the same time as land claims negotiations. The government, at that time, was not prepared to negotiate self-government concurrently with land claims because it feared that this would lead to de facto constitutional protection for self-government. While a clause allowing for the opportunity to explore a public government was inserted into the final agreement, this option was never explored until much later in 1992, when the Gwich'in of the region entered into their own land claims negotiations. During this process the Inuvialuit saw an opportunity to work together in concert with the Gwich'in to establish a public government for their citizens. While the idea initially appealed to the Gwich'in (the two groups signed an Agreement-In-Principle (AIP) with the federal government in 2003), eventually the Gwich'in withdrew to pursue their own self-government agreement.

With a framework in place, however, the Inuvialuit persisted and converted the combined AIP into a more Inuvialuit-specific self-government agreement. The original goal of the agreement remained to provide a public-type government for the Inuvialuit people that would essentially set up an autonomous region where the public government would serve everyone in the communities and the region, and where the Inuvialuit would comprise the majority of the population. However, given the many Aboriginal groups that live in the NWT, and taking into account the number of Gwich'in sharing Inuvialuit communities (including Aklavik and Inuvik), the Inuvialuit are on a new path to creating a more Aboriginal-specific government that would only serve the Inuvialuit community. As an Inuvialuit Self-Government Representative suggested in an interview on 28 September 2007, there are greater limitations placed on a government like that. For instance, a public-style government would require a pooling of resources and produce economies of scale, but the further you reduce the government, the more you break apart the available funds and the less the government is able to do. Since 2003, the Tli Cho government has provided community services and operationalized self-government. Indeed, securing sustainable and long-term development is a real key to self-government, as it is critical to creating programmes that can serve the Inuvialuit community better.

Despite the absence of self-government, the Inuvialuit have been very successful in generating economic opportunities for its members within the region. They have, to date, focused on business and the generation of wealth. Building on the potential of the Mackenzie Valley Pipeline project, in the context of a massive natural resource extraction proposal and pipeline project, the Inuvialuit have proven to be very aggressive, using the provisions of the land claims agreement to pursue new avenues for development, and demonstrated their ability to generate profit from those pursuits. One of the main intentions of the Inuvialuit land claims process was to provide for economic development opportunities as well as resource management. The consequence was that the Inuvialuit focused on establishing a stable, long-term economic base that would enable them to contribute to the regional and national economy. To achieve this goal they chose to invest in regional services to generate economic growth. There is no doubt that the IFA provided positive economic benefits in the region, but the beneficiaries wanted more: they wanted improved socio-economic conditions. Despite the economic success of the Inuvialuit and its corporations, many Inuvialuit people continue to live in poverty and experience a lower standard of living. As profits are reinvested into corporate initiatives, some members suggest the money might better be spent on the people in the communities. As one Inuvialuit member puts it in an interview on 2 October 2007, 'My way to measure success is, sure you can make some money, you got to make money, but you also have to reinvest it into the communities, in the people that put you there in the first place.' This, the member suggested, means building programmes that serve the community, and to that end suggested the community would benefit from having a local women's shelter or homeless shelter available to its citizens. With self-government, this type of development would be more likely because the community would be less concerned about spending corporate profits on social programming, as federal financing combined with tax revenues would ensure the Inuvialuit could operate its government effectively without drawing on its corporate side.

So, even as land claims provide compensation money that funds economic development projects and ventures, the federal and territorial governments remain primarily responsible for the provision of basic social services. With a new self-government regime, decisions regarding programmes and services such as education or health care could be transferred into the hands of the Inuvialuit government. While they wait for the self-government regime to be finalized and established, the Inuvialuit have been very successful in building a community development division that delivers government programmes and services. However, what this means is that they are basically administering someone else's programmes. As one Inuvialuit member put it, in an interview on 2 October 2007, 'I think there is a big leap between that and self-government where you are going to shape your own programmes.' In short, self-government is more than just being a good corporate citizen and donating money to support local community initiatives and programmes (like Christmas hampers or donating funds to the youth centre). In the end it is about social programming designed to increase the well-being and security of citizens.

Like other Aboriginal groups that are not officially self-governing but that have assumed more responsibility for governance in the region or community, the Inuvialuit have accumulated a lot of experience in administering social programmes, which is critical to preparing them for the next step. According to the current IRC chairperson and Chief Executive Officer, Nellie Cournoyea, 'the Inuvialuit have always had a vision for governing our region and people', and this is why they have been busy negotiating a self-government agreement that would see the federal, territorial and Inuvialuit governments respecting and recognizing each other's role in governance in the Inuvialuit Settlement Region (Department of Indian and Northern Affairs Canada 2007). Powers of self-government cover a range of topics but typically include control over issues of culture, language and heritage as well as child and family services, social services, early childhood development, education, training, health, income support, marriage, justice and policing, financial planning, housing and economic development. More specifically, for the Inuvialuit, self-government means gaining more direct control over programmes and services; being able to protect the Inuvialuit language and culture; establishing new relationships between the people and their governments; and building stronger communities in the region. Certainly a key feature of self-government involves the Inuvialuit deciding upon the structure and form that their government will take. It also includes outlining its jurisdictional or governing responsibilities to its citizens, determining the scope of its governance and establishing new standards for service delivery. Given the potential for greater Inuvialuit control, self-government will increase representation and participation of Inuvialuit beneficiaries who live outside the region as well.

Despite the potential transformation self-government might provide, there is significant apprehension surrounding it. For some, it is a matter of levels of jurisdiction and the uncertainty as to who will do what. As one Inuvialuit member explained in an interview on 2 October 2007,

> [self-government] is going to be a really touchy issue because there's already talk about, OK, what's going to happen to the Inuvialuit Regional Corporation? And what's going to happen to the Hamlet. In other words, there is a lot of uncertainty as to how institutions will be organized and how they will serve people. Will they be better off? Or, will they lose out in the end? Yet at the same time the potential ability to improve programmes like education in communities is very appealing.

The same member went on to say:

> Know if we had like self-government we oversee it and say 'OK, this is where we are lacking'. And we'd have a say in that and make sure it was changed. Right now for lack of words, you bitch and complain, but the government doesn't listen.

The member even went so far as to articulate a vision for education: 'I always say that maybe from K to Grade 3 its all day everyday in our language.' Despite concerns that have been raised, it appears that the vision for a better future fuels the desire to negotiate because, in the end, self-government coupled with land claims means Aboriginal groups can do more, moving beyond economics, profit and jobs to programming and social well-being.

Vuntut Gwitchin

Right next door in the Yukon, in February 1995, the VGFN of Old Crow signed its comprehensive land claim agreement, awarding its 300 citizens 7,744sq. kilometres with territory and CAD$19 million in cash compensation.[3] Although there were not huge resource pressures at the time, there was always the threat of natural resource development as well as the issues surrounding it. As one VGFN official remarked in an interview on 20 September 2007, 'There was a sense in that the north was gonna boom in some way, the pipeline, you know there were all these developments that were not quite in the Yukon but that could have an impact on the Yukon.'

He elaborates:

> So with the Alaska highway gas pipeline you had a major construction project that was going to impact a huge chunk of the population, you also had the Dempster lateral as an add-on project that would capture the other Yukon communities. So there wasn't a resource extraction issue but it was a resource-related issue that was a bit of a driver.

Because First Nations' interests often collide with strong economic forces supporting resource development, issues of political security are as important as, if not more than, economic security (Deiter and Rude 2005:21). In some cases, where Aboriginal groups have been pressed to concede either cash or land to achieve a final agreement with the state, they have given up cash because, as Joe Linklater, Vuntut Gwitchin Chief, explained in an interview on 5 September 2000, 'you can always make money grow. So you can earn it back. But you can't make land grow.' This is because the land is not just something to be protected, it is their home and the basis for their survival.

While the cash compensation element of land claims is critical to the transformation of material relations between Aboriginal peoples and the state, it is not the sole motivating factor. Rather, the overall thrust of land claims is to improve the socio-economic condition of Aboriginal peoples. And as the state searches for ways in which to alleviate the financial burden posed by Aboriginal dependency (poverty is expensive!), land claims offer an excellent opportunity to redistribute financial responsibility for economic development, shifting it from the government to the private sector and to Aboriginal groups. Indeed, land claims are as much (if not more) about stimulating regional economic development, from clarifying title to infusing the region with money, as they are about settling claims to land (see Slowey 2008).

In contrast to the Inuvialuit, the Vuntut Gwitchin walked away from an initial settlement with the territory because it lacked enough self-government. While self-government was not an option in 1984, by 1995 a federal policy was in place and it was clear to the VGFN that powers of self-government had to be included in any negotiated agreement. They not only wanted more land but they also wanted more responsibility to ensure they would be able to govern that land. To that end, the VGFN sought self-government as a way to protect land, devolve hierarchical decision-making processes and empower community members. The new VGFN government's goal is to protect their ancient cultural heritage and ancestral homeland, exercise its inherent right to govern, assume responsibility for the welfare of its citizens and provide responsible governance for the communities, lands and resources. In addition to the economic opportunities provided by a completed claim, self-government transforms the relationship between the Aboriginal group at hand and the state. For instance, since finalizing their land claim and self-government agreements in 1995, the VGFN no longer functions as an 'Indian Act' band. Instead, the VGFN is a legal entity and has the capacity, rights and powers of a natural person. This means that unlike a conventional First Nation or Aboriginal group operating under the confines of the federal government of Canada, the VGFN are able to enter into contracts, acquire and hold property, and raise, invest and borrow capital. Moreover, as a self-governing community, the VGFN has exclusive power to legislate in areas pertaining to the administration of VGFN affairs and operations and the management of the rights and benefits emerging out of the finalized land claim agreement. It also holds power to enact laws in relation to the provision of programmes and services for its citizens living on settlement land.

In the beginning, the VGFN government admits it fell flat on its face, lacking any sort of idea as to how to construct a government. They were thrown in the deep end and they sank like a stone. The doors and windows of the band office were boarded up and former Chief Linklater admits they had no idea where to begin in terms of operating a government. However, given their tenacity, relative geographic isolation and desire for a better future, they were forced to learn how to swim quickly and managed to remedy their situation. A lot of their success is attributed to the skills and vision of their chief, Joe Linklater. To date the VGFN government has been busy developing its governance structure, building capacity, drafting policies and even preparing a constitution. To achieve its goal of operating well as a government, the VGFN has drawn on many resources including at one point hiring as CEO a VGFN member who was also a former member of the territorial legislature, and who infused the VGFN organization with a bureaucratic model of operation designed to streamline the decision-making process.

Further tied to security and political power is capacity. That is, self-government is also about capacity-building: providing Aboriginal peoples the tools to govern. Indeed Deiter and Rude (2005) urge the Canadian government to provide training and direction to First Nations engaged in self-government to ensure among other things that they possess the necessary tools to negotiate policing agreements, develop their own constitutions, devise policy and implement land management regimes.

The VGFN government plays a supporting role in creating an environment conducive to economic development as opposed to initiating economic development per se. It is a separate entity from its investment corporation, which is assigned the task of using compensation dollars to generate economic wealth. Instead, the VGFN government is engaged in developing housing, health care and educational initiatives, as well as dealing with issues of infrastructure and intergovernmental relations. It was active in a territorial curriculum review (see the Yukon Education Reform Project 2007) and also in devising a new land use planning model. In short, it is busy tackling the day-to-day business of governance in the region and in its traditional territory as a government. As former Chief Linklater explains it:

> In the past, affected First Nations had very little, if any, say in how land and environmental management worked in the traditional territories. These practices and the ties to the land had a very important role in First Nation life and still do today. The First Nations' traditional practices of land management were largely ignored by government and exploration companies, who instead chose to impose the Western European values of the time for land management through ownership and exploitation for profit. Exploration companies would come into First Nation areas with Federal Land Use Permits allowing them to explore for various resources and the affected First Nation would have no voice when it came to expressing concern about the socio-economic impacts these activities had on their lifestyle.
>
> (Linklater 2000:1)

Linklater points out that, because of self-government, the VGFN are no longer at the mercy of the federal government in deciding what issues they will address or how they will be addressed.

As more and more Aboriginal groups regain political control over settled lands, they gain an important and strategic voice in resource development. Nowhere is this more necessary than in the community of Old Crow, which lies in close proximity to the Arctic National Wildlife Refuge (ANWR) located in Alaska. Although the potential area for resource extraction lies in another country, the Vuntut Gwitchin have used their position as a self-governing First Nation to argue in both Washington and Ottawa that the industrial development of the ANWR would negatively affect their security, as it would inevitably disrupt the migration of the Porcupine Caribou and disturb their calving grounds. This is critical because the community continues to depend on the caribou for cultural and practical subsistence. Hence, the community stands in opposition to the opening up of any part of the ANWR region, as it would seriously undermine their security.

First and foremost, the Vuntut Gwitchin stand opposed to the ANWR development for social, cultural and environmental reasons. They are still suffering the effects of the last oil boom, which saw a dramatic rise in foetal alcohol syndrome and the abuse of drugs. The Vuntut Gwitchin are also concerned about the serious environmental threats that any development would pose. Already in other parts of Canada, like in northern Alberta, oil and gas development is having an adverse effect, not only on wildlife and the land but also on the Aboriginal peoples who live in the region (Urquhart 2010).

The VGFN government experience draws attention to one of the key wildlife management issues of the day, which remains the protection of the Porcupine Caribou herd. In terms of its orientation, its self-government is intimately tied to the dynamics between non-renewable resource development, and resource development and its role in managing environmental issues, and being involved in decisions that have environmental impacts on their people. From the insight of a former Territorial official interviewed on 20 September 2007, it can be understood that whether it is about instituting a new management structure for the Porcupine Caribou herd, lobbying in Washington and in Ottawa around the preservation of the Porcupine Caribou herd, or being part of a national northern lobbying effort, all these issues are tied to fundamentals around tradition and culture and were the driving force behind VGFN land claim negotiations. While socio-economic impact control, economic participation and environmental monitoring can be effectively realized where comprehensive claim settlement agreements are reached and implemented, self-government takes it to another level (Notzke 1994:220). The former Territorial official also remarked in their interview (on the topic of the VGFN):

> I think the essence of First Nations' self-government agreements being part and parcel of the claims process is that it established inter-governmental relationships. It wasn't a community working with government, it wasn't a collective of individuals or it wasn't individuals working with government, it was governments working with governments.

Put differently, self-government means indigenous peoples are recognized as governments and relate to other orders of government as governments. They are no longer simply wards of the state or stakeholders in development but participants in the decision-making process that affects their people, in their communities, on their land, across the Canadian Arctic.

Conclusion

It can be argued that the power of self-government agreements, like the power of human security and all that it entails, lie in their potential to transform the lives of Aboriginal peoples. While each of these examples highlight agreements that provided land and cash compensation, what differentiates them is one key element: self-government. Self-government therefore is critical not just for the land it provides as a basis for governance but also for the community and social well-being that can be constructed. In order to improve Aboriginal well-being, a community has to have more than just jobs and job security. Governance enhances the authority and position of an Aboriginal community and gives them the power to make decisions affecting their own security. The Yukon land claims agreement is also accompanied by a self-government agreement that gives the VGFN the right to enact legislation in a number of areas.

Given the recent resurgence in oil and gas exploration in the Canadian Arctic, does self-government improve human security? The utility of linking human security to self-government is clear. Security is more than the absence of war or military threats. Like self-government, security includes non-military threats arising from socio-economic inequalities, environmental degradation and a lack of social justice. It is about environmental protection and decision-making power. Indeed, this approach is not necessarily new. Deiter and Rude suggest it is 'echoed in the analysis of women's groups around the globe that have called for redefining security in terms of human and environmental needs instead of national sovereignty' (2005:4). Like the women's groups Deiter and Rude (2005) identify, Aboriginal peoples cannot be secure until their economic *and* social, personal and community needs as human beings are fulfilled and they are full and equal participants in political institutions. The study of self-government thus lends a practical dimension to the study of human security as well, and provides a lens through which we can unravel the relationship between self-government and the development of oil and gas in the Arctic. While the language of human security per se is not something one would find or discuss within Aboriginal communities, the ideas inherent to human security – of protecting a way of life, promoting governance and building economies – are everywhere. Like self-government, human security is a vague, often an undefined, concept that is best understood in terms of its principles and explained through examples.

When asking the specific question 'Does self-government improve the human security of Aboriginal peoples living across Canada's Arctic?', the answer is a resounding yes. Looking at the question from the community perspectives of Old Crow (Yukon) and Tuktoyaktuk (NWT), it is apparent that when land claims are settled without self-government then there is a transformation of economic opportunities but there is not necessarily any change in the ability of communities to make important political decisions or chart a specific political course. In contrast, with the concurrent negotiation of land claims and self-government, Aboriginal groups are not only given control over land development, but also important powers to regulate and legislate, which provides them with a voice in development as well as the ability to control it. Thus, not only do land claims promise to transform the economic condition of Aboriginal peoples but they can also increase security by returning political power to the communities. While it might be that self-government improves the human security of Aboriginal peoples, it also serves the security interests of the state as well by solidifying Canada's relationship with its Arctic peoples and over Arctic lands.

Notes

1 This project is supported by a Social Sciences and Humanities Research Council (SSHRC) Northern Development Research Grant (2006–2009) and forms part of a larger IPY (International Polar Year) Grant, Project 310: The Impacts of Oil and Gas Activity on Peoples in the Arctic Using a Multiple Securities Perspective (GAPS). For more details please see http://www.ipygaps.org (accessed 23 March 2013).

2 It is important to point out that the spelling of Gwich'in/Gwitchin varies and that the former is used in the NWT while the latter is used by the VGFN in the Yukon.

3 This number is for Settlement Lands only (which is important for self-government purposes) and thereby excludes Vuntut National Park, Fishing Branch Ecological Reserve and Old Crow Flats which are designated 'Special Management Areas', which means they are owned by Canada but co-managed with the Vuntut Gwitchin.

References

CBCNews Canada (2005) 'Canada ranked low in UN native report'. Online. Available at: http://www.cbc.ca/canada/story/2005/04/11/UNNatives-050411.html (accessed 11 April 2010).

—— (2013) 'Canada drops out of top 10 most developed countries list'. Online. Available at: http://www.cbc.ca/news/canada/story/2013/03/14/canada-slips-.html (accessed 20 March 2013).

Deiter, C. and Rude, D. (2005) *Human Security and Aboriginal Women in Canada*, Ottawa: Status of Women Canada.

Department of Foreign Affairs and International Trade (2009) 'The Arctic: Questions and resources'. Online. Available at: http://www.international.gc.ca (accessed 30 January 2011).

Department of Indian and Northern Affairs Canada (2007) 'Canada, Northwest Territories, and the Inuvialuit move ahead on Inuvialuit self-government process and schedule agreement'. Online. Available at: http://ainc-inac.gc.ca/ai/mr/nr/m-a2007/2-2890-eng.asp (accessed 30 May 2010).

Linklater, J. (2000) 'The perspective of a settled aboriginal land claim "stakeholder": Realities to date in terms of oil and gas operations and exploration', presentation to the 'Extending Oil & Gas into the Yukon and the NWT' conference, Calgary, September, p. 1.

Notzke, C. (1994) *Aboriginal Peoples and Natural Resources in Canada*, Toronto: Captus Press.

Owen, T. (2008) 'The critique that dog doesn't bite: A response to David Chandler's "human security: the dog that didn't bark"', *Security Dialogue*, 4(39):445–54.

Slowey, G. A. (2008) *Navigating Neoliberalism: Self-Determination and the Mikisew Cree First Nation*, Vancouver: UBC Press.

The Canadian Press (2007) 'Full text of speech from the throne'. Online. Available at: http://www.parl.gc.ca/Parlinfo/compilations/parliament/ThroneSpeech.aspx?Language=E (accessed 16 October 2007).

'The Responsibility to Protect' (2001) (December) *Report of the International Commission on Intervention and State Sovereignty*, Ottawa: International Development Research Centre.

Urquhart, I. (2010) 'Between the sands and a hard place? Aboriginal peoples and the oil sands'. Online. Available at: http://www.bcics.northwestern.edu/documents/workingpapers/Energy_10-005_Urquhart.pdf (accessed 20 June 2013).

White, G. (2002) 'Treaty federalism in northern Canada: Aboriginal-government land claims boards', *Publius: The Journal of Federalism*, 3(32):89–133.

Yukon Education Reform Project (2007) 'Education reform project final report'. Online. Available at: http://www.education.gov.yk.ca/pdf/ed_reform_report.pdf (accessed 23 March 2013).

Chapter 12

Women's participation in decision making

Human security in the Canadian Arctic

Stephanie Irlbacher-Fox, Jackie Price and Elana Wilson Rowe

Human security in the Arctic: a new dialogue?

The concept of human security was developed primarily to assess and ameliorate dramatic challenges to life and livelihood in the 'Third World'. In this chapter, we explore the relevance of the concept of human security to political participation of women in the Northwest Territories (NWT) and Nunavut in Canada. We start with the assumption that functioning and appropriate political systems and institutions are fundamental to human security. We also assume that legitimate and effective Indigenous representation and the creation of political institutions appropriate to Indigenous peoples' senses of identity, community and culture are essential to the well-being of Indigenous individuals and communities.

In this introductory section, we explore briefly the vocabulary of rights that is presently active in the Canadian North and hypothesize about the ways in which human security may provide a useful way of thinking, through the rigid juxtaposition of individual versus collective rights that so often characterizes discussions of Indigenous women's rights. Subsequently, we draw upon and combine our individual research in the NWT and Nunavut to further explore these issues. First, Stephanie Irlbacher-Fox describes broadly some ongoing tensions between colonial and Indigenous political systems in these areas of the Canadian Arctic and the influence this conflict exerts over the participation of Indigenous women. Second, Elana Wilson Rowe presents a case study of how institutional reform aimed at ensuring the participation of Inuit women in formal politics in Nunavut has failed. Finally, Jackie Price proposes an alternative approach to address the lack of voice that Inuit women have in politics – a model that inverts more conventional approaches to involving Indigenous persons in politics. Throughout, we examine how the concept of human security can contribute to thinking about the issues of political participation and the appropriateness of political institutions.

Indigenous and/or human rights discourse is one of the primary political discourses generated by Indigenous leaders engaged in activism and negotiations directed towards the Canadian state.[1] This political activism is based upon the idea that Indigenous peoples, as peoples colonized by settler states, have human rights – structured as 'aboriginal rights' by colonizing states – that existed before the imposition of the colonial state, and continue to exist. In many cases, particularly in the Americas where colonial governments and peoples became permanent, the refusal of the state to recognize Indigenous peoples' rights have resulted in an unresolved relationship with the state, wherein Indigenous peoples remain colonized (e.g. dispossessed and in conditions of economic, political and social marginalization).[2] In general, human rights are understood as protecting an *individual's* fundamental human interests, such as protection of the home and bodily safety, and are believed to be universally applicable to all persons in all societies. Canadian political philosopher Will Kymlicka has been influential in broadening the conception of human rights to describe how Indigenous rights differ from the dominant understanding of individually held human rights developed by Western liberal political thinkers and in positing a place for collective rights in the human rights discourse. His argument is based on the notion that Indigenous peoples require rights that exceed individual human rights, such as a collective right to the traditional lands that form the basis of continuing group livelihood and coherence, or a right to meaningful political engagement, in order to ensure access to a secure base of Indigenous culture. Kymlicka argues that, in the absence of such a firm societal basis, individual rights and the attending concepts of individual autonomy and freedom cannot be attained (Kymlicka 1998).

Yet, Kymlicka ascribes the title of 'national minorities' to Indigenous peoples within the Canadian state. As the classification of minority requires there to be an established authority (i.e. Canadian Federal Government), this designation represents another attempt to assimilate Indigenous peoples within the colonial structure by placing Indigenous peoples in a fixed relationship with a particular political standard, as Kymlicka does not engage with an Indigenous understanding of nationhood.[3] These points highlight the need to further explore the tension in discussions surrounding Indigenous rights as human rights. If Indigenous peoples have a collective right to maintain practices and institutions that differ from the dominant settler state, would their social and political institutions be exempt from basic human rights legislation and charters of rights? This issue becomes particularly prominent in terms of thinking about the rights of 'sub-groups' of vulnerable persons, often depicted as Indigenous women, within an Indigenous community, and is often posed as an irreconcilable conflict between group, or Indigenous, rights and individual liberal rights.

Indigenous feminist scholars recognize the vulnerability of Indigenous women in particular communities and in particular situations, but ascribe disparities in the power accorded to men and women in indigenous cultures to the influence of Western beliefs (Turpel-Lafond 1997; McIvor 1999; Kafarowski 2002). Thus, the protection and empowerment of women in Indigenous communities would not hinge upon the enforcement of Western individualist liberal rights specifically for women, but rather a restoration of the Indigenous decision-making processes and political institutions that have been undermined by constant colonial interference (Alfred 1999). It must be recognized that colonialism has rendered issues of human security for Indigenous women inextricably intertwined with the questions of self-determination and freedom which transcend gender lines. Colonialism has attacked the very basis of Indigenous cultural practices, which underpin the equality and freedom of Indigenous women vis-à-vis Indigenous men and non-Indigenous people. In the words of one Indigenous feminist, anti-colonial perspectives encompass 'a theory and movement that wants to fight all forms of oppression, including racism and colonialism ... we could see it as a struggle for unity among all oppressed men and women' (Sunseri 2000:144).

This approach, which articulates the necessity to interrogate broader social and political processes bent on the destruction and de-legitimization of Indigenous cultural and political practices, guides the following discussion of the implications of two colonial political systems for human security in the Arctic, with respect to Indigenous women and peoples generally.

Western political theorists are now also working to think beyond the individual versus group rights discourse that characterizes much of political theory's approach to Indigenous claims. Bern and Dodds propose that the individual versus group rights paradigm conceals that there may be

> a diversity of interests within a group that are not so much opposed to one another as they are to all aspects of the rich complexity of a shared way of life. As such, it is not a case of an oppressed subgroup's interests against the interests of the wider group, but rather an array of partially overlapping but different interests that together form the full array of group interests.
>
> (2000:169)

In the same volume, Ivison *et al.* (2000:11) propose that moving beyond this dichotomy requires abandoning the idea that rights are universal and unchanging, and call for an understanding of the historicity of rights, human rights discourse, and their implementation mechanisms and institutions. However, they do not propose to abandon

> the language of rights completely ... [but rather] the moderation of our desire to translate every claim into one that can be classified as an individual or group right ... A postcolonial political theory needs to focus as much on these processes as it does on the language(s) of rights.

Perhaps the concept of human security is another way of overcoming the group versus individual rights debate. As the United Nations Development Programme notes, human security is an 'integrative' as opposed to a 'defensive' concept. It is this integrative aspect that recognizes the interconnected nature of all facets of life in achieving a sense of security, which could help us see beyond the perennial juxtaposition of Indigenous women's human rights versus the group rights of the peoples to which they belong. Human security was initially defined as freedom from danger, poverty and apprehension, but both in theory and in practice today it encompasses political, economic, health and environmental concerns. In opposition to the language of rights, in which an individual's rights are either respected or not, human security invites us to think of the embodied and situational experience of feeling more or less 'secure' – a spectrum as opposed to absolute possession or dispossession. Also, this more faithfully reflects the concrete issues of felt security or insecurity which are closer to the everyday lived experience of individuals in communities. This stands in contrast to an abstract concept of being a rights-bearer and what that entails. Thus, the vocabulary of human security may be a more empowering one as it could provide the tools for individuals and/or peoples to make assessments on the state of human security in their home territories and to generate local solutions.

Indigenous political process and colonial political systems

Literature on political development in the NWT and in Nunavut, Canada, focuses largely on emerging political institutions, and the 'indigenization' of systems predicated on Western political institutions. In particular, this has taken the form of the inclusion of 'TK' (Traditional Knowledge), 'TEK' (Traditional Ecological Knowledge) and 'IQ' (Inuit Qaujimajatuqangit), which has received much attention by scholars as a groundbreaking and appropriate way for Indigenous values, ways and practices to be 'integrated' with Western institutions. It is argued that this 'integration' is essential to proper governmental administration of programmes, and political articulation between Indigenous institutions and those of dominant society.[4] Sadly, nothing could be further from the truth. Since the inception of the original Government of the Northwest Territories (GNWT) Traditional Knowledge Policy in the early 1990s, and proven by literature detailing problems of integrating TK and 'scientific' knowledge in the context of co-management institutions created by land claim institutions (for example, see Morrow and Hensel 1992; Cruikshank 1998; Nadasdy 1999, 2004), it is evident that efforts of hybridization of culturally based governance practices is nothing more than a replay of colonial dominant–subordinate relations characterizing the political development of the NWT and Nunavut generally. Part of the reason is that the only political institutions recognized in the North by Canada (and, importantly, by its funding policies) as being legitimate are the ones sanctioned through the laws of non-Indigenous governments, namely the territorial and federal ones. This section of the chapter attempts to describe the different political institutions extant in the NWT and Nunavut, broadly demarcated as 'colonial' and 'Indigenous', with attention to the often interactive relations constituting each. Through this description, we attempt to draw attention to the different political institutions and forces at play in decision-making in the North in order to explore the potential implications of the interrelationships and tensions between these institutions for human security in the Arctic and to provide background for the more detailed case studies of and proposals for women's involvement in the political life of Nunavut that follow.

Characteristics and institutions of Indigenous political processes

Generally, Indigenous political processes are embedded within the social and political cultures and processes that have historical continuity with pre-colonial practice, characterized by Indigenous autonomy.[5] Generally, these are 'institutions' conceived broadly as practices unrecognized by colonial governments. Indigenous political processes exist without reference to colonial systems and thus are not designed to correspond with colonial political institutions. Generally, Canada has chosen to bypass Indigenous political practices and institutions in favour of creating institutions most suited to its own purposes. For example, following the 'Indian Act', Bands replaced Indigenous institutions (often referred to as 'traditional leadership') in First Nation communities of the NWT. These Bands were designed and mandated by the Aboriginal Affairs and Northern Development Canada (AANDC) to serve specific purposes in relation to the responsibilities and policies of Canada, rather than to provide legitimate representation to Indigenous peoples.[6] In many communities in the NWT, such Indigenous political institutions are predicated on community recognition versus election, and, as one leader described to Irlbacher-Fox, Indigenous leadership predicated on Indigenous practices consists of people who are 'respected, not elected'. This recognition as a leader stems from a person's abilities, behaviours and characteristics in light of specific expectations, requirements and conventions of cultural and social practices and rules. For example, an Elder for purposes of a government pension is a person having achieved a specific age, such as sixty-five years old. In Indigenous communities, the title of Elder is assigned to persons who may have reached a certain age, but is equally, if not more, attached to a person's cultural expertise, personal attributes, behaviours and participation within a community and within a set of specific cultural and social expectations.

Indigenous political institutions reflect variances of Indigenous political cultures with those of the dominant society. In the NWT, Indigenous peoples' 'Assemblies' – community, tribal, regional and national – are regular gatherings of community representatives established to provide general direction to leaders and make joint decisions, mainly on a consensus basis. Elders

– sometimes recognized collectively as Elders committees – are involved in decision-making through mandated consultation and through regular individual consultation with leaders and decision-makers. Leaders chosen to head specific organizations or to speak on specific issues are bound to the authority of assemblies and ongoing consultation and consensus building among Elders and respected leaders with communities and regions represented.

Women's participation within Indigenous political institutions is far more prevalent than in colonially based political institutions. The emphasis on consensus within an entire community rather than a majority-rule approach, the necessity for ongoing consultation and consensus building throughout a leader's mandate and the recognition of individuals as uniquely qualified for asserting authority and legitimacy in certain situations based upon the specific knowledge held by these individuals ensures that women participate as vocally as men. This is not to justify the exclusion of Indigenous women from colonially based institutions.[7] Rather it provides a basis for recognizing the dual nature of political institutional decision-making within Indigenous communities, as described in greater detail below.

Characteristics and institutions of colonial political systems

Land claim and self-government negotiations in the NWT and Nunavut have resulted in dramatic changes in governance institutions and practices since the 1980s. While the 1980s and 1990s marked a period of optimism about the power of negotiated agreements for Northern decolonization, recent experiences in agreement implementation have resulted in renewed attention to the necessity for increased economic and political independence – rather than inter-dependence – as an essential outcome for land claim and self-government agreements.

Colonial political systems are ones that are legally created or recognized by non-Indigenous institutions (federal and territorial legislation, for example), and based on non-Indigenous institutional practices and designs. Generally, these are institutions with a leadership elected in first-past-the-post elections by majority vote at regular intervals in which consensus and consensus building is replaced by delegated authority to make decisions, and are characterized by specific legal responsibilities to correspond with regional, territorial and national non-Indigenous political institutions. Perhaps most importantly, these institutions are recognized as 'legitimate' according to the values and practices of non-Indigenous institutions and, as a result, receive ongoing funding and political recognition for discharging responsibilities.

Examples of such institutions include both 'public governments' – those that represent all citizens of a particular geographic area – and 'aboriginal governments' – those institutions representing Indigenous peoples' rights and interests vis-à-vis the state, which are recognized through land claim, self-government or other negotiated agreements involving devolution or delegation of authority from other governments. Examples of such institutions include the legislative assemblies and bureaucracies of the GNWT and the Government of Nunavut (GN); community or municipal governments; NWT First Nation Band Councils created by the 'Indian Act'; regional and community Inuit/Inuvialuit organizations created through land claim agreements; and public/aboriginal community governments created through self-government agreements.

Some will object to the description of 'aboriginal' governments as colonial. However, as the analysis below attempts to show, it is the colonial origin of the design and functions of such institutions that result in their inherent instability, due to the inability of colonial-based institutions to accommodate the reality of Indigenous practices. In other words, the Canadian government's insistence that Indigenous peoples adopt political institutional structures based on alien practices and political culture results in an inevitable and ongoing friction between Indigenous practice and imposed political structure. Instability and dependence on other governments (for funding, implementation obligations) are some characterizations of these governments; the predominant exclusion of Indigenous political processes in their functioning is another.

Consequently, several aspects of these systems' architecture result in the decreased participation of women in the decision-making and political process as compared with the Indigenous political process, as is discussed in greater detail below. The election process significantly decreases the need for leaders to seek advice from 'recognized' Indigenous leadership, among which women are prominent. Further, representatives are not bound to seek consensus on critical issues within communities. The nature of the system itself is predicated on the delegated authority of community to elected individuals acting in decision-making roles. From the election process to forums, such as the legislative assembly sessions and tribal council meetings where Robert's Rules of Order rather than consensus-building political techniques are seen as authoritative, the divergence of decision-making from community practice is evident.

Implications of the Indigenous process for colonial systems and vice versa: implementation stories

In many NWT and Nunavut institutions, the influence of Indigenous political culture is unmistakable. In both legislatures, there are no political parties, and Cabinet and regular members sit facing each other in a semi-circle to debate and discuss Cabinet policy and direction. This is touted as 'consensus-based' decision-making, stemming from Indigenous political practices. Within land claim and self-government agreements, decades of legislated and de facto exclusion of Indigenous peoples from decision-making bodies in their traditional territories has been replaced by a 'co-management' approach, which guarantees Indigenous peoples' representation in public institutions. For example, in the NWT and Nunavut land claim agreements, public resource management bodies guarantee Indigenous peoples' equal representation with the GNWT and Canada, and mandate that Indigenous TK or TEK is incorporated in decision-making.[8] Within self-government agreements currently being negotiated in the NWT, similar provisions are being made in some regions that community councils guarantee 50 per cent Indigenous representation. In exchange for this guarantee, the councils will exercise 'self-government powers' attached to the exercise of Indigenous peoples' rights of self-government.[9] Certainly, such arrangements represent steps forward in the struggle for Indigenous peoples' participation in, and indigenization of, non-Indigenous political institutions.

But experiences of implementing such innovative approaches tell a different story. Land claim and self-government agreements are not receiving enough funding to achieve the objectives of their design. Instead of creating coherently Indigenous institutions, individual obligations are assessed and funded by Canada in isolation of each other. The result has provided for the implementation of the 'letter rather than the spirit' of land claim agreements, according to the Canadian Auditor General, the national independent audit agency (Auditor General of Canada 2003:1). The same is true for public co-management institutions, where often the question of 'what is traditional knowledge (TK)?' is as difficult and controversial as receiving enough funding for its collection, recording and use alongside 'Western science' by resource managers illiterate in the cultural and experiential genesis of such knowledge (for example, Nadasdy 2004). Echoing the predictions and observations of political philosophers, such as Iris Marion Young (1998), James Tully (2000) and Raymond Geuss (2001), institutions created through land claim and self-government agreements are rendered ineffective through their inadequate or lack of implementation, inattention to the provision of mechanisms to monitor and conceive of implementation as an essential part of the agreements themselves, and the propensity of the dominant state's institutions to marginalize Indigenous voices and aspirations in the very institutional structures set up to accommodate Indigenous voices and culture. During 2004, this inspired a coalition of land claim institutions, including all land claim-based governments in the NWT and Nunavut, to propose a national effort towards effective land claim implementation, based on a common experience of land claim failure and shortcomings.[10]

Perhaps just as interesting as the problems posed by bureaucratic versus Indigenous interpretation of what constitutes appropriate implementation measures is the interactions between Indigenous and colonial institutions within the NWT and Nunavut. The persistence of Indigenous institutions alongside colonial ones creates an inherently destabilizing current within Indigenous communities. Band council elections premised on first-past-the-post majority voting sometimes subsequently result in chiefs and councillors being unceremoniously ejected from office by community petitions. Often these disputes are more a result of the council system's inability to accommodate cultural approaches to governance rather than being about the performance of the chief and council. In many communities, public meetings may include complaints of the failure of elected politicians to 'communicate effectively' with constituents. Often such instability might be rectified by clan or traditional 'family' representation in political institutions rather than majority vote contests, or the use of 'assemblies' to canvass community members on significant issues.

Some colonial institutions work towards achieving accommodation with Indigenous institutions. For example, one tribal council has an Elders committee that has developed the convention of choosing a Grand Chief before delegate voting takes place. This practice, though often criticized by proponents of individual rights, has not yet been overruled by voting delegates. Often elected leaders of colonial institutions practise Indigenous political culture, regularly meeting with community leaders and recognized Elders to seek advice and so build consensus and cohesion in accordance with Indigenous political practice. While at the level of political decision-making the colonial and Indigenous are not mutually exclusive, the technologized and bureaucratic structure of implementation institutions render indigenization efforts much more problematic.

As a result, the focus on political systems in the North has progressively turned to the issue of indigenizing foreign institutional structures required by self-government and land claim agreements. Public belief in the political legitimacy of these foreign institutions often rests upon the success or failure of a government to be perceived to be indigenized. In Nunavut, the GN has come under increasing criticism for its inability to reach its target of having Inuit as 85 per cent of public service employees. Land claim governments in the NWT have run into the same problem, with the Inuvialuit Regional Corporation, for example, recently being criticized for having over 85 per cent non-Inuvialuit employees in its various organizations and businesses, twenty years after the land claim creating it was initially signed. Other land claim governments face similar situations as they sign on to agreements predicated on their ongoing financial dependence on Canada to implement agreements and underwrite their economic development. This is compounded by the fact that land claim agreements, far from restoring dispossessed lands and resources, instead extinguish rights and ownership to lands and resources that could form the basis for Indigenous economic independence and self-sufficiency. Land claims restore small measures of political and economic control to Indigenous peoples, but they have not successfully addressed underlying sources of poverty and the attending social suffering that continue to plague communities in all regions of the NWT and Nunavut. One needs only to look at the social statistics in any land claim region to see that this is true. The failure of land claim agreements to 'fix things' in communities underscores an intransigent and embarrassing truth to the Canadian state: if the state dispossesses a people and continuously controls their political, social and economic realities, that people's social suffering will persist and require constant attention and resources. Unfortunately, all indications are that Canada and most Canadians are content to subsidize Indigenous suffering and the suffering industries spawned by the operations of the AANDC, rather than restore economic and social well-being in Indigenous communities. The imposition of colonial institutions, and massive – and failing – efforts to indigenize these institutions, may be read as a central aspect of creating suffering – a hallmark of human insecurity.

Implications for Indigenous women and human security in the Arctic

Insofar as Indigenous political processes continue to operate, both Indigenous women and men hold equally important responsibilities; however, women's participation in colonial institutions lags far behind the participation of men. Within colonial institutions representing exclusively Indigenous interests (such as Tribal Councils and land claim governments), women fare better than in public government institutions. For example, both the NWT and Nunavut legislatures, which have nineteen members each, have two women representatives (including the Premier of Nunavut) (current to 2013). Regional Inuit associations, Inuvialuit organizations and Band and Metis councils include several women as presidents, chairpersons, executive members and councillors.

If human security is constituted, in part, by the participation of all citizens in decision-making processes, what are the implications of a lack of women's representation in colonial institutions, and the broader problem of the lack of Indigenous participation and control of institutional implementation of 'indigenous inclusive' institutions, such as land claim created bodies and responsibilities? What are the implications for human security that a significant aspect of Indigenous political practices – Indigenous institutions which are respected, not elected – are unrecognized by colonial governments that have dispossessed Indigenous people and subsequently fund colonial institutions of their own choosing as the only legitimate avenues for Indigenous participation and consultation?

Clearly, human security, in terms of access to and control of the political process, is deeply undermined by current colonial practices in the Arctic, with a doubly damaging blow to women in particular, who participate more freely and in greater numbers within Indigenous rather than colonial institutions. To rectify this problem, governments could take a first step by expanding their notion of what constitutes 'legitimate' Indigenous governments and 'legitimate' methods of consultation. They must move beyond the perception that only a few elected Band councillors or land claim government representatives speak for the people. They must recognize that Indigenous peoples practise expansive political cultural ways, in contrast to Western norms. Therefore, requests for additional funding for community consultation and longer timelines to enable communities to conduct consultation and decision-making processes on their own terms must be heeded. What is labelled as 'consultation' by colonial systems, thereby signalling a lesser form of democratic power than 'decision-making', forms a central political practice of many Indigenous peoples. This is because it is the foundation for consensus building within Indigenous communities. As detailed by Irlbacher-Fox (2002) elsewhere, the Canadian government's continued ignorance of the reality for political consensus building as an essential practice of Indigenous political institutions results in an inherent instability in the negotiation and creation of self-government and land claim agreements. Put plainly, Indigenous communities do not operate politically similarly to non-Indigenous ones; everyone in the community has a say and people are generally unwilling to delegate their voice and opinion to one person through a majority vote held every four years.

In the following sections, we continue to discuss the questions of participation and institutional appropriateness outlined above, but focus solely on Nunavut as a more detailed case study and explore two concrete models designed to increase the participation of Inuit women in the politics of Nunavut. One model, outlined by Elana Wilson Rowe immediately below, was actually proposed and voted upon in 1997, and the debate surrounding the proposal highlights some of the difficulties in rights-based institutional reform, particularly as related to bringing Inuit women into formal territorial-level politics. In the next section, Jackie Price explores her own model of institutional reform and illustrates an approach to increase the participation of Inuit women in politics that emphasizes the need to bring the consultation process in line with traditional Inuit governance, rather than attempting to bring Inuit women in line with Western politics and institutions.

Addressing the political access of Inuit women in Nunavut: a case study of the gender parity proposal

On 1 April 1999, one-fifth of Canada's landmass became Nunavut, a new Arctic territory in which 85 per cent of the population of 28,000 is Inuit.[11] The gender parity proposal,[12] discussed prior to the territory's official establishment, was an attempt to address the under-participation of Inuit women in formal politics by guaranteeing gender parity in the Nunavut Legislative Assembly through an electoral system in which two representatives would be chosen: one man and one woman. To locate the debate, the politics that led to the establishment of Nunavut are sketched out briefly, focusing specifically on how changes in leadership and power structures affected the political participation of women. Looking closely at the gender parity debate, which extended from the proposal's inception in 1994 to a public plebiscite in

1997, ways in which different political actors described the role of women in political institutions, in terms of traditional Inuit culture and Western colonial political institutions, are highlighted. Finally, in light of the failure of the gender parity proposal, the concept of human security is analysed as a potentially useful tool for developing other proposals related to the political participation of Indigenous women in decision-making.

Interviewees argued that the concept of a gender-equal legislature emerged from an awareness that traditional modes of Inuit gender relationships, leadership and the nature and structure of political power and organization have changed over time.[13] Despite the leadership of several high-profile Inuit women, overall levels of participation of Inuit women in formal colonial political structures remain relatively low, particularly in Nunavut. The assertion that the Nunavut government cannot adequately represent the interests of all Inuit if it consists almost entirely of males was one of the guiding principles behind the gender parity proposal. The idea of a gender-balanced legislature was discussed first in 1994 by the Nunavut Implementation Commission (NIC), which was mandated by the 1993 'Nunavut Act' to provide advice on the establishment of Nunavut. The majority of the nine-person commission came to see the idea developed by the NIC staff – an electoral system that would 'build' gender balance into the very structure of the legislative assembly through a voting system in which one man and one woman would be chosen from each electoral district – as a combination of practicality and innovation, and a rational step towards overcoming a history of Inuit women's voicelessness and non-participation in territorial-level politics.[14]

Prior to contact with Europeans and Euro-Canadian bureaucratic structures, the basis of Inuit identity was the extended family unit, which was usually led by the oldest male who took decisions 'informally, gently and ... in consultation with members of his extended family' (Duffy 1988:196). Ethnographic analyses of indigenous northerners' gender relations highlight that men and women were highly interdependent and that there was a mutual awareness and appreciation of this complementary relationship (Guemple 1986; Dorais 1988; Bodenhorn 1990; Reimer 1996). However, the Euro-Canadian traders, merchants, missionaries and government officials who controlled the North favoured the participation and leadership of Inuit men, a tradition that caused Inuit women to feel that their experiences and knowledge were not applicable to this new Western political process (Thomsen 1988; Reimer 1996). This was followed by the sedentarization of the formerly nomadic Inuit by the Canadian state in the 1960s, which caused the Inuit to interact more regularly with those outside their own clan or kinship groups and reinvent and reorganize their societal patterns, including what could be described as more formal, or more accurately more Western, political organization in the shape of settlement councils (Honigmann and Honigmann 1965; Vallee 1967). The 1970s marked the beginning of a campaign, led by Inuit politicians and largely in keeping with the policies of the Canadian state, to settle land claims and to create the GN, a process that greatly emphasized the right to exercise authority over land, a traditionally male domain of activity (Cassidy 1993; Damas 2002). The economic and political emphasis on natural resources and the centrality of the image of the hunter (Dybbroe 1988; Thomsen 1988) may have positioned masculine concerns at the centre of Inuit nation-building.

Further, the emphasis on Inuit, and other Indigenous, women as 'givers of life, custodians of culture and language and caretakers of children' (RCAP 1996:64) positions domesticity at the centre of Inuit womanhood, creating a situation in which the mobility of political office is problematic and women are distanced from the public sphere – the space of Western political decision-making and the space upon which political institutions in Nunavut are modelled. Nancy Karetak Lindell, a Member of Parliament for Nunavut, cited the criticism faced by female politicians for leaving their families behind when travelling as a disincentive for women to participate in formal territorial politics (*Nunatsiaq News* 2001). Other reasons cited in interviews for women's low participation in elected politics include the difficulty of balancing family, career and community obligations and the challenge of gaining access to the funds and supportive networks required for political campaigns.[15]

Prior to the public plebiscite on 26 May 1997, political leaders in favour of gender parity toured Nunavut's communities as part of the 'Yes' campaign and published articles and made radio appearances. Subsequently, a smaller 'No' campaign, championed by Manitok Thompson, a prominent female politician who had served as a Member of the Legislative Assembly in the GNWT, was established in opposition. However, all meetings about gender parity had relatively low attendance, and turnout for the final vote, in which gender parity was rejected by 57 per cent of those casting ballots, was only 39 per cent of all eligible voters (Dahl 1997). Regardless, the debate amongst the politically engaged was heated and clearly elucidates some of the major discourses active in conceptualizing gender relationships and women's political participation in Nunavut, a debate that drew upon both traditional Inuit and modern Western ideas of politics and who belongs in spaces of power.

Arguments based on competing perspectives of traditional Inuit culture

Although the idea of gender parity in the Nunavut legislative assembly was not initially presented or discussed in relationship to Inuit tradition, the idea of traditional and inherent gender equality within Inuit society came to be used by both supporters and opponents of gender parity, who framed their arguments with competing views on the continuity between traditional gender relationships and those of the present-day. Supporters of gender parity argued that the proposal would restore a tradition of respect and equality that had been lost, whereas those against the proposal hearkened to traditional Inuit gender relationships as a firm foundation assuring mutual respect, which made the proposal itself superfluous. James Arreak, in a letter to the editor of *Nunatsiaq News*,[16] argued that 'women do not need to earn respect because they already have our respect' (1997). Acknowledging men and women differently was also seen to undermine an Inuit 'spirit', which is, in the words of Arreak (1997), 'communally based and individualism is second to it'. As Paul Quassa (1997) argued, '[our ancestors] did things collectively in order to survive ... I believe that this [idea of gender parity] will only make the Inuit think and act as if there are two distinctive groups rather than viewing us all as one.' Also, a relatively conservative interpretation of Christianity, which has in some ways been incorporated into Inuit 'tradition' across the North, was invoked by a vocal minority who, particularly over community radio, used religion as grounds for opposition to the gender parity proposal (Dahl 1997; Gombay 2000; Hicks and White 2000).[17]

The conceptualization of women as the carriers of tradition, responsible for the care of the national family, is central to Inuit nationalist discourse and echoes throughout understandings of Inuit womanhood. As carriers of knowledge about the communities and the home, supporters of gender parity argued that the presence of more women in formal politics would help the government deal with social problems, like those relating to health and education. One Igloolik woman said that social issues would have a higher priority if the Nunavut legislature had gender parity. She commented that 'if we don't make a change, I don't see these types of problems going away ... Nobody talks about them in the present government' (Bourgeois 1997a). The importance of the family and women's role within it was also used as a reason to reject the proposal, based upon the notion that the absence of women in the home while participating in territorial politics could lead to a deepening of the social problems afflicting many communities in Nunavut. One elder in Pond Inlet voiced his concern about women being elected as members of the legislative assembly, as children would be left at home. 'I see kids who are left alone at home ... I think they're the ones committing suicide when they get older. What's going to happen to those kids when their mothers are at the capital?' (Bourgeois 1997b).

Arguments based on competing perspectives of formal political structures

Those against the gender parity proposal argued from a position of belief in the ability of existing forms of representation and rights to ensure the fair treatment and equal participation of all citizens, an idea that feminist scholars argue has led to the suppression of difference from public discourse in the name of objectivity – a practice that reinforces the privileges of already dominant groups (Pateman 1989; Young 1998). Paul Quassa (1997) argued that 'each and every able Canadian elector has the right to run for office ... we don't elect people because they are men or women, but because they have experience and have proven their ability to constituents'. In light of the Canadian Charter of Rights and Freedoms, gender parity was seen to be superfluous, as a legal requirement for and commitment to equality already existed.

Supporters of gender parity attempted to point to the reality of women's under-participation and the institutional barriers inherent to governance structures (Nunavut Implementation Commission 1995). Martha Flaherty saw Nunavut as a unique opportunity to reverse this trend:

> We can avoid some of the problems with existing governments, one of the most significant problems being the under participation of women ... in the old days, Inuit survived in the harsh environment through cooperation, and now NIC is proposing to carry on this long-standing tradition of working together.
>
> (1994)

In attempting to incorporate an acknowledgement of gender difference into the structure of the legislative assembly, supporters of the proposal were challenging powerful notions about the nature of representation in the public sphere. The public sphere was, in many ways, constructed largely in contrast to the assumed particularity and subjectivity of power relations within the home (Phillips 1998). Feminist scholars argue that this contrast resulted in two concepts that powerfully characterize modern political thought: the public/private divide and the related notion of women's responsibility for the spheres of domesticity and reproduction and men's obligation to the public world of economic and political life (Okin 1998). This created a situation in which both women and women's issues are excluded from the public sphere of political life (Pateman 1989; Okin 1998; Phillips 1998; Young 1998). While it is important to exercise caution in relating Western literature on the public/private divide to non-Western societies, much of the evidence outlined in this case study points to the influence of Euro-Canadian political ideas and institutions in shaping Nunavut's politics.

The gender parity proposal can be seen as a 'rights-based' approach, despite the fact it could be argued that the proposal's detailed implementation mechanisms and practicality overcomes the criticism frequently levelled at the use of rights language, namely that rights are meaningless without the mechanisms that allow rights to be realized. Regardless, the proposal is based in the idea that women have a right to be involved in politics and seeks to implement this right by moving Inuit women into formal political institutions. The failure of the gender parity proposal essentially ended productive dialogue about how to involve more Inuit women in political activity and, in the end, it was an all-or-nothing game – the gender parity proposal was to be accepted or rejected in referendum, a right to participation as a way of overcoming voicelessness to be implemented or not. Perhaps the idea of human security, which calls attention to the particular historical, cultural and social circumstances of individuals and collectives in a certain place and forces us to think about the individual and collective embodied experience of security, is a more empowering and useful concept that can measure a spectrum of experience from insecure to secure. Perhaps those experiencing insecurity can describe and perhaps even address the roots of insecurity more simply and accurately than they could describe the more abstract and all-or-nothing state of being a rights-bearer. Therefore, the concept of human security, rather than just an idea, might be a speech-giving practice that allows people to talk about their own experience on their own terms and to point to areas requiring improvement and change. In a society of speakers, however, there would need to be institutions required and appropriately designed to listen, and such a model is outlined in the subsequent section.

Moving beyond 'our rights' in 'our land': Inuit governance in Nunavut

Human security discourse is applicable to a full range of Inuit experience across the Arctic. Issues such as climate change and food security, particularly in regards to persistent organic pollutants (POPs), present real threats to Inuit livelihood. These threats have obtained international recognition within academia and politics, and have supported Inuit in their political work. For a discourse on human security to create roots within the circumpolar, discussions surrounding the security of Indigenous governance must be included. One way to evaluate a possible compatibility is to consider whether human security discourse can support the revitalization of Indigenous governance processes. Jackie Price considers this question by examining the various requirements of public dialogue in Nunavut.

In Nunavut, political processes and the dialogue it supports are in a current state of conflict. As highlighted by Irlbacher-Fox, this conflict is a result of the continued presence of colonial political structures in the Inuit homeland. In Nunavut, the effectiveness of colonial structures, such as the Nunavut Land Claims Agreement (NLCA), to support or represent Indigenous governance practices must be critically assessed in order to move beyond the cyclical nature of this conflict.

Diluting the call for critical engagement are the very political discussions on Inuit Qaujimajatuqangit, which seeks Inuit support within the NLCA political systems.[18] In the previous section, Wilson Rowe outlined the complexity of negotiating a direct relationship between these two systems with her overview of the gender parity debate. This debate represents an attempt of the Nunavut political systems to be reflective of Inuit Qaujimajatuqangit, and it inspired fascinating discussions amongst Inuit in Nunavut. Equipped with political interpretations on what Inuit governance 'means', Inuit publicly debated gender parity. In reviewing these debates, it appears that Inuit governance was both in favour and against the gender parity proposal. I believe this conflict is an organic result of the conflict that emerges when colonial institutions and Inuit governance are forcefully managed simultaneously without a critical understanding of their fundamental differences. Such debates result in an inability to understand the needs of Inuit governance beyond institutional structure.

This section considers possible requirements for a return to Inuit governing processes. Specifically, it explores the need for the governing process to be practised outside colonial institutions. I advocate for a new political consultation process called the 'Kitchen Consultation Model'. The foundation of this model recognizes the ability and potential of communities to initiate, lead and support community discussions and recognizes the need for communities to maintain, and defend, their connection to the perspectives held by their members.

Introduction to Inuit governance

> Words have always been cleansing. Words are what has made us Inuit. Words are very important.
>
> (Aalasi Joamie in Therrien and Laugrand 1999:251)

The security and strength of words emerges from a collective structure that guides people in their talk. I learnt this lesson in reading the pages of the *Interviewing Inuit Elders* book series.[19] This book series contains constant and consistent lessons on how Inuit Elders speak and share words. As a passionate reader of the complete series, I was made aware, and reminded of, many foundational lessons in sharing words. I was reminded of the direct connection between words, experience and knowledge. Since knowledge is acquired through personal experience, which is localized and influenced by family teaching, the sharing of words was a meaningful, deliberate action and the group reinforced the necessity of discipline when sharing words. This discipline was reinforced in the respect afforded to individuals with specialized knowledge, as it was rooted in an appreciation that such knowledge was accumulated through various experiences and the result of an ability to respect the diversity of opinion and knowledge on a particular topic.

I was also consistently made aware of the direct connection between sharing words and sharing food. Perhaps the greatest structure and support for supporting dialogue is food sharing. Food sharing was a primary forum for future planning, as explained by Noah Piugaattuk, an Elder from Iglulik, Nunavut:

> Inuit would be called to one place to eat and they would make plans as to who is going to hunt the day after ... if the weather co-operated. They would gather at night and tell their stories of the day's hunt and make plans for the next day.
>
> (Piugaattuk and Angiliq 1986:15)

Food sharing is also reflective of social responsibility:

> Food sharing was necessary for the physical and social welfare of the entire group. It made people feel that they were part of the community. Someone you shared food with was someone you could turn to for help at a future date ... Selfishness was not tolerated, and young hunters were taught to take care of all camp members.
>
> (Bennett and Rowley 2004:87)

Food sharing is also a time of celebration, by welcoming gesture to the community, well articulated by Isapee Qanguq:

> When the dog teams returned they had big loads of meat ... After they completed the task of storing the meat, my grandfather Inuutiq went outdoors; as he got to the entrance he started to shout in a loud voice: 'Tamuattuaaq, tamuattuaaq, avagusukkama, tamuattuaaq.' [Something to chew, something to chew, it is my wish to share it, something to chew.] Everyone came so that our home was packed with people.
>
> (Bennett and Rowley 2004:87)

The culmination of celebration, social responsibility and practicality is central in conceptualizing Nunavut's political reality beyond its present colonial systems to a political system really rooted in the Inuit way.

Moving beyond a continued contradiction in Nunavut

Nunavut is Inuit homeland, and Nunavut's political system stems from the signing of the NLCA, a legal contract between the Government of Canada and the Inuit of Nunavut. The NLCA secured the creation of an Inuit majority territory within the Canadian federation supported by institutions that function at the standard of the Government of Canada. The contradiction that Inuit in Nunavut surrendered their aboriginal rights to secure the creation of a government that functions according to the standard of a colonial mechanism within Inuit homeland is the forum in which Inuit governance must re-emerge from. How can this be done?

There are many ways to answer this question. I have chosen to focus on political consultation as it is the bureaucratic equivalent of communication and interaction, and it is the central interaction between 'government', 'community' and 'Inuit'. In Nunavut, political consultation is a legal requirement and a strategy utilized within the political process to secure some level of accountability. The current process of political consultation is straightforward. Political institutions decide what issues will be discussed, which communities will be visited and when, and what decisions will be made from these discussions.[20] Political consultation is a strategy for political institutions to record and acquire a community perspective, thereby creating an imagined community history within its political structure and identity.

The practice of consultation has a short but poignant history in Nunavut. R. Quinn Duffy introduced me to this history in describing a meeting held in Ottawa in May 1952. The meeting, which focused on the strategies to bridge the continual presence of the Canadian Federal Government in the North while protecting the well-being of Inuit, brought together representatives of federal departments conducting business in the North. There were no Inuit at this meeting. The justification for this absence was:

> 'The only reason why Eskimos were not invited to the meeting,' the director explained, 'was, apart from the difficulties of transportation and language, that it was felt that few, if any, of them have yet reached the stage where they could take a responsible part in such discussions.'
>
> (Duffy 1988:198)

Duffy goes on to argue: 'Thus the government recognized the difficulty of involving local Inuit in planning policy at levels higher than the community ... So it was left to those southern Canadians to settle the future of the people of the north' (Duffy 1988:199).

Inuit in Nunavut have consistently been aware of this 'well-intended' invisibility, and have successfully rallied against it as a means to provide Inuit with some access to political authority. Prior to 1999, Nunavut was the eastern section of the NWT. The isolation and distance experienced by 'the East' acted as a springboard for the campaign calling for a government that would better represent and address Inuit needs. The argument was that, to do this, Inuit needed to be politically and physically separate. The vision was informed by an understanding that a political system located within Inuit homeland would inspire Inuit to participate in the political decision-making processes, a belief expressed by Emile Imaruittuq: 'We don't put our knowledge to use. Maybe when we get Nunavut we will stand up for ourselves' (Oosten *et al.* 1999:93). This campaign was effective, and resulted in the creation of Nunavut. Many Inuit believed the GN would allow Inuit Qaujimajatuqangit to direct political and government decisions and processes. Achieving this goal has proved challenging. In its 2003 final report titled 'Our Words Must Come Back to Us', Inungni Sapujjijiit, the Nunavut Taskforce on Suicide Prevention and Community Healing, expressed its concern over the continued practice of southern Canadian experience determining the direction of

Northern programmes and services. Within the early pages of this report, one can still sense the isolation experienced by Inuit vis-à-vis the Federal Government in the 1950s:

> Many programs in our communities are imported from the south or other aboriginal groups. We must seek our own solutions to improve our solutions. Our ability to make our own decisions was taken away from us; we must now take it back.
>
> (Inungni Sapujjijiit 2003:10)

The recognition that Inuit Qaujimajatuqangit is not the basis from which decisions in Nunavut are made represents the inability of Inuit governance to secure a solid place within colonial structures. If Inuit systems are to have a rightful place in Nunavut, they must untangle themselves from institutional systems. Nunavut communities can be an important support in this work, by the action generated from their own dialogue.

Developing a Kitchen Consultation Model

The logistics of the Kitchen Consultation Model are straightforward. Under this model, interested community groups, organizations or individuals would lead the administration of the consultation. Ideally, the community group would already have a presence in the community as a non-governmental organization in order to draw from the strength and experience already gained, such as having an established presence in the community,[21] and having the experience of addressing financial and reporting requirements.[22] Once a consultation mandate was accepted by a community organization, the organization would then be responsible for determining what additional supports would be required to successfully represent the diverse perspectives of the community. This model works from the assumption that different modes of interaction would be used a number of times. Examples of interaction strategies would include public meetings, radio call-in shows, group meetings and individual interviews. The identification and selection of support staff would also have to occur; specific roles that would need to be filled include facilitators, translators, interpreters, transcribers and final proofreaders. Organization or community members may fill these roles, and no one person may necessarily be assigned to the same role as a new consultation process is initiated within the community.

Currently, the majority of consultations occur by holding public meetings. Public meetings assemble community members into one location to publicly discuss a topic with political officials and/or representatives. I believe public meetings are ineffective at supporting in-depth dialogue around a variety of complexities due to logistical difficulties and unrealistic timelines.[23] Under the Kitchen Consultation Model, public meetings would be given a more specific responsibility, one that respects the strengths of its practice. Under this model, public meetings would be a time when consultation intent is introduced and explained to the community. Not only would an overview be provided, but there would also be a discussion on the possible scope such discussions may yield. Also, if consultation is being done to provide direction on bureaucratic technicalities, such as legislation or budgeting allocation, public meetings provide the opportunities for community members to learn about these technicalities, thereby providing a public and collective forum where their limitations can be articulated. Providing this technical information early in the process would alleviate the continual need to hint at logistical constraints, ideally leaving more time and energy to talk about the topic at hand. This can occur quickly, as communities in Nunavut already have a variety of public infrastructure and collective experience to support community dialogue. Public meetings currently occur in existing community halls, and the majority of communities have community radio stations.

This model also respects the complexity of community dialogue, and respects the responsibility of accurately portraying the dimensions of a community's perspective by requiring appropriate and in-depth support and securing the opportunity for various perspectives to be heard. Under this new consultation model, public meetings would be supplemented with individual and group meetings. Such meetings will allow different perspectives to interact at different scales, as they would include various meetings of a small group of people knowledgeable on a particular subject or representative of a particular demographic. These meetings could happen once or a number of times. Alongside these meetings would be individual meetings or interviews. Again, the strength of these meetings is to provide an additional dimension in the community's understanding of a particular topic. Once these meetings were completed, another public meeting may occur to provide the community with an update, allow a general approval/disapproval to be given, allow for further recommendations to be brought forward or provide a general discussion for closure.

To ensure continual movement in this process, a team of community supports are required. Facilitators would be responsible for supporting community dialogue by facilitating community interaction. In order to encourage different perspectives, more than one facilitator may be required. Facilitators would be responsible for ensuring that the consultation process is executed in a way that respects the form and function of the community, and would work with translators to respect the various language responsibilities in the community level (Inuktitut and English) and the requirements of political institutions (largely English). Methods of recording would also reflect these ongoing dialogues, and make them accessible in different languages.

Although this model represents a territory-wide vision, the autonomy of the community will be respected.[24] Therefore, the success of the Kitchen Consultation Model will develop as communities gain experience in supporting consultation. To support this success, it is necessary for both communities and political institutions to acknowledge and appreciate the history of dependency resulting from a colonial past. More specifically, communities require support, cooperation and space from these external systems in their present-day interaction with political institutions and their direction, authority and assumptions.

The practice of dialogue as a social responsibility is central to the Inuit governing process, but, placed within a colonial context, this responsibility has been translated into a 'fly in and fly out' consultation process that prevents interactive dialogue. The example of consultation, though conceptually equivalent between the two political systems, is fundamentally different in theory and practice. Although there are constant attempts to synergize the two political processes, experience has proven that these efforts do not work, regardless of how much space is created for Indigenous processes within the colonial system. This inability to do so expresses itself by way of Indigenous apathy in the colonial political systems, apathy that portrays itself as, and is the result of, realities of social conditions in Indigenous communities.

In establishing experience, communities create space for individuals, organizations and political institutions at the community, regional and territorial level, to support community development through education with and for the community, region and/or territory. Recognizing this, Indigenous populations can no longer commit the time and energy to support a failed process that increases the tension within their own populations. The Indigenous population must return its focus to revitalizing their own political processes. Strategies such as the Kitchen Consultation Model are an example of how this goal can be accomplished. This model breaks away from the fluidity of colonial processes by returning the necessity of experience to political processes. Returning to an experienced-based political process will acknowledge the contribution both women and men make within the political processes. Respecting that knowledge is obtained directly through experience, or indirectly through family teachings, it becomes clear that both women and men have a responsibility in all realms of political reality and, therefore, must be present within the system to do so. This will require Indigenous communities and colonial structures to work through the complexity of moving beyond the colonial process. This will be difficult, as financial sources, colonial political mentalities and apathetic communities are difficult realities to untangle, but it must occur in order to create stable political systems. Once a space between colonial and Indigenous processes is established, education on how to navigate political processes can occur within each system separately. Perhaps thinking in terms of human security is one mode of encouraging and respecting this necessary space.

Conclusion

The introductory section of *Interviewing Inuit Elders: Perspectives on Traditional Law* states: 'Inuit elders were not concerned with theory, but with practice' (Oosten *et al.* 1999:6). Throughout this chapter, we have attempted to take up that directive and to determine how human security may be a useful tool for thinking and, more importantly, for action, especially in relationship to issues of governance and Indigenous women's involvement in governance structures. In our definition of human security, we emphasize that the concept must include recognition of the fact that Indigenous representation and Indigenous political institutions are connected to a people's identity, community and culture. They are essential to Indigenous individuals' and communities' well-being. Drawing upon this definition, we have argued throughout this chapter that human security is more conducive to achieving an integrated understanding of what it means to live well and live freely – a concept more open to participation and discussion than the defensive and theoretical discourse of rights. Further, thinking in terms of human security allows us to move beyond the prevalent and hindering juxtaposition of individual rights versus collective rights that has characterized so much of the debate about Indigenous women's rights within their collectives. The concept of human security acknowledges and must continue to acknowledge that individual and collective security are intertwined and, as highlighted in this chapter, the security of Indigenous women is deeply engaged with the security of Indigenous peoples, both men and women. Further, we have found that the dominant rights language, which was developed by Enlightenment thinkers and is at odds with much Indigenous political philosophy, intrinsically favours Western colonial systems, particularly in terms of the question of political participation. Through the domination of non-Indigenous understanding of rights, the right to participate in present political systems for Indigenous peoples amounts to Indigenous peoples having a right to take part in the system that has perpetrated oppression, rather than opening a space that recognizes and accepts the existence and functioning of alternative Indigenous political systems. Finally, the idea of 'security' is one that speaks to the embodied experience of individuals and of groups. Discussion of human security invites those experiencing insecurity to identify, describe and define the roots of their vulnerability and to develop solutions. Through such an exercise, discussions of Indigenous suffering and the direction of an Indigenous future could move beyond the confines of colonial systems that both perpetuate suffering and limit the future of Indigenous governance systems that are in consonance with Indigenous realities.

Notes

1 This summary does not include Indigenous political theories and practices that operate at the level of Indigenous communities and peoples – such theories and practices vary from people to people and are difficult to make generalizations about and will be discussed in more detail in the section on Inuit governance in Nunavut. By contrast, strategies directed towards the Canadian state for achieving Indigenous political goals are more homogenous in that key practices and discourses are set by the Canadian state itself and by politically savvy Indigenous leaders.

2 In Canada, the state has pursued the extinguishment of aboriginal rights through treaties and through the so-called 'modern treaties' – the land claims and self-government negotiations process, which are discussed in detail in the section on the Indigenous political process and colonial political systems.

3 See Taiaiake Alfred's *Peace, Power, Righteousness* (1999) for a discussion on Indigenous understandings of nationhood.

4 For example, see the extensive and excellent discussion of challenges and difficulties inherent in an 'integration' approach in Nadasdy (2004).

5 The observations in this section are based on Irlbacher-Fox's participation in political development processes through over close to a decade of professional work in the NWT and her anthropological fieldwork conducted in the NWT during 2000/2001.

6 AANDC administers the Indian Act, and various legislative and policy-based responsibilities for Indigenous peoples in the NWT. These responsibilities are structured and discharged within a national policy, legislative and political context that promotes uniform treatment of Indigenous peoples in Canada as much as possible.

7 Indigenous women who serve as elected leaders have emphasized to me that given the fact of colonial institutions within Indigenous communities, it is important to continually advocate for women's representation in these and other elected institutions. As stated elsewhere, this distinction between colonial and Indigenous systems is not intended as an apology for the exclusion of Indigenous women from electoral politics nor advocacy for Indigenous women to settle for 'whispering in the ear of power' rather than asserting their own. Instead, it is to distinguish between the existence of the two types of systems, and make the case that power and authority is exercised differently within Indigenous political processes than in colonial ones.

8 See the respective land claim agreements and, in the NWT, the *Mackenzie Valley Resource Management Act* mandates the use of 'aboriginal traditional knowledge' in its decision-making bodies.

9 See for example the Inuvialuit and Gwich'in Self-Government Agreement in Principle (Government of Canada 2001).

10 Resulting from a November 2003 conference on land claim implementation, the coalition issued a major policy paper in 2004 and initiated the establishment of a coalition secretariat based in Ottawa.

11 Prior to Nunavut's establishment, the Eastern Arctic was part of the NWT. Although the GN is a public one, in which Inuit and non-Inuit alike can participate, the founding principle of Nunavut was that it was meant to be an Inuit homeland with governmental structures and political processes reflecting the values and interests of Inuit society. It is debatable the extent to which the GN is an 'Inuit' government and the effects of integrating 'Inuit values' into Euro-Canadian governance institutions.

12 See Wilson (2005) for further discussion of the gender parity proposal and debate.

13 This observation stems from interviews with Jack Hicks and John Merritt, which were conducted by Wilson Rowe in Iqaluit, autumn 2003. Jack Hicks served as a staff member of the NIC and John Merritt was legal counsel to the NIC from December 1993 to January 1998.

14 This is an observation drawn from interviews with individuals (Hicks, Merritt and Kenn Harper) associated with the NIC conducted by Wilson Rowe in Iqaluit, autumn 2003.
15 At the time of interviews with Wilson Rowe (August and September 2003), Veronica Dewar was President of Pauktuutit, the Inuit Women's Organization, Jennifer Dickson was Executive Director of Pauktuutit and Maureen Doherty was Executive Director of Qullit (Status of Women Council, Nunavut).
16 Articles and commentary in *Nunatsiaq News*, an English-Inuktitut weekly newspaper that has the largest circulation of any newspaper in Nunavut, were the primary texts analysed for this chapter and supplement the interviews I conducted with politicians and policymakers in Nunavut from July to December 2003.
17 Observation also drawn from an interview (October 2003) with Natsiq Alainga-Kango, who served as Secretary of Nunavut Tunngavik Incorporated, the Inuit land claims organization, during the gender parity plebiscite.
18 For example, Article 23 of the Nunavut Land Claims Agreement 'Inuit Employment in Government' represents a legal obligation placed on political institutions and Inuit to work within the colonial system.
19 This five book series, published by Nunavut Arctic College, is available in English and Inuktitut and covers a range of topics, including health, child care, law and cosmology. This series is complemented by other series, including *Inuit Perspectives on the 20th Century*, *Memory & History* and *Life Histories of Northern Leaders*. Online. Available at: http://arcticcollege.ca/en/shop-on-line/books/life-stories-of-northern-leaders-series (accessed 4 June 2013).
20 Although this equation is experiencing creative additions, and success stories, such as the consultation surrounding the Nunavut *Wildlife Act*, its reality does not stray far from this description, and time will tell what, if any, legacy results from this wide-ranging and expensive consultation tour.
21 Organizations that fulfil these requirements already exist in Nunavut communities. For example, in Clyde River, the Ilisaqsivik Centre filled gaps in social services programmes delivered in their community by creating their own wellness centre which provides healing sessions, addictions counselling, suicide prevention and prenatal care; in Iglulik, the Inullariit society is an active presence in the community, with this Elder's society not only supporting youth camps, but also supporting community language activities that are communicated over the community radio. Finally, in Arviat, the Arviat Youth Piliriqatiqiit is a strong force that is working towards establishing a youth centre in the community, and has initiated various suicide prevention campaigns.
22 Funding stresses is a recognized reality. In an article by *Nunatisaq News*, the director of the Clyde River Ilisaqsivik Centre, Tracey Thompson, was quoted as stating: 'As the coordinator, now I spend all my time scrambling for money. We have great projects, but they could be greater. I don't have time for quality control' (George 2004).
23 Logistical difficulties include the time and energy required for simultaneous translation and dialogue. Timelines for consultation often only allot one community meeting within a three-day visit, and often public meetings go on past five hours in duration. The patterns associated with the political consultation process is based on my own personal interaction with consultation growing up in Nunavut, as a former employee of the GN, and in my discussions with Nunavummiut on their own experiences with consultation.
24 While all communities have the potential to support their own consultation, some communities may not be prepared to take on this responsibility immediately. I strongly believe that experiences from communities will inspire communities Nunavut wide. To support this mobilization, lessons learnt from the Iglulik Oral Histories Project, as well as the methodology used in the 'Interviewing Inuit Elders' book series, will be utilized. Prior to the initiation of a consultation process, it would be recommended that potential consultation organizations first establish themselves in the community by interviewing

community members' kinship and local hunting routes/place names. This is the necessary first step in supporting a community understanding of its own history and family. Once this is started, the organization will utilize this information to empower the community. The information collected on family and kinship could be used by the community organization to support 'membership' dialogues occurring within the community, a role currently initiated and resolved by land claim organizations. Also, the discussion on hunting areas would revitalize the knowledge on traditional place names, strengthening community ties to the land, and could be used in campaigns such as changing community names or changes to electoral districts.

References

Alfred, T. (1999) *Peace, Power, Righteousness: An Indigenous Manifesto*, Don Mills, ON: Oxford University Press.

Arreak, J. (1997) 'Letter to the editor', *Nunatsiaq News*, 4 April.

Auditor General of Canada (2003) *Report of the Auditor-General of Canada*, Chapter 8: Indian and Northern Affairs Canada – Transferring Federal Responsibilities to the North, Supply and Services Canada, Ottawa.

Bennett, J. and Rowley, S. (2004) *Uqalurait: An Oral History of Nunavut*, Montreal: McGill-Queen's University Press.

Bern, J. and Dodds, S. (2000) 'On the plurality of interests: Aboriginal self-government and land claims', in D. Ivison, P. Patton and S. Sanders (eds) *Political Theory and the Rights of Indigenous Peoples*, Cambridge: Cambridge University Press.

Bodenhorn, B. (1990) '"I'm not the great hunter, my wife is": Iñupiat and anthropological models of gender', *Etudes/Inuit/Studies*, 14(1–2):55–74.

Bourgeois, A. (1997a) 'Yes tour gets tepid support in communities', *Nunatsiaq News*, 16 May.

—— (1997b) 'Small turnout for yes tour's rankin inlet stop', *Nunatsiaq News*, 16 May.

Cassidy, F. (1993) 'Troubled hearts: Indigenous people and the crown in Canada', in T. Brantenberg, J. Hansen and H. Minde (eds) *Becoming Visible: Indigenous Politics and Self-Government*, Tromsø: University of Tromsø, Centre for Sami Studies.

Cruikshank, J. (1998) *The Social Life of Stories*, Lincoln, NE: University of Nebraska Press.

Dahl, J. (1997) 'Gender parity in Nunavut?', *Indigenous Affairs*, 3(4):42–7.

Damas, D. (2002) *Arctic Migrants/Arctic Villagers: The Transformation of Inuit Settlement in the Central Arctic*, Montreal: McGill-Queen's University Press.

Dorais, L.-J. (1988) 'Inuit Identity in Canada', *Folk*, 30:23–33.

Duffy, R. (1988) *The Road to Nunavut: The Progress of the Eastern Arctic Inuit Since the Second World War*, Kingston: McGill-Queen's University Press.

Dybbroe, S. (1988) 'Participation and control: Issues in the debate on women and development, a Greenlandic example', *Folk*, 30:111–32.

Flaherty, M. (1994) Pauktuutit Press Release, 'Statement by Pauktuutit President Martha Flaherty on the release of the NIC Discussion Paper on Gender Equality in the Nunavut Legislative Assembly' (6 December 1994), in J. Merritt (1996) *Memo to John and Simon; Date: November 18, 1996; Re: Legislative Assembly Report* (first cut draft). Finding Information: Government of Nunavut Archives, Iqaluit, Nunavut, volume 11, folder 1.

George, J. (2004) 'Clyde River Wellness Centre hits cash crunch', *Nunatsiaq News*, 5 March.

Geuss, R. (2001) *History and Illusion in Politics*, Cambridge: Cambridge University Press.

Gombay, N. (2000) 'The politics of culture: Gender parity in the legislative assembly', *Etudes/Inuit/Studies*, 24(1):125–48.

Government of Canada (2001) *Gwich'in and Inuvialuit Self-Government Agreement-in-Principle for the Beaufort-Delta Region*, Ottawa: Department of Public Works Canada. Online. Available at: http://www.aadnc-aandc.gc.ca/eng/1100100032131/1100100032133 (accessed 13 June 2013).

Guemple, L. (1986) 'Men and women, husbands and wives: The role of gender in traditional inuit society', *Etudes/Inuit/Studies*, 10(1–2):9–24.

Hicks, J. and White, G. (2000) 'Nunavut: Inuit self-determination through a land-claim and a public government?', in J. Dahl, J. Hicks and P. Jull (eds) *Nunavut: Inuit Regain Control of Their Lands and Their Lives*, Copenhagen: IWGIA.

Honigmann, J. and Honigmann, I. (1965) *Eskimo Townsmen*, Ottawa: Canadian Research Centre for Anthropology.

Inungni Sapujjijiit Task Force on Suicide Prevention and Community Healing (2003) *Our Words Must Come Back to Us: Report of the Inungni Sapujjijiit Task Force on Suicide Prevention and Community Healing*, Iqaluit, Nunavut: Department of Health and Social Services.

Irlbacher-Fox, S. (2002) 'Participation of indigenous women in self government negotiations in the Northwest Territories, Canada', in *Taking Wing: Conference on Women and the Arctic: Proceedings*, Reykjavik: Ministry of Social Services and Health, Government of Iceland.

Ivison, D., Patton, P. and Sanders, W. (2000) 'Introduction', in D. Ivison, P. Patton and S. Sanders (eds) *Political Theory and the Rights of Indigenous Peoples*, Cambridge: Cambridge University Press.

Kafarowski, J. (2002) 'Women and natural resources in the Circumpolar North: Striving for sustainable development through leadership', in *Taking Wing Conference Report: Conference on Gender Equality and Women in the Arctic*, Helsinki: Ministry of Social Services and Health.

Kymlicka, W. (1998) *Finding Our Way: Rethinking Ethnocultural Relations in Canada*, Don Mills: Oxford University Press.

McIvor, S. (1999) 'Self-government and aboriginal women', in E. Dua and A. Robertson (eds) *Scratching the Surface: Canadian Anti-Racist Feminist Thought*, Toronto: Women's Press.

Morrow, P. and Hensel, C. (1992) 'Hidden dissension: Minority–majority relationships and the use of contested terminology', *Arctic Anthropology*, 29(1):38–52.

Nadasdy, P. (1999) 'The politics of TEK: Power and the "integration of knowledge"', *Arctic Anthropology*, 36(1–2):1–18.

—— (2004) *Hunters and Bureaucrats*, Vancouver: UBC Press.

Nunatsiaq News (2001) 'A seat of one's own', *Nunatsiaq News*, 9 November.

Nunavut Implementation Commission (1995) *Footprints in New Snow*, Ottawa: Nunavut Constitutional Forum.

Okin, S. (1998) 'Gender, the public and the private', in A. Phillips (ed.) *Feminism and Politics*, Oxford: Oxford University Press.

Oosten, J., Laugrand, F. and Rasing, W. (eds) (1999) *Interviewing Inuit Elders: Perspectives on Traditional Law*, Language and Cultural Program of Nunavut, Iqaluit: Arctic College.

Pateman, C. (1989) *The Disorder of Women: Democracy, Feminism and Political Theory*, Cambridge: Polity.

Phillips, A. (1998) 'Introduction', in A. Phillips (ed.) *Feminism and Politics*, Oxford: Oxford University Press.

Piugaattuk, N. and Angiliq, J. (1986) *Iglulik Oral Histories Project* [interview], Iglulik Research Institute.

Quassa, P. (1997) 'Letter to the editor: Former NTI President opposes gender parity', *Nunatsiaq News*, 28 March.

Reimer, G. (1996) 'Female consciousness: An interpretation of interviews with Inuit women', *Etudes/Inuit/Studies*, 20(2):77–100.

Royal Commission on Aboriginal Peoples (RCAP) (1996) *Report of the Royal Commission on Aboriginal Peoples. Volume 4: Perspectives and Realities*, Ottawa: Ottawa Communication Group.

Sunseri, L. (2000) 'Moving beyond the feminism versus nationalism dichotomy: An anti-colonial feminist perspective on aboriginal liberation struggles', *Canadian Women's Studies*, 20(2):143–8.

Therrien, M. and Laugrand, F. (eds) (1999) *Interviewing Inuit Elders: Perspectives on Traditional Health*, Language and Cultural Program of Nunavut, Iqaluit: Arctic College.

Thomsen, M. (1988) 'The role of inuit women with regard to the development of self-government in the Canadian Arctic', *Folk*, 30:85–110.

Tully, J. (2000) 'A just relationship between aboriginal and non-aboriginal peoples of Canada', in C. Cook and J. Lindau (eds) *Aboriginal Rights and Self Government*, Montreal: McGill-Queen's.

Turpel-Lafond, M. E. (1997) 'Patriarchy and paternalism: The legacy of the Canadian state for First Nations women', in C. Andrew and S. Rogers (eds) *Women and the Canadian State*, Kingston: Queen's University Press.

Vallee, F. (1967) *Kabloona and Eskimo in the Central Keewatin*, Ottawa: Saint Paul University.

Wilson, E. (2005) 'Gender and nationalism in Nunavut's territorial 'House': A case study of the gender parity debate.' *Arctic Anthropology*, 42(2): 82–94.

Young, I. (1998) 'Polity and group difference: A critique of the ideal of universal citizenship', in A. Phillips (ed.) *Feminisms and Politics*, Oxford: Oxford University Press.

Chapter 13

Human security and women's security reality in Northwest Russia

Kirsti Stuvøy

Human security encapsulates a wide variety of security realities and experiences. In the 1994 report of the United Nations Development Programme, 'Human Security Now' (UNDP 1994), the term was introduced to draw attention to people's experiences of security, which were often ignored due to the emphasis on state security. In the post-cold war context of the early 1990s, therefore, the traditional focus was supplemented with a concern for humans and their experiences of insecurities. State-defined security, revolving around political and military sectors, was conceived as too narrow and insufficient (Hoogensen 2004:8). Despite some current analyses that suggest human security is a tool used to support practices for increased state security, such as (Western) military interventions (in states of the global South, e.g. Afghanistan) (Duffield and Waddell 2006; Chandler 2007; Duffield 2007), human security entails a normative concern with people and security. It is noted however that the people to whom human security matters are mostly passive subjects in the policy discourse (Krause 2005:6). At the same time, it is claimed that directing the focus of research towards people's experiences can contribute to altering the understandings of security (Darby 2006:467). To achieve that, a bottom-up and people-centred perspective of human security research is called for (Krause and Jütersonke 2005; Ewan 2007:187).

Such a people-centred understanding of security has inspired my interest in the work of non-state crisis centres in Northwest Russia. As a region free of large-scale conflict, Northwest Russia has not attracted much attention from human security researchers. In addition, women's groups that work on violence against women in Northwest Russia represent an agent of security that is rarely paid attention to in international security studies. Within a narrow, state-centred conceptualization of security, violence against women is not even a security issue. Violence against women affects women around the world, however, and this has spurred initiatives by the international community to address this issue, visible in United Nations (UN) resolutions and documents. For example, in 1993, the UN General Assembly adopted resolution 48/104, 'Declaration on the Elimination of Violence Against Women', and, in 2000, UN Security Council resolution 1325 on women, peace and security also identified women's security concerns. These initiatives indicate that security today is conceptualized broadly, and that the struggle for acknowledgement of the issue of violence against women has led to results. Nonetheless, as one Russian NGO activist on violence against women explained, other security issues mostly gain more attention:

> Today here [in Russia], [and] equally in the whole world, questions of terrorism etc. are bigger. That's probably the case in many societies. The thing is that preventive measures are more difficult to [present] than reactive. That is, reaction is easier. Or at least, the problem becomes more visible when you react. And preventive [action] is invisible.
>
> (Informant 10/2006)

In popular discourse, security continues to be a special domain associated with the state. The choice of addressing security on the basis of non-state women's groups and their work to remedy violence against women reveals a concern with analysing security based on accounts of other actors than state representatives. Such an actor perspective has been advocated and applied by feminist international relations scholars since the 1990s (see Enloe 1996; Sylvester 2004; Ackerly and True 2006), which have strongly inspired my work. The concern with people's own capacities or empowerment to enhance human security has been a further inspiration, and my research interest has been to examine how security is created by non-state crisis centres in Northwest Russia. In Russia, non-state crisis centres were established in the 1990s on the initiative of local women's groups and funded and encouraged by foreigner donors. Women's groups represent an active civil society actor that is acknowledged as strong and stable (Sperling 1999; Richter 2002; Sundstrom 2002; Johnson 2006). I illustrate the inventive, networked and *ad hoc* character of their work, which explains their successes in security production as well as its fragility (Stuvøy 2010a). This contributes to understanding the role of local crisis centres in Northwest Russia and also intervenes in debates on human security because the analysis exposes the struggle over the understanding and conceptual clarification of security.

To begin with, I address a core question of the human security debate: 'What is security?' Two other questions, security by whom and for whom, are addressed by analysing crisis centres as a local security actor and their security practices.

What is security?

Over the course of my work, I have experienced that a security analysis of violence against women is often met with bewilderment. My informants were puzzled by my interest in the concept of human security. They did, however, also talk about violence against women as security: 'Security in the family. Time to act' was for example the title of a recommendation prepared by the Russian Association of Crisis Centres (RACC) in 2003 that aimed to inform legislative and executive powers on how to improve the efforts to prevent violence in the family in Russia. The concept of human security is not theorized or discussed among these women activists, however, which suggests thereby that the realization of human security can be localized but its theorization less so. In international politics, among core stakeholders and elites, the human security concept is a means to legitimating international responses to complex emergencies and conflicts, but the local level, such as women's security from violence in Northwest Russia, is less relevant to the thinking of human security in terms of policy. Thus a question is raised, a hesitancy expressed, regarding my project and how the case of violence against women fits in with the concept of human security. In contrast, Caroline Bunch, for example, emphasizes that violence against women is the epitome of a human security issue: 'Yet, if you look at definitions of human security, there is no better paradigm for human insecurity than violence against women, which directly and indirectly affects a vast number of people' (Bunch 2004:32).

She suggests that the human security concept is a means of examining fundamental threats to security inherent in the everyday violence of domestic abuse. Moving away from the tendency to ignore women's lives in security analyses will, it is argued, contribute to an improved understanding and consideration of a comprehensive human security reality (Bunch 2004).

In conceptual debates, the scholarly interest in learning about the complex reality of human security competes against demands for a delimitation of the concept, in terms of issues, causal inferences and values (Buzan 2004; Krause 2004; Mack 2004; Paris 2004). The empirical descriptiveness of the concept, the number of issues that it can potentially describe and incorporate into security thinking, is identified as a problem.[1] Specifically concerned with interconnectedness of threats in the human security concept, Andrew Mack (2004:367) notes that this is an 'unhelpful' jargon because '[a]ny definition that conflates dependent and independent variables renders causal analysis virtually impossible'. Conceptual clarification is a focal point of critique against the concept of human security. Sabina Alkire (2004) remarks that there are thirty or more definitions of human security, and that the key conceptual challenge is to name priority issues that represent the vital core of security. This character of the concept is the basis for a critical view of human security as a framework of analysis because the concept is considered too vague to generate specific research questions (Paris 2001). Kyle Grayson importantly notes however that the politics involved in delimiting and defining human security need to be openly and critically assessed:

> [I]t is imperative that 'the aspiration to power' that is inherent in any definitional claim be exposed and debated in terms of both what is being

> positively affirmed as comprising human security and what is concurrently disqualified; there must be analytical sensitivity given to the people, places and things that are marginalized when an 'expert' claims to be providing a precise/scientific/workable definition of human security that is of practical use.
>
> (Grayson 2004:357)

According to this view, any delimitation of human security aimed at making the concept analytically useful therefore entails a political positioning.

One way in which many have dealt with the problem of conceptual clarification is by defining human security in terms of a limited set of issues. For example, in the Human Security Report 2005, the first annual report of the Human Security Centre at the University of British Columbia in Canada, human security is defined in terms of political violence. Their aim is to map 'the incidence, severity, and consequences of political violence around the world' and thereby to provide 'data and analysis that is essential to evidence-based security policy'.[2] The measuring of political violence is linked to a threshold of twenty-five deaths (Human Security Centre 2005:67).[3] In order for human security to be quantifiable, it must be defined and operationalized, and, thus, a narrow conceptualization is needed in order to employ quantitative measurements. Furthermore, the focus of the Human Security Report has rightly been recognized as being 'broadly within the realist rubric' (Roberts 2006:257) with a focus on 'hard security' in terms of political violence. Alternatively, Booysen (2002) defines human security by including all the seven components listed by UNDP (economic, food, health, environment, political, community, personal) and calculates an Inefficiency Ratio. This expresses 'the extent to which efforts at human security are translated into actual achievement' (Booysen 2002:274).[4] Other attempts at composite indexing, e.g. Steve Lonergan *et al.* (2000) and Gary King and Christopher J. L. Murray (2001/2002), also display the difficulties of choosing dimensions to include and, importantly, problems in regard to the availability of reliable data (Tadjbakhsh and Chenoy 2007:241).[5] The indexing is a task of the researcher(s), as seen in King and Murray's (2001/2002:592) technical definition of human security: 'We define an individual's human security as his or her expectation of years of life without experiencing the state of generalized poverty. Population human security is then an aggregation of individuals' human security.' Further suggestions for mapping human security link the specification of human security to a certain threshold (Owen 2004a, b) or to a scale of severity (Roberts 2006). The rationale behind the threshold-based definition put forward by Taylor Owen is that openness to all threats that people possibly experience is crucial. Instead of listing specific threats to human security and thereby narrowing down the possible threats to humans, an open approach 'allows all possible harms to be considered' (Owen 2004b:381) where the scale of the threat is what matters. Consequently, all threats should be considered potential human security issues, and it is their indexing in terms of whether they cumulate to a certain threshold that decides whether they are acknowledged as human security issues. '[O]nly the most serious [threats], those that take or seriously threatens lives, are included' (Owen 2004b:383). For the purpose of establishing the truth about human security, rules such as a specific threshold to be observed empirically are suggested and employed. These delimitations of the human security concept are analytically motivated and explained. Threshold-based definitions thus allow twenty-five deaths to be relevant for consideration as a human security issue, but as noted by Grayson (2004), these suggestions of a 'precise/scientific/workable' definition of human security exclude twenty-four civilian deaths. The exclusion of these deaths determines the understanding and knowledge production of the comprehensive human security reality when such criteria defined from above and operationalized in indexes dominate empirical human security research: a policy maker, or an academic, decides what is worthy of attention for human security. Security is then examined from the top down. Such a focus on indexing as exemplified with the Human Security Report and the work of Booysen and Owen, among others, reveals an objectivist bias that is not uncommon in security research (Pouliot 2007:377).

Annick Wibben's (2008) call for an 'opening' of the security debate addresses an analytical sensitivity to people, places and issues in regard to definitions of security. Understood as an alternative epistemological outlook (Hoogensen and Stuvøy 2006; Burgess *et al.* 2007:12), human security, through its focus on non-state actors like individuals, by its nature implies openness to meanings of security. Phillip Darby (2006:467) argued that 'groundedness in lived experience is a vital corrective to the faceless, placeless narratives so characteristic of security texts'. This requires reduced control over the interpretation of security but also acts as a basic positioning for beginning to examine the comprehensive human security reality. The aim of examining local practices of security, and thereby contributing to the comprehension of people's human security reality and a bottom-up perspective, inspired my work on women's security in regard to violence against women in Northwest Russia. Empirical groundedness and a focus on local viewpoints do, as noted above, imply openness to the interpretation and understanding of what security is and, in the following, I will explain the fragile security production characteristic of the context of Northwest Russia and women's security.

Non-state crisis centres as a local security actor

Although violence against women is recognized to be a global problem, media and political focus on specific countries and regions, like Afghanistan, ensures there are certain parts of the world that get less attention, but which can possibly provide new insights into the issue. As the chapters in Part I of this book illustrate, one can probably safely argue that the combination of the terms 'Russia' and 'security' seldom generates thoughts about violence against women. Instead, Russian territorial security (traditional), and ethno-national disputes in Chechnya and elsewhere in the Russian Caucasus, are dominant Russian security images, closely followed by energy security and protection of energy resources. In Russia, Moscow defines the agenda, and less is known about both the regional and local experiences within the diverse peripheries of the world's largest country – or this is not considered relevant to the dominant image or narrative of Russian security internationally. To push the argument – this focus on top-down, central perspectives (Moscow) – means that definitions of security are derived from incomplete information and experience.

The work of women's groups in Northwest Russia was initiated during the 1990s, and their experience is about a struggle for the institutionalization and recognition of their work as contributing to the security of women and stability of the overall society. In my work I have focused on three regions of the Russian Northwest: Murmansk oblast, Archangelsk oblast, and the Republic of Karelia (see Stuvøy 2009, 2010a, b). The first non-state crisis centre in the region was established in Murmansk in 1997, and institutional responses to the issue of violence against women have therefore been established only during the last 15 years in Northwest Russia. In an interview with a representative of one of the most recently established crisis centres in the region, the 'Severodvinsk city societal organization for women – Cabinet of psychological help for women in crisis situations' (Cabinet), the interviewee spoke about this recent recognition of the issue of violence against women:

> When we [in 2001] began to work on this programme [on the improvement of women's situation], we were the first in the city to address the problem of violence against women. Because, in principle, earlier no one ever talked about it.
>
> (Informant 7/2006)

Although the first crisis centre in the Northwest region was established in 1997, the Cabinet in Severodvinsk exemplifies that the work on violence against women in the region was established even more recently. This does not mean that the problem is unacknowledged. Citing a Russian study of violence in the family and society, Janet E. Johnson (2007:40) noted that, according to a 2001–2002 survey, 87 per cent of men and 93 per cent of women agreed that violence against women in the family is a problem.[6] Recognition of the issue is, however, not synonymous with institutionalization of an assistance structure for women. The establishment of non-state crisis centres in Russia in the 1990s changed the field of security for women because they gave women a place to seek help and support (beyond family and friends). They therefore represent a new agent within the field of security for women in Russia, an agent of change.

Since non-state crisis centres began to emerge in Russia in the early 1990s, they have achieved a great deal of international attention; and became recognized as a vital agent of civil society development in Russia (Sperling 1999; Richter 2002; Sundstrom 2002; Johnson 2006). Amid concerns of weak civil society development in Russia, this agreement of the solid standing of women's groups and the level of institutionalization and organization of the Russian women's movement in the mid-2000s made them particularly useful for a study of a civil voice of security as they are rather well established. This rather solid standing of the organization allowed a focus on the research method to understand human security from the bottom up (Stuvøy 2010b). The development of non-state crisis centres in Russia was initially furthered by money from foreign sources (Richter 2002; Sundstrom 2006). The international gender cooperation that emerged with the end of the Communist era and as part of Western support of civil society in the 'third wave of democracy' (Ottaway and Carothers 2000) strongly affected the kind of groups that developed in Russia. International benefactors had specific motives with their funding, pertaining to political and social agendas reflected in their programmes that aimed at promoting democracy, civil society and a free market (Kay 2004:250). Further, many funding bodies assumed the flow of knowledge to be unidirectional from the West to the East (Kay 2004). Thus the demands of the foreign funders regarding the development of organizational structure and requirements of grant applications significantly impacted the development of Russian women's groups. The previous informal entities women engaged in and with during perestroika (*neformal'nye*) were suddenly required to formalize, register as NGOs and professionalize (Hemment 2004b:316). Julie Hemment identifies frustration among women activists in regard to the agendas of international collaborators. For example, the foreigners emphasized the importance of violence against women, but local women activists were sceptical to crisis centres because 'they did not think gendered violence was the most pressing issue facing Russian women and expressed concern that so many resources were put into it' (Hemment 2004a:823). For example, crisis centres are premised upon a Western assumption about women's economic dependence on men and that women are stuck in the home. Hemment emphasizes that '[t]his was not true for Soviet women, who were brought into the workforce and guaranteed formal equality by the socialist "paternalist" or "parent" state' (Hemment 2004a:823). Also, the notion of the nuclear family was hardly realized in Soviet times, not least due to housing shortages, and the extended family was common. 'For all these reasons,' explains Hemment, 'domestic conflict most commonly expressed itself in the form of tension over rights to living space, interpersonal strife, or alcoholism' (Hemment 2004a:823). For Russian women, therefore, it was a complex and diverse set of issues that comprised the 'crisis' they found themselves in. When crisis centres began to emerge in Russia, Russian women approached the centres with all these crisis situations, not only domestic violence.

Thus, although the crisis centres were set up to address violence against women, a wide array of issues were raised on their phone hotline service. Violence against women was clearly a local, human security issue, but could not be divorced from, or reject, other issues of insecurity experienced by women. Local activists became critical of the effects of the international intervention in their work. Their practical experience with immediate service provision created a space for critique and invention. As practical experience broadened, also regionally, the impact of the international agendas and restrictions was challenged. James Richter noted that the expansion of donor funding beyond Moscow and St Petersburg facilitated the emergence of local crisis centres that developed forms of assistance and activism beyond donor restrictions (Richter 2002:39). Although non-state crisis centres in the regions were not divorced from international trends of women's activism and agendas, their immediate and concrete action was oriented towards practical challenges and grassroots needs. Their concerns were mainly local challenges. While keeping international donors pleased, local crisis centres were committed to objectives of altering or improving the immediate life-situation of vulnerable victims, as defined by the victims themselves. The tensions between the objectives of the Western NGO bureaucracy and the local livelihoods inspired critical reflection by crisis centre representatives on their own work. Drawing upon extensive research in post-Soviet Russia, Julie Hemment acknowledged the influence of the international agendas on local concerns but advocated a local lens:

> [I]f we pay close attention to what is happening on the ground, we will see that this [the dominance of the international agendas] is not the whole story. First, NGOs are diverse in status and form and their members are not solely elite. Second, although international aid certainly exerts a powerful influence on group activities, it does not determine them. In the nongovernmental sphere, Russian women draw on international aid to strategize and make sense of the realignments in their lives and to articulate a way forward. What comes into view if we shift the lens?
>
> (Hemment 2004b:316)

Hemment shifts the lens in order to examine conceptions of democracy development on the local level and violence against women. For my focus on security, I thus learned from Hemment's work, but also from Richter's (2002), that this lens can be particularly fruitful because local activists develop their own strategies independently of international demands and donor requirements.

As a basis for exploring local security production, I therefore examined how non-state Russian crisis centres strategized in the local context.[7] The women's groups formulated the objective of addressing violence against women and established and provided assistance to victims with the aim of institutionalizing a dependable system of response. To this end they interacted with victims and local stakeholders, and in this work they were inventive in adjusting to local challenges while facing economic constraints and not least continuing difficulties of gaining support and legitimacy of their work.

Local security practices on violence against women

The emergence of non-state crisis centres in Russia in the 1990s represented a challenge to traditional practices of silencing the issue of domestic violence. The state, with its power to define legitimate social concerns and problems, dominates decisions relevant to addressing women's security situations, and this position was challenged by the emergence of non-state crisis centres. Women's groups have played an important role as key narrators of the problem of violence against women in Russia, as they publically described and mapped knowledge of anonymous victims' experiences, and also registered how public servants responded to the victims. Notwithstanding these important tasks, non-state crisis centres in Northwest Russia were preoccupied with their meagre financial situation. To begin their work and set up the required infrastructure, the funding that crisis centres received in the initial phase from international donors but also infrastructural support from local authorities was essential to their establishment. In contrast to the more generous financial situation in the late 1990s and early 2000s, the work of these local crisis centres in Northwest Russia during the 2000s became characterized by a lack of resources that impeded the predictability in their work. Some crisis centre representatives even expressed reservation regarding resources from local authorities. They couched their reservation in an attitude of social responsibility towards the budget limits of the municipality while at the same time claiming how pressing the needs of the crisis centre were. Despite the difficult economic situation, where the women's groups continued their work, their position, which rested on prestige, was a possibility for affecting views on the issue of violence against women – to establish recognition of their work and the suffering of the victims. The logic of their work – their security practice – was embedded in a struggle for the acceptance of their claim: that violence against women is an issue of women's personal security.

There is nevertheless a lack of crisis centres and shelters. In Northwest Russia only two crisis centres offer a shelter, a protected place for their clients to stay in emergency situations. Such shelters represented a primary structure of crisis centres in Western Europe and the USA when they were first established four decades ago. In Russia, the lack of shelters has led to the characterization of crisis centres as a 'shelter without a shelter':[8] their knowledge, experience and information contributes to a discursive space in which victims can seek advice and make choices pertaining to their insecurity. Crisis centres have little help to provide, however, in terms of physical and material support, and also few alternatives to suggest, which limits their contribution to women's security. A vulnerable economic situation and housing difficulties are the most important issues for many women. A main security strategy that the crisis centres offered was an invitation to discuss alternative ways of living. This comprised a set of choices that the victims can make. In regard to the economic challenges that must be surmounted in such a process, the crisis centres thus offer little tangible help. Practising security in this field basically involves a re-description of self and, contributing to this process, the crisis centres offer their resources. This also involves advice and suggestions on how to solve legal and economic problems. These challenges also demand specific action by the victims as they are the ones who need to take action to change the situation. The crisis centres described how they point out to the victims the various options they have available, but choices and actions need to be taken by the victim, who in this process produces the new self.

The process of understanding and re-describing the self is one in which the subject is confronted with a range of immaterial constraints. Empirically these can be seen in the form of, for example, the social pressure that intimidates the victims. They fear consequences such as social judgement and loss of status. The actions of victims are embedded within processes that give symbolic meaning to people's practices. Interviews with victims of violence demonstrate this, as they face already a challenge in the naming of their situation, i.e. the explicit recognition that they live in a situation of violence. Local crisis centres experience that the naming of violence against women is contested, which can be linked to a continuing tendency to treat men's violence carefully, with understanding, almost with justification. This raises the question as to what extent this careful treatment of men's violence affects the individual victims and their way of understanding their own situation and the context in which choices are made. Such understanding is supported by traditional views expressed by women, emphasizing a woman's need to be patient. The analysis has showed how such gender roles are also predominant in public discourse (Stuvøy 2010a). Thus, their impact on the individual level is highly probable, though difficult to predict in regard to how the impacts will manifest themselves in visions of the self.

Action taken by a woman to actively change her life-situation to one where violence and insecurity does not define her life-world any more is, drawing upon the above experiences, not always associated with symbolic meanings such as prestige and social recognition. The evidence demonstrated that due to subjective constraints of these meanings, women's individual actions are not necessarily interpreted as reconstructive or positive. When crisis centre representatives encouraged women to act, to consider the choices outlined before them, they confronted a difficult situation. In interviews, crisis centre representatives were keen to emphasize the importance of individuals making the decisions themselves, highlighting the 'freedom of choice' for the victims. The material and immaterial constraints on action are forceful, however, and constrains women's freedom to choose. The victims do not necessarily possess the symbolic capital that ensures a socially positive recognition and acceptance of their action. In this the women who have been victims of violence differ from the crisis centres and the latter's possession of symbolic capital. Thus the security practices of crisis centres are enabled by a social recognition of their practices, whereas the individual victims are exposed to traditional norms and habits that do not support or applaud the recommended actions provided by the crisis centres. Thus, both crisis centres as well as their clients are challenged by significant material and social constraints.

The work of non-state crisis centres nevertheless functioned despite weak material capital, because of the various ways and means of interacting with local stakeholders. In addition to the interaction with victims, interactions with local authorities, law enforcement, other activists and international collaborators contributed to the security production of non-state crisis centres. These activities were conducted on the basis mainly of their specific social capital, through which the continuation of such interactions is continuously reproduced, which establishes trust, an important key to the further establishment of security (Hoogensen Gjørv 2012). Crisis centre representatives have had different perceptions, assessments and experiences with local law enforcement, for example, but trust has been acknowledged as a key to collaboration, demonstrating empirically that social capital defines the work of non-state crisis centres in Northwest Russia. This is, as we will see below, not without difficulties.

Crisis centre representatives are keen to emphasize that there has been a change in the recognition that they receive from local authorities. This process has developed from a situation in which crisis centres were neglected, to a situation in which they are both listened and talked to. Crisis centre representatives have noted, however, that there are a number of differences in how the issue of domestic violence is approached in their interactions with local authorities. From the perspective of crisis centre representatives, local stakeholders understand the issue of violence against women more in terms of numbers and administrative instructions, evoking a threshold-based understanding of security whereby acknowledgement of a violent incident is defined on the basis of specific parameters or criteria. In other words, not all incidents will qualify. This contrasts with the approach of NGO crisis centres, who are keen to stress that they provide assistance to those who contact them and, in their own words, are motivated by open-mindedness and empathy – an inclusive rather than exclusionary approach to security.

Crisis centre representatives claim that they are focused on informing the public about their work and that they endeavour to establish broad networks. For a brief comparison, one can look at the assessments of Norwegian collaborators in the Network of Crisis Centres in the Barents Region (NCRB), as they perceive this work differently, and they compare their own (Norwegian) approaches to that of the Russian crisis centres, emphasizing a lack of rules in the Russian system. The local crisis centres in Northwest Russia operated differently from what the Norwegian counterparts expected, as personal networks, contacts in police and health departments, etc., were key nodes in their work. Such relationships are used to find solutions on an *ad hoc* basis for the victims and, in this work, trust is the key to the functioning of the assistance system, more than rules. Additional differences are relevant if one compares the Norwegian and Russian experiences. According to the Norwegian crisis centre representatives, their engagement with victims is with a long-term perspective. They engage victims' experiences and assist in immediate recovery as well as in long-term aftercare. In contrast, the emphasis in the Russian context is recognized to be at the acute phase. The aim is to calm the situation, and a short-term perspective defines their approach. Russian actors have far less opportunity to engage the social context in which victims live in Russia (in comparison to Norwegian crisis centre actors), and are unable to address the issues of housing and finances, as these issues are less accessible for Russian crisis centres because of the lack of transparent rules. This is why Russian crisis centre representatives in their accounts emphasized their contact with the victims and how they are empathetic and open-minded in these interactions – those are the resources they have in abundance.

An idea of being a mediator between people and power was also reflected among the non-state crisis centres in Northwest Russia: 'We are kind of the information provider or guide between the people and the power' (Informant 8/2006).

This description as a guide also defines the crisis centres as local security producers. As interlocutors between people and power, crisis centre representatives draw upon their understanding of the victims and their situations, as well as their experience with local stakeholders. In their role as a guide, therefore, crisis centre representatives use their social capital. The knowledge they have established on the basis of their experiences both with victims and with stakeholders is the key resource in their security production. Material capital constrains their activities, and also their interaction with their networks. Since social capital is their main resource for assisting the victims in finding a new way of organizing their lives, non-state crisis centres are vulnerable as security producers because they work on a case-to-case basis. They individually adjust their security practices to the needs of the victims but also need to strategize in regard to how their personal connections can be used to acquire help from stakeholders in individual cases. But their work might also remain concentrated on the victims and personal therapy. Although this implies instability within their practices, over time the crisis centres gain routine. In regard to victims, crisis centres are a cornerstone. Whether short term or long term, the crisis centres are there for therapeutic work. The crisis centres gathered knowledge on the basis of the stories they heard from victims and became experienced in working with victims. This experience is a key resource for crisis centres. They are aware of the suffering and the struggles of individual women, and convey their understanding of these difficulties to a broader public. This is the essence of their involvement in the symbolic struggle for power to produce a vision of the world of violence against women. By using their experiences to promote advocacy of measures against violence towards women, they also use this knowledge and experience to the benefit of other women that may experience violence.

Describing how security practices are conducted does not necessarily explain what kind of security is produced. A few quotes from the interviews suggest what sort of key issues are relevant to the kind of security established by crisis centres. For example, it is said that 'We talk about abuse, make words, phrases – what it is that we are doing, that is, building' (Informant 16/2006). This implies that a key activity is to make sense of the kind of insecurity experienced by people. The importance of such meaning-making is emphasized through the experience of Russian crisis centres regarding women who do not acknowledge that they live in a situation of violence. Making words to describe women's insecurities in situations that were previously naturalized was therefore a key aspect in the work of non-state crisis centres in Russia. Closely connected with this is access to information. In one interview it was stated that: 'If you are informed, it means you're protected' (Informant 9/2006).

This includes the information needed to acknowledge that you are a victim of violence. Beyond that, however, information is security because it implies that a victim knows where help can be found and whom to contact and through which methods. To act, however, requires a belief that the system functions. In one interview it was commented that: 'They [Norwegian women] are more confident [than Russian women] that they will receive help' (Informant 16/2006). The kind of security produced by non-state crisis centres in Northwest Russia was tied to processes of meaning-making and information provision that required confidence in those processes.

Human security and women's (in)security in the North

Bringing human security studies to a local, people-centred setting makes contingencies in security production visible. In regard to women's security in Northwest Russia, security is addressed institutionally, i.e. by crisis centres, but is still subject to personal connections and *ad hoc* organizations, which implies that security for women is an ambiguous project in this context. Therefore, the analysis of the work of non-state crisis centres in Northwest Russia provides an empirical case that can explain in some detail what local actors are doing in regard to women's security. But is this a re-conceptualization of (human) security?

Marlies Glasius (2008) has pointed out that human security involves a shift in epistemological perspective from objectivism to an engagement with subjective views on security. Being concerned with subjectivity and security evokes different opinions, interests and values. What is at stake here is a struggle between a positivist approach to human security that demands measurements and substantial delimitation of the concept, and an epistemology that questions what and who defines security in the life-world of individuals. In terms of knowledge of security, human security addressed from the bottom up therefore is the attempt to keep a technical or technocratic understanding of security at a distance (Büger 2006; Wibben 2008).

By prioritizing the role of *ad hoc* interactions based on personal network connections in security as well as the role of the victims' process of re-describing self, the complexity of local security production has been emphasized. Each individual woman may choose differently in this process. This points, however, to another important, but under-reflected, aspect in regard to security and security policies that is implicit in the human security debate: de-militarized security practices. In regard to empirical studies on human security, a particular concern is the identification of a security issue. Who identifies what human security is in a specific empirical setting? The deliberative definition of an issue as a matter of security is a practice considered particularly problematic by a group of human security critics termed post-positivist by Ewan (2007). The argument is that processes of securitization elevate issues on to an emergency agenda, which implies exceptional politics, in particular by military means (Wæver 1995). Human security issues are best kept off this agenda, argues Buzan (2004), because they are more effectively addressed by normal politics. Such a perspective acknowledges military security policies and neglects the existence of alternative security strategies. In regard to women's security in Northwest Russia, such a perspective disregards the security practices of not only non-state crisis centres, but a plethora of actors that engage in security practices pertaining to individual and community concerns. However, there are also differing views on the issue, and we see this in the specific context of Northwest Russia:

> The understanding of social problems, and this also characterizes the problem of women, is completely different among state and societal [non-governmental] organizations. The state understanding [of the problem] is [based on] strict instructions. For example, a woman receives help from the state if she has a small income or if she is raising a child. That is, there are specific parameters on the basis of which they help her. Of course that is not the way it ought to be. And non-governmental organizations, they don't do that. For them it is not important how much she earns. For them it is important what kind of problem she has, that she contacted them [the non-governmental organization/crisis centre] with that problem [and that] she needs help.
>
> (Informant 9/2006)

The state has a very different understanding of security for women, it is argued, and the problem-oriented focus reflected in the work of the crisis centres allows for pragmatic recognition of various security concerns. Importantly, in this context, it opens up the thinking about what and who produces security for people locally. Ewan (2007:187) associates alternative security strategies with ethical dialogue, empathy and self-restraint, while Fiona Robinson (2008) has emphasized the importance of care in human security. These are but two attempts to identify non-military security practices.

Human security research is not substantial in its specification of what human security practices are in regard to a people-centred and not necessarily military-centred approach. Over the course of my research on women's security in Northwest Russia, the concern with defining what security 'really is' has been an issue that I have continuously been confronted with, during field work, at conferences and in other academic contexts of feedback and critique. In international documents, violence against women is acknowledged as a security concern but, in practice, the presentation of my research project under the heading of human security seldom if ever generated responses of recognition by the respondents. The interviewees did, both in the interviews and in written material, refer to violence against women as a security matter, but other annotations, in particular 'human rights issue', is just as important, if not more than the term human security, to their work in practice. In academic contexts, the bringing together of violence against women and human security in my project has also been questioned. One concern has been why the human is included, as security on its own would suffice, it is suggested. Other suggestions have been to exchange the term human security with other terms, such as safety for example. Throughout this project these debates have showed that delimiting security is a contingent task that interconnects theory and practice, the local and the international. The analysis of the work of non-state crisis centres in Northwest Russia exemplifies the complex constraints that define people's security reality. The interest in women's security in Northwest Russia has displayed specific and contextual constraints on (de-militarized) security production. By studying ongoing practices of security production that define people's comprehensive security reality and making this visible, pre-conceived conceptions of security and their connection to the use of military force are challenged. Regarding people and their security experiences, a militarized security concept may not be what is reflected on the ground. Particularly in the Circumpolar North, the human security concept contributes to a much needed questioning and reflection upon diverse understandings of security, from whose perspective, when and in which context.

Notes

1 In regard to the descriptiveness of the concept, Andrew Mack in fact argues that '[i]f the term "insecurity" embraces almost all forms of harm to individuals – from affronts to dignity to genocide – it loses any real descriptive power' (Mack 2004:367).
2 Quoted from the website of the Human Security Report. Online. Available at: http://www.humansecurityreport.info/content/view/25/60/ (accessed 26 September 2006).
3 For the purpose of this global mapping of political violence, a new dataset was set up in cooperation between the Human Security Centre and the Conflict Data Program of Uppsala University, Sweden. This dataset is based on the conflict data that Uppsala University has developed together with the Peace Research Institute Oslo (PRIO) that counts inter-state wars and armed conflicts in which the state takes part on at least one side. The new dataset has been broadened to include previously excluded forms of political violence, such as armed conflicts in which the government is not party to the conflict, or genocide. In the first Human Security Report in 2005, numbers from 2002 and 2003 were reported.
4 The indices measure the performance of one country relative to the other countries in the study, but this is admittedly a crude measure of human security (Booysen 2002:288). A crude overall conclusion is that 'nations that are ethnically more homogeneous are more likely to make better efforts at and to achieve higher levels of human security' (Booysen 2002:292).
5 Lonergan *et al.* explain that 'most countries did not have complete time series [1970–1995] for most indicators'. King and Murray (2001/2002:601) indicate difficulties related to data collection when noting: 'Measuring population human security requires as its starting point good information on the current and past levels of income, health, education, democracy and political freedom.'
6 Sundstrom refers to a 2003 survey of about 2,000 respondents around the country that found that 43 per cent still felt that a man beating his wife was a private matter, but also notes that, on this specific aspect, change has been observed (Sundstrom 2005:441).
7 Empirical material comprised interviews with representatives of seven crisis centres in Northwest Russia, documents and reports, as well as interviews with women's organizations in Moscow. Interviews were conducted in 2006 and 2008.
8 This characterization originates from Janet Elise Johnson's presentation 'Women's Crisis Centers in Russia: How Transnational Collaborations Can Work', Eighth Annual Aleksanteri Conference, Helsinki University, Finland, 10–12 December 2008.

References

Ackerly, B. A. and True, A. (2006) 'Studying the struggles and wishes of the age: Feminist theoretical methodology and feminist theoretical methods', in B. A. Ackerly, M. Stern and J. True (eds) *Feminist Methodologies for International Relations*, Cambridge: Cambridge University Press, pp. 241–60.

Alkire, S. (2004) 'A vital core that must be treated with the same gravitas as traditional security threats', *Security Dialogue*, 35(3):359–60.

Booysen, F. (2002) 'The extent of and explanations for international disparities in human security', *Journal of Human Development*, 3(2):273–300.

Büger, C. (2006) *Better Together! Human Security as a Boundary Object and the Fallacies of Technocratic Reasoning – Or an Ecological Defence of Human Security*. Second COST Training School, 'Constructing Insecurity and the Political'. Tampere, Finland. Online. Available at: http://www.eui.eu/Personal/Researchers/bueger/Documents/Tampere%20-%20Better%20Together.pdf (accessed 16 January 2009).

Bunch, C. (2004) 'A feminist human rights lens', *Peace Review*, 156:29–34.

Burgess, P., Amicelle, A., Bartels, E., Bellanova, R., Cerami, A., Eggum, E., Hoogensen, G., Kittelsen, S., Knibbe, K., de Koning, M., Koser, K., Krause, K. and Salemink, O. (2007) *Promoting Human Security: Ethical, Normative and Educational Frameworks in Western Europe*, Paris: UNESCO.

Buzan, B. (2004) 'A reductionist, idealistic notion that adds little analytical value', *Security Dialogue*, 35(3):369–70.

Chandler, D. (2007) 'The security development nexus and the rise of "anti-foreign policy"', *Journal of International Relations and Development*, 10:362–86.

Darby, P. (2006) 'Security, spatiality, and social suffering', *Alternatives: Global, Local, Political*, 31(4):453–73.

Duffield, M. (2007) *Development, Security and Unending Wars: Governing the World of Peoples*, Cambridge: Polity Press.

Duffield, M. and Waddell, N. (2006) 'Securing humans in a dangerous world', *International Politics*, 43:1–23.

Enloe, C. (1996) 'Margins, silences and bottom rungs: How to overcome the underestimation of power in the study of international relations', in S. Smith, K. Booth and M. Zalewski (eds) *International Theory: Positivism and Beyond*, Cambridge: Cambridge University Press, pp. 186–202.

Ewan, P. (2007) 'Deepening the human security debate: Beyond the politics of conceptual clarification', *Politics*, 27(3):182–9.

Glasius, M. (2008) 'Human security from paradigm shift to operationalization: Job description for a human security worker', *Security Dialogue*, 39(1):31–54.

Grayson, K. (2004) 'A challenge to the power over knowledge of traditional security studies', *Security Dialogue*, 35(3): 357.

Hemment, J. (2004a) 'Global civil society and the local cost of belonging: Defining violence against women in Russia', *Signs: Journal of Women in Culture and Society*, 29(3):815–40.

—— (2004b) 'Strategizing gender and development: Action research and enthnographic responsibility in the Russian Provinces', in K. Kuehnast and C. Nechemias (eds) *Post-Soviet Women Encountering Transition: Nation-Building, Economic Survival, and Civic Activism*, Baltimore and London: Johns Hopkins University Press, pp. 313–33.

Hoogensen, G. (2004) 'Human security in the North: Is it relevant?', *IHDP Update. Newsletter of International Human Dimensions Programme on Environmental Change*, 3:8–9.

Hoogensen, Gjørv, G. (2012) 'Security by any other name: Negative security, positive security, and a multi-actor security approach', *Review of International Studies*, 38(4):835–59.

Hoogensen, G. and Stuvøy, K. (2006) 'Human security, gender and resistance', *Security Dialogue*, 37(2):207–28.

Human Security Centre (2005) *The Human Security Report 2005: War and Peace in the 21st Century*, Lieu Institute of Global Issues, University of British Columbia, Canada.

Johnson, J. E. (2006) 'Public–private permutations: Domestic violence crisis centers in Barnaul', in A. B. Evans, L. A. Henry and L. M. Sundstrom (eds) *Russian Civil Society: A Critical Assessment*, New York and London: Armonk, pp. 266–83.

—— (2007) 'Contesting violence, contesting gender: Crisis centers encountering local government in Barnaul, Russia', in J. E. Johnson and J. C. Robinson (eds) *Living Gender After Communism*, Bloomington, IN: Indiana University Press, pp. 40–62.

Kay, R. (2004) 'Meeting the challenge together? Russian grassroots women's organizations and the shortcomings of Western aid', in K. Kuehnast and C. Nechemias (eds) *Post-Soviet Women Encountering Transition: Nation Building, Economic Survival and Civic Activism*, Baltimore and London: Johns Hopkins University Press, pp. 241–61.

King, G. and Murray, C. J. L. (2001/2002) 'Rethinking human security', *Political Science Quarterly*, 116(4):585–610.

Krause, K. (2004) 'The key to a powerful agenda, if properly delimited', *Security Dialogue*, 35(3):367–8.

—— (2005) 'Human security: An idea whose time has come?', *S+F, Sicherheit und Frieden/ Security and Peace*, 23(1):1–6.

Krause, K. and Jütersonke, O. (2005) 'Peace, security and development in post-conflict environments', *Security Dialogue*, 36(4):447–62.

Lonergan, S., Gustavson, K. and Carter, B. (2000) 'The index of human insecurity', GECHS Project, 6, AVISO series. Online. Available at: http://unpan1.un.org/intradoc/groups/public/documents/apcity/unpan037033.pdf (accessed 12 August 2013).

Mack, A. (2004) 'A signifier of shared values', *Security Dialogue*, 35(3):366–7.

Ottaway, M. and Carothers, T. (2000) *Funding Virtue: Civil Society Aid and Democracy Promotion*, Carnegie Endowment for International Peace, Washington, DC.

Owen, T. (2004a) 'Challenges and opportunities for defining and measuring human security', *Disarmament Forum*, 3:15–24.

—— (2004b) 'Human security – conflict, critique and consensus: Colloquium remarks and a proposal for a threshold-based definition', *Security Dialogue*, 35(3):373–87.

Paris, R. (2001) 'Human security: Paradigm shift or hot air?', *International Security*, 26(2):87–102.

—— (2004) 'Still an inscrutable concept', *Security Dialogue*, 35(3):370–1.

Pouliot, V. (2007) '"Subjectivism": Towards a constructivist methodology', *International Studies Quarterly*, 51(2):359–84.

Richter, J. (2002) 'Promoting civil society? Democracy assistance and Russian women's organizations', *Problems of Post-Communism*, 49(1):30–41.

Roberts, D. (2006) 'Human security or human insecurity? Moving the debate forward', *Security Dialogue*, 37(2):249–61.

Robinson, F. (2008) *The Importance of Care in the Theory and Practice of Human Security*, Annual Conference of the International Studies Association, San Francisco, 26–29 March.

Sperling, V. (1999) *Organizing Women in Contemporary Russia: Engendering Transition*, Cambridge: Cambridge University Press.

Stuvøy, K. (2009) *Security Under Construction: A Bourdieusian Approach to Non-State Crisis Centres in Northwest Russia*, PhD dissertation: Universitetet i Tromsø.

—— (2010a) 'Symbolic power and (in)security: The marginalization of women's security in Russia', *International Political Sociology*, 4:401–18.

—— (2010b) 'Human security research practices: The case of crisis centres for women in Russia', *Security Dialogue*, 42(3):279–99.

Sundstrom, L. M. (2002) 'Women's NGOs in Russia: Struggling from the margins', *Demokratizatsija*, 10(2):207–29.

—— (2005) 'Foreign assistance, international norms, and NGO development: Lessons from the Russian campaign', *International Organization*, 59:419–49.

—— (2006) *Funding Civil Society: Foreign Assistance and NGO Development in Russia*, Stanford, CA: Stanford University Press.

Sylvester, C. (2004) 'Woe or whoa! International relations where it's not supposed to be', *Brown Journal of World Affairs*, 10(2):57–68.

Tadjbakhsh, S. and Chenoy, A. M. (2007) *Human Security: Concepts and Implications*, London and New York: Routledge.

UNDP (1994) *Human Development Report 1994: New Dimensions of Human Security*, New York: United Nations Development Programme.

Wæver, O. (1995) 'Securitization and desecuritization', in R. D. Lipschutz (ed.) *On Security*, New York: Columbia University Press, pp. 46–86.

Wibben, A. T. R. (2008) 'Human security: Toward an opening', *Security Dialogue*, 39(4):455–62.

Chapter 14

The political exclusion and commodification of women

Maria Lvova

Introduction

As the Russian Federation evolved from the ashes of the Soviet Union, an emerging phenomenon appeared to take shape in the form of an increasing tendency to tolerate the 'sex industry', 'trafficking' and 'prostitution' and to accept them as essential parts of post-Soviet society. At the same time, Russian society has also appeared to reject, if not ignore, such notions as 'violence', 'discrimination', 'inequality' and 'sexual harassment'. Attitudes towards sex since the fall of the Soviet Union provides just one example of how women have been commodified in post-communist Russia. At the same time, women's participation in democratic processes, as well as attitudes towards such participation, has been very poor. In combination, the democratization processes taking place in Russia have not had a positive effect on women and women's security. Russian democratization has been operationalized largely through an assumption that periodic elections combined with an unregulated market economy leads to democracy. In this chapter I wish to raise the conflicting issues around democratization processes and how this has affected women, in particular with regards to attitudes towards the sex trade. The prognosis does not look promising for Russian women and their overall security. I argue here that we need to look into Russia's past, and the attitudes inherited from the USSR, to understand how attitudes towards women have developed over time. I then examine this issue through an analysis of the relationship between economics and the concept of 'democracy' occurring at both the state and individual levels, as well as through concepts within liberalism (negative and positive freedom). Thereafter I discuss current state policy, what values are reflected in these policies towards prostitution and trafficking, and how this relates to what democracy means in Russia.

The two terms 'trafficking' and 'prostitution' are often closely linked. As stated in the study commissioned by the Norwegian Ministries of Local Government and Regional Development, and of Justice and the Police, carried out by the Fafo Institute for Applied International Studies: 'the terms prostitution and trafficking are not necessarily self-evident in their meanings, and to operate with a clear line of demarcation between the two may be problematic' (Brunovskis and Tyldum 2004:17). This problem of differentiation has been a topic of methodological discussion for the last fifty-five years. I agree with the words of Askola who posits that 'approaches to trafficking in women greatly draw from attitudes to prostitution in general, and these vary greatly from one country to another' (Askola 2001:17).

In the era of the Soviet Union, the state declared that trafficking and prostitution would not be a problem. Women were deemed to be liberated and free. They were socially active and independent.

> The new socialist government of Russia believed that prostitutes were the vestige of capitalism, so they became the first in the line of those for whom the government was going to do a lot of good. It was officially proclaimed that communism was the grave of prostitution.[1]

Elizabeth Waters has noted that the dominant discourses surrounding an understanding of the problem of prostitution in Russia are explicitly different from Western ones. After the October Revolution there was a period where two coexisting points of view on prostitution existed: prostitutes as villains and prostitutes as victims. The latter was reflected in the words of Lenin, who wrote: 'prostitution is a great misfortune for humanity' (Lenin quoted by Waters 1992:160). This understanding of prostitution did not hold the day, however. At the beginning of the twentieth century, the West (understood primarily as Europe and North America) appeared to address the issue of prostitution quite differently than the Soviet Union. The Western perspective developed as a result of the scientific and public debates on sexuality, and the 'moral' side of the problem was reduced. In the Soviet Union the morality of the issue became the crux of the problem. In the West 'rescue work has been replaced by welfare, and prostitution redefined as a medical problem ... In the Soviet Union, this process was interrupted by the repressive policies of the early thirties' (Waters 1992:174). Hence, in the former Soviet Union prostitution was criminalized and prostitutes were categorized as villains, not victims. As such, it is difficult to find any evidence that prostitution and trafficking actually existed during the time of the Soviet Union. While prostitution may not have been a mass or open phenomenon, this does not mean that it did not exist at all, however, but it was well hidden. Prostitution and pimping were called a 'parasitic lifestyle'. The issue of trafficking was not taken into consideration by the USSR at all. For example, in 1966 there was a discussion within the United Nations about establishing a committee to address the problem of slavery and trafficking in women. The response of the Soviet Union was as follows: 'the Soviet delegate condemned slavery but insisted that apartheid, racialism and colonialism were even more pressing problems and only these should be regarded as manifestations of modern slavery' (Barry 1984:63). When, many years later, the political situation drastically (and quickly) changed and the Soviet Union ceased to exist, the society was not prepared to face the societal sexual problems, including those of trafficking in women and prostitution.

Post USSR: the emergence of the problem with respect to state policy

Since the democratic reforms starting in the early 1990s, Russia has become a country where the transit, export and import of women is common, and ever increasing (Tverdova 2011). After the fall of the Soviet Union, many people regarded the opportunity to express their sexuality as the freedom to make a personal choice and as a newly obtained right. They considered it to be an essential aspect of the implementation of democracy: we are free to do what we want to do. They accepted primarily a freedom as a right to be independent from the state and to be free of any imposed values (either democratic or communist) in their actions. 'The liberalisation of moral attitudes, following a period of relative Puritanism, may find women as well as men enjoying what they see as newly won personal freedom' (Bridger *et al.* 1996:167).

This attitude of the vast majority of the population left almost no room for addressing and solving the problem of trafficking in women. Rapid declines in support, the impoverishment of the population, chaos and confusion in ideals and no clear guidelines for appropriate behaviour have led to changes in priorities. In January 1993, a report on young women and prostitution appeared in the media: 'According to sociologists, in answer to the question, "Which is the most attractive profession?" 60 per cent of Moscow female school-leavers replied, "prostitution"' (Beliaev 1992, quoted by Bridger *et al.* 1996:174). Prostitution, especially that motivated by the desire to gain hard currency, became an essential part of life for some women, and was viewed by a society as a good possibility to make a lot of money. 'As the social climate in Russia has become ever more oriented towards making money, by whatever means present themselves, becoming a prostitute could be viewed as an entirely rational choice for a young woman to make' (Bridger *et al.* 1996:174). Prostitution has become a widespread profession. Boris Panteleev,[2] in his interview to the Russian BBC office, mentioned that, in post-reform Russia, a new generation has grown up that does not consider prostitution to be a crime or immoral action.[3]

Authorities and bureaucrats have not grasped the magnitude of the problem as it has emerged. The 1999 survey by the MiraMed Institute stated that because of the lack of interest from state authorities in the issue of trafficking, the number of women trafficked from the CIS countries (Armenia, Azerbaijan, Moldova, Kyrgyzstan, Tajikistan, etc.) had exceeded the numbers of women that were being trafficked from Asia (Preliminary survey report on sexual trafficking in the CIS 1999). Three years later, a United States-based researcher defined the same situation in the Russian Federation:

> Although trafficking in women from Russia has been going on for at least a decade, with increasing severity, the Russian Government has taken little notice of the problem ... One of the largest problems is the lack of political will to recognize and address the problem of trafficking in women.
>
> (Hughes 2002:56)

In terms of what has been done, we can look to the 'Protocol to Prevent, Suppress and Punish Trafficking in Persons, Especially Women and Children, supplementing the United Nations Convention against Transnational Organized Crime' (UN 2000) which was signed by Russia in December 2000. It was ratified only on 26 May 2004 after the implementation of amendments to the Criminal Code of the Russian Federation for criminalizing trafficking. The list of crimes in the code was increased, through such articles as 1271 (trafficking in persons), 1272 (use of slave labour) and 2421 (production and circulation of materials or goods with pornographic images of non-adults).

The Federal Law of the Russian Federation No. 163-F3, 'On introduction of changes and addenda in the Criminal Code of the Russian Federation' (2003), is actually the first official state condemnation of trafficking.

Although the amendments to the Criminal Code were passed so that trafficking in women was finally criminalized, there is still an acute need for a mechanism to realize Russia's anti-trafficking goals. There is neither consensus among different levels of government (federal, regional, local) nor between branches (e.g. executive and judicial) at the same level about where responsibility for and implementation of amendments to the Criminal Code reside. Therefore, experts agree that the Russian Federation needs a special law against trafficking, which will empower the existing executive system (Gavrilov 2003; Mattar 2004; Holschevnikova 2005). Since trafficking is criminalized under the Criminal Code, another enabling law is required. This law would establish contact between state and other actors, not least non-state actors such as NGOs, identifying the responsibilities of various power structures, and providing social and legal guarantees to the victims of trafficking. The name of the proposed law is 'On the status of victim of trafficking in women'. Mohova stresses: 'If the law would be accepted it would be the first and unique precedent in Russia when the state takes the responsibility for the victims. Usually the state only punishes the offender' (Mohova quoted in Lvova 2005:61). The project of the Federal Law 'On the status of victim' originated through the efforts of a working group under the instruction of Elena Mizulina, Deputy of the State Duma. The discussions around this law are ongoing and it is unclear when the proposed principles will be accepted.

Human (personal) and state security through democracy and economic freedom

The term 'democracy' has a long history that can be viewed as being a result or outcome of the development of Western civilization. It developed simultaneously with the formation and evolution of civil society and law-governed state (rule of law). The word first appeared in Ancient Greece and it means 'rule of people' ('demos' – people; 'kratos' – power, rule, govern). With reference to Robert Dahl's observation that 'there is no single theory of democracy – only theories' (Terchek and Conte 2001:xiii), it is possible to conclude that different authors have ascribed different meanings to this term.

Huntington defines various approaches to the definition of democracy: empirical, descriptive, institutional and procedural definitions (Huntington 1991:6). He states that the procedures for constituting government are the most important aspect of democratic institutions. From his point of view, 'the definition of democracy in terms of elections is a minimal definition' (Huntington 1991:9). Huntington traces the origin of the procedural definition to the work of Schumpeter. Arguing with the 'classical theory of democracy', Schumpeter shows that the source of democracy, 'the will of the people', and the aim of it, 'the common good', are not sufficient enough, because 'none of them seriously considered any substantial change in the economic framework and the habits of bourgeois society' (Schumpeter 1994:252). Therefore, Schumpeter suggests another theory of democracy, the so-called 'democratic method', which 'is that institutional arrangement for arriving at political decisions in which individuals acquire the power to decide by means of competitive struggle for the people's vote' (Schumpeter 1994:269). I can extract several consequences that follow on from Schumpeter's position, which I argue are useful in an examination of prostitution and trafficking in Russia.

First, there is an essential connection between democracy and the freedom of the individual. The definition of democracy 'involves the two dimensions – contestation and participation – that Robert Dahl saw as critical to his realistic democracy or poliarchy' (Huntington 1991:7). It is linked to the state such that 'it is ... impossible to imagine a democracy that is not based on the definition of a polity, and therefore a territory' (Touraine 1997:27). Democracy is embodied in the territory of the sovereign state, where political rights are supplemented with civil liberties and economic freedom. Civil liberties

> comprise freedom of the press and the independence of the major media from government dictation; the protection of the individual by the courts; the freedom to express individual opinions without fear of imprisonment; the respect for private rights and desires in religion, occupation, residence, education, and the like; and the individual's ability to engage in rational political activities without fear for his or her life.
>
> (Rummel 1983:31)

Additionally, Rummel links economics and democracy, where he posits that economic freedom is measured by independence from governmental ownership, control and regulations (Rummel 1983). In contrast, a critique of Rummel's theory of freedom, by Erich Weede, states: 'economic freedom is usually not part of definitions of democracy' (Weede 1984:655). The lack of agreement on the relationship between democracy and economy is an important one, not least when the activities under examination here have implications economically for both the women themselves as well as for those controlling prostitution industries, as well as trafficking rackets. Thus the role of economic freedom in ideas about democracy has been a much-debated and tenuous element. We must examine these linkages in relation to the international context and ongoing political processes. 'The rapid spread of the language of democracy ... cannot be studied without reflecting upon the changed international political context of the late twentieth century' (Rai 2000:1). Weede's article was published in 1984, and since that time the political map of the world has drastically changed. Therefore, Schweller fairly notes: 'current democratization of Eastern Europe and the Soviet Union seems to demonstrate that in addition to political freedom, a market and private property economy may be an indispensable ingredient of democracy' (Schweller 1992:240).

If we acknowledge that democracy and the economy of a sovereign state are interrelated, it becomes of paramount importance to take into consideration the positive and negative outcomes of the free market economy. Dahl points out that usually market capitalism leads to economic growth:

> [T]o begin with, by cutting acute poverty and improving living standards, economic growth helps to reduce social and political conflicts. Furthermore, when economic conflicts do arise, growth provides more resources that are available for a mutually satisfactory settlement in which each side gains something.
>
> (Dahl 1998:168)

This also has social and political outcomes. People become more active, they search for education and gain personal freedom, property rights and civil rights. At the same time, Dahl stresses such negative consequences of market economy as the emergence of economic inequalities and, as a result, inequities in the distribution of political resources. But since 'democracy and market-capitalism are locked in a persistent conflict in which each modifies and limits the other' (Dahl 1998:173), there is an opportunity for the creation of a certain set of rules[4] that set boundaries on institutions of market-capitalism (property rights, monopolies, contracts, ownership, etc.). Following this, it is possible to argue that nature of democracy is strongly bound to the process of development, particularly economic development. Democracy allows for the possibility of implementing change. Thus, there is another term that is intimately related to the term 'democracy': 'democratization'. Huntington identifies tensions involved in processes of 'democratization' which include both obstacles as well as forces for democracy, such as politics (particularly when there is a lack of experience with democracy in the past), economics (such factors as poverty and economic development) and culture (attitudes, values, beliefs and patterns of behaviour existing within the society before the process of democratization) (Huntington 1991). Huntington's words reflect the insights of other scholars who examined the transformation of the Soviet Union as it moved towards democracy: 'after all, democracy is more than a set of political institutions; it requires sustenance from a myriad of political, social, legal, and economic values resident in the hearts and minds of the ordinary members of the polity' (Gibson *et al.* 1992:330). Consequently, it is logical to say that 'democracy' is an ongoing project that is actualized by the process of formation and is based on 'the rule for law, the separation of powers, an independent judiciary, and respect for individual initiative and human rights' (Lukin 1999:39).

Liberalism

At this point it is valuable to draw on a broad tradition of theory, that of liberalism. Liberalism and theories of democracy have much in common, although it is incorrect to equate the two. Processes of liberalism that will be elucidated below can be seen to interact with processes of democracy, which I argue is relevant to our understanding of the development of attitudes and political action that has taken place in Russia since the fall of the Soviet Union. As previously mentioned, democracy provides the necessary conditions for personal freedom, rights and liberties. Inclusion of all citizens gives them an opportunity to 'shape your life in accordance with your own goals, preferences, tastes, values, commitments, beliefs' (Dahl 1998:52). Within the body of liberal-democratic theory, the free individual is acknowledged as an independent actor. Therefore, there may be a tendency to use 'liberalism' as a synonym for 'democracy'. Dahl stressed that although liberalism has contributed to the development of democratic theory, it would be wrong to substitute the notion of democracy with that of liberalism, confirmed by Carter and Stokes:

> There is a historic tension between the ideals of liberalism and those of democracy. Classical liberalism stresses the importance of individual and minority rights, the rule of law and constitutional limits on government power. Democracy, on the other hand, places more emphasis upon popular participation and popular sovereignty.
>
> (1998:2–3)

Liberalism is defined as a particular set of social and political attitudes and values. Doyle writes: 'there is no canonical description of Liberalism' (Doyle 1986:1152, quoted by Keohane 1990:173). For the purpose of the argument I am developing here, it is useful to view liberalism as embedded within the principle of the freedom of the individual (Carter and Stokes 1998). With a departure point in the freedom of the individual, it is important to explore related concepts that help us understand what this means, in this case 'negative' and 'positive' freedom. 'Negative freedom' implies the necessity of protecting the individual from the interference of the state in personal life. 'Positive freedom' means that an individual can be subordinate to a whole community.

> The link between negative and positive liberty is the democratic will to enable those who are subordinate and dependent to act freely and to discuss rights and guarantees on equal terms with those who possess economic, political, and cultural resources.
>
> (Touraine 1997:11)

It also means that both freedoms are interconnected and that it is not possible to prioritize one or the other. There are areas of social action where public authority is needed, and *in addition* the individual should have the possibility to act freely and to contribute to the design of legislation, which might be needed for the protection of individual rights.

With the collapse of the Soviet Union, that negative perception of freedom became most important. This impacted the perceptions and practices surrounding prostitution. Suddenly it was out in the open, as opposed to hidden like it was during the period of the Soviet Union. Using the argument that the right for self-expression is essential, prostitution was liberated from any moral connotations. It became 'just a profession' equal to any other, and was explained in terms of self-expression, liberation from the state suppression and triumph of sexuality in the post-Soviet space. The necessity for any public control and state action in regard to the situation was not present in the dominant discourses.

As Doyle states, there is one important aspect that is often missing from theories of democracy, in that while debating the importance of the freedom of the individual, it is necessary to remember that there is 'the importance of moral freedom, of the right to be treated and a duty to treat others as ethical subjects, and not as objects or means only' (Doyle 1996:4). Thus, the focus should be shifted to not just liberty and rights but also duties. The principle of responsibility, or duty, was developed by such philosophers as Immanuel Kant and Adam Smith. Kant noted that the human being is essentially able to freely make rational and morally good choices because man is led by the concept of duty: 'When we talk about duty we must presuppose that the required action is possible. It is illogical to say that something is obligatory according to ethical theory, but yet impossible to realize in action ... Ought implies Can' (Airaksinen and Siitonen 2004:317). Adam Smith developed the idea of the role of commerce for democratic development and promotion of peace, stating that commerce can give an opportunity to exercise liberty in a civil society.

> The progress of society in both dimensions, social and moral, was the key. Adam Smith set out to discover the social foundations of moral behavior. Commerce and armies together could, he argued, push societies from servile dependency and a continual state of international war to something better, a civil society enjoying natural freedom and prospect of peace.
>
> (Doyle 1997:241)

The fundamental nature of human beings has led founders and proponents of liberalism to believe that not all people and nations are free just as a matter of course. There are two basic requirements of liberation: 'First, persons or nations must be themselves enlightened, aware of their interests and how they should be secured. Second, people must live under enlightened political institutions which allow their true interest to shape politics' (Owen 1996:123). Consequently, there is a distinct and important relationship between the state and the individual. The state holds the possibility of liberating the individual, and provides the necessary conditions for freedom through the use of democratic institutions and systems of legislation. It is difficult to generalize about liberalism as individual freedom within the body of democracy, and it is important to take into consideration the distinct understanding of liberalism within the national context. 'Typically, liberals are associated with the values of freedom, tolerance and democracy in the world. Liberalism, however, may be shaped differently in different social and cultural contexts' (Tsygankov and Tsygankov 2004:54). Russian individuals and communities had a long history of a particularly oppressive experience with, and relationship to, the state. This, among other things, has contributed to the formation of democracy after the fall of the Soviet Union, and to subsequent attitudes towards prostitution and trafficking.

Democracy in Russia

There is a significant amount of literature discussing the challenges with the development of democracy in Russia. Ross has written that: 'Many authors have alluded to the unique nature of Russia's dual transition and its difficult task of simultaneously reforming its economy and polity' (Ross 2000:405). He suggests that the literature on the problem of democratization in Russia may be divided into several schools. The first one pays attention to the preconditions that are important for the emergence of democracy (such as industrialization, capitalism, class structure, institutional factors, political culture and civil society); the second focuses on the transition processes; the third addresses the period following the collapse of the old regime and the emergence of current problems (Ross 2000).

In order to avoid going too deeply into the discussion of the nature of democracy in Russia, I will adopt the following approach. The situation in Russia is complicated and ambiguous. There is no clear answer to whether Russia has a democracy or not. This question remains disputed in the existing literature on democratization. Defining democracy in terms of elections (procedural approach) leads us to conclude that Russia is a democratic country. 'Governments produced by elections may be inefficient, corrupt, shortsighted, irresponsible, dominated by special interests, and incapable of adopting policies demanded by public good. These qualities may make such governments undesirable but they do not make them undemocratic' (Huntington 1991:10). The changes that have happened in Russia for the past fifteen years are significant and drastic, and researchers agree on this. But Russia must follow its own road to democracy. It is largely impossible in a short period of time to get rid of all the old problems and to successfully solve new ones. It is in this context that Russian attitudes towards prostitution and trafficking have developed. These developments have also had a significant impact upon women.

Women's issues: from the USSR to contemporary Russia

In the Soviet Union, women were in an ambiguous position: equality between men and women was formally proclaimed, but in reality women had a secondary position compared with men. Thus, the Russian researcher on gender issues, Voronina, states: 'formal equal rights for men and women, enshrined in the constitution and in a number of other laws, certainly do not ensure equal rights in practice' (Voronina 1994:39). The issue of the 'double burden' of domestic duties and public life was largely ignored by the state. 'Feminism was denounced as a luxury of bourgeois Western women' (Sundstrom 2002:224). Soviet women were hardly aware of their problems of inequity. There was no active women's movement to aid in articulating these difficulties. Zhensovety (Women's Councils) were the only women's organizations during the Soviet Union time. They were set up only in 1986 with the aim of establishing a contact between the state and women in order to help them to carry this 'double burden'. There was otherwise no effort to facilitate women's political consciousness and to develop grassroots movements.[5] At the same time, however, the situation was not completely catastrophic. The level of unemployment was low, and basic social needs were being met. Existing inequality was not taken into consideration by communist ideology since the society was implementing the goal of stability and happiness.

Since the fall of the Soviet Union, changes in Russia have significantly modified the situation. Democratization began with the move to the market economy. The need for economic development dominated all spheres of life: political, social and cultural. The impact of the economic reforms on the society was high. In mixing the terms 'market economy' and 'democracy', people started to substitute economic principles for everyday rules.

> When the Soviet Union lost its empire and its regime and then finally ceased to exist, the reconstruction of social life began, not with popular movement or new ideas, but with economic management ... Given that the Soviet regime had been totalitarian, economic freedom seemed to many people to mark an immediate transition to democracy and even prosperity.
>
> (Touraine 1997:175)

From one point of view, the adoption of the market model was a necessary step. The market economy allowed the integration of the Russian economy into the world economic system. It gave an opportunity for Russia to develop free competition in its internal market. After empty shelves in the shops and a period of deficit, people were looking forward to change for good. But from another perspective, in addition to ill-planned reforms and mass confusion, phenomena such as a growing sex industry and criminal activity were left without any attention from a highly corrupted state. The transition to capitalism brought Russia the possibility of participating in the global market as a free and equal partner. But there were other less welcome outcomes such as involvement in the global sex industry and labour migration. Trying to establish democracy in a form of market economy, the government did not notice that a so-called 'market approach' started to operate in all spheres of Russian life. The commercialization and privatization of the education and healthcare systems was rationalized as an essential step. The state that controlled so much previously now neglected a number of social issues: closed kindergartens, poor municipal services, increased prices for travel and unchecked pyramid schemes, for example. Lack of state regulation facilitated increasing discrimination against women in terms of inequity in wages, reduced access to jobs, reduced career promotion and increased redundancies. This played well into the sexual objectification of women, who became increasingly linked to the domestic or entertainment spheres as creatures defined by sexual capacity. The vast majority of advertisements were based on the exploitation of women's bodies and their sexuality. The sex industry developed rapidly. Women were increasingly objectified as a possession, with sexuality as a profitable object. Democracy understood as market economy had a distinct effect upon women's bodies and lives.

The substitution of the term 'democracy' with the term 'market economy' led to the transformation of mass consciousness. Young Russian capitalistic society easily perceived sexuality as a commodity. Interpersonal relationships became grounded in the 'market approach' or 'business imperative' (Klimenkova 1994:14) and became the norm. The market economy brought freedom and liberalization and reinforced the notion that everything could be bought and sold. Thus, exploitation of woman's sexuality is just a business. Economic reforms and steps towards market economy 'brought about a new social orientation whose essence was the commercialization of all areas of our social life' (Klimenkova 1994:15). The prioritization of the economy above social needs, the rapid stratification of society and drastic changes in social attitudes played important roles in increasing society's level of tolerance towards the phenomenon of trafficking in women.

Thus, the breakdown in the functioning of almost all spheres of society, of industry, along with inflation, empty shelves in the shops and problems with salary were reflected in the gender issue and vice versa. Women were regarded as unprofitable employees. The privileged place belonged to men who were considered the breadwinners. The alternative open to women was either a low-paid, routine job in bad conditions or sitting at home and fostering children. The competition on the labour market was different according to age, depending on if a woman was young or middle aged; older women trying to find work relied on a combination of luck, skill and wits, whereas young women were sought out with regard to their bodies, being able to work in modelling businesses, to marrying rich men or working as prostitutes. Internationally, women had 'better' chances regardless of age due to the popularity of the bride market and entertainment services in different countries outside the former Soviet Union.

The report of the International Helsinki Federation for Human Rights in 2000 stated that many of the problems experienced in modern Russia had originated during the transition period of the 1990s: 'Illegal dismissals and jobs that have no social guarantees and are frequently without any official labour contract have become the norm' (International Helsinki Federation for Human Rights 2001:375). Thus, the political indifference towards women resulted in a situation where women were not socially and legally protected by the state. The state also did not provide security for women in the new working (like the modelling business) and social conditions (trafficking, domestic violence, sexual harassment).

The exclusion of women as a result of an incorrect understanding of democracy

The prevailing political attitude towards Russian women in the 1990s led to their exclusion from the process of democratization. Politicians actively promoted the traditionalist view of women:

> From about the spring of 1989 it became quite clear that perestroika was basically a male project and that not only the Communists, but also the democrats had turned their backs on women ... Most of them had no women's programme, and their personal views amounted to the same old

theory about the need to 'return women to the home and give them a rest from socialism'.

(Posadskaya 1994:3)

Six years later the 'Women 2000' report stated that 'among the political parties and movements represented in the State Duma, only the Liberal Democratic Party's program included a special section addressing women's issues' (International Helsinki Federation for Human Rights 2001:379).[6] This attitude resulted in an unwritten false assumption that democracy can be established without women's participation; women just have to relax at home and do their best carrying out the role of wives and mothers. The 'Woman question' was reduced to everyday life and domestic problems and was considered to be trivial and uninteresting for the state. This political attitude continues to guide the public perception that women's issues are a secondary problem, which can be addressed once the situation is better, not least economically (according to a non-inclusive model of a successful market economy).

Another consequence of the prevailing attitude is that women are regarded as alien to politics. After a little over a decade of 'democratization', an analysis of the weekly poll carried out by the Public Opinion Foundation in 2001 examined data for the three previous years. The aim of this work was to clarify the gender dimension of the developing democracy and civil society in Russia. Differences in value orientation and political attitude between men and women were identified. Russian women were less willing than men to participate in solving municipal problems. The question about the wish to participate in municipal government was positively answered by 38 per cent of men and 30 per cent of women, although 51 per cent of men and 59 per cent of women supported the statement 'women have to be more active in politics'. A clear one-third of respondents were sure that politics is not women's business. Statistics showed that fewer Russian women supported such institutional elements of democracy as freedom of speech, political pluralism, market economy and freedom of conscience, and were often ready to refuse them in favour of state care and regulation of social relationships. Existing differences in attitudes between genders in modern Russian society may explain these data. Aivazova and Ketman (2001) concluded that gender inequality resulted in women not perceiving civil and political rights as having a significant institutional source.[7] Only forty-five women were elected to the State Duma in 2003 (10 per cent of members of parliament). Parties such as 'Women of Russia' and the 'Russian party of women's protection' got 2.04 per cent and 0.80 per cent of votes respectively and did not achieve representation in the Duma. The lack of representation of women in political processes must also be seen in light of the developments in the market economy, of the commodification of sex (and the concurrent attitudes towards sex), to be able to get a sense of the position of women in Russia today.

Conclusions

Thus, we can state that there is a mixture of approaches towards women as child bearers and mothers through what I would argue is a distorted understanding of democratic values and liberalism. It became widely accepted that every 'real man' has a right (as a result of liberalism) to have sex and to be entertained by women whether she is his wife or a prostitute. At the same time a double standard was evoked towards women. As a response to decades of socialism that included (at least formally) women's equality, women now could take a break from the demands of equality by becoming good housewives, devoting their lives to family (time to remember the famous 'kitchen, church, children'). Democratization early on, then, was articulated through a classic patriarchal framework. At the same time, however, one was suddenly 'free', where such freedom was articulated through market principles. Thus one had the freedom to choose the things one wanted to do, including working as a prostitute or buying sex. 'Prostitution', 'sex' and 'pornography' had been banned under socialism; in the era of democracy, they became an expression of new-found freedom. If prostitution was then a free choice, a 'career option' so to speak, and something that could develop unregulated in a market economy, the newly emerging phenomenon of trafficking that involved complicated, similar but different issues with prostitution was not taken very seriously, and definitely was not considered being relevant to 'big politics'.

The failure to understand or see this situation unfolding, and the resulting threat to the personal security of women, can be traced to a public/private dichotomy exercised by the state. 'The family sphere of life was identified with all that was female, and was counterpoised to general, social and state issues' (Voronina 1994:40). Thus those issues that had to do with women were private matters, and the state could hardly be concerned about these private matters since they were not important for public policy. Askola posits that this attitude can be seen in the different ways smuggling and trafficking have been treated. Smuggling is considered to be a crime against the state, because it violates state security and the state's boundaries. However, trafficking 'is a crime against the individual, and only secondary a violation of state sovereignty' (Askola 2001:18). Since Russian politicians continue to consider women as an attachment to family, they are not inclined to treat women as independent individuals in need of state support or protection. If a woman takes a risk (makes a choice, in the free market) to improve her earnings by working abroad, she has to bear the responsibility for the consequences. This logic of exclusion is rather simple: if a woman does not function as mother or wife and dares to challenge the state (and societal) position towards women, then she does not deserve to be protected. Unfortunately, individual rights are frequently violated in Russia. And the situation is even more complicated when it comes to women, who are often viewed as second-class citizens.

This current attitude can be explained with reference to the time of the USSR. At that time it was claimed that the 'Russian mentality' was centred and built upon ideas of collectivism, commonality and confidence in the power of the state (e.g. Mjasnikov 1998; Shulyndin 1998). The Soviet ideology embraced these values, and the unique model of the identity of the Soviet people was created. Personality and individualism were neglected by the state. Values of collectivism and state interest were promoted as significant. Shulman, in analysing the problem of the individual versus the collective in the Soviet Union, points out: '[privacy] is a seldom used noun' (Shulman 1977:381). Because of this governing ideology that was promoted by communists, the rights of individuals were suppressed (Bakacsi *et al.* 2002). This mentality remained, at least within state structures. In the aftermath of the collapse of the USSR, with the economic and political chaos that ensued, the state's position towards promoting individual and human rights was not prioritized, or even really recognized as being relevant.

Weiler claims that the Russian authorities do not consider human beings to be elements that are endowed with value in and of themselves, despite any formal commitment to the contrary (Weiler 2002). Additionally, there is a contradiction between the formal commitment of the state towards upholding individual human rights and its actual actions. Weiler states: 'the interaction of declining state resources with the lack of accountability of state authority for its actions and inaction has created a deadly combination for socially vulnerable individuals' (Weiler 2002:260). The latter include such vulnerable groups as the poor, minorities, criminals (and criminal suspects) and women (especially in cases of domestic violence and trafficking). When Russia adopted the liberal notion that 'private matters should not be of public concern' (Weiler 2002:274), this view was transformed through the lens of the past in a country where the right of the individual was considered to be unimportant. Suddenly a lack of concern for the individual was further justified by a perception of democratization, associated with liberal ideals regarding the private and public divide and the drive for a free market.

Democracy provides an opportunity to develop at both the individual and the state level, including the opportunity for individuals to express themselves (freedom of opinion and expression) and to make one's own choices. In order to achieve the latter, special attention must be paid to human rights and the social values within the population and entrenched within state institutions. The effort to liberate the individual requires that attention be paid to the gender issue and related problems: 'all attempts at liberating the individual without liberating the female sex is just as pointless as the socialist project of liberating the sex without liberating the individual' (Posadskaya 1994:4).

Notes

1 *Three centuries of the Russian prostitution*. Online. Available at: http://english.pravda.ru/news/society/sex/30-04-2002/42121-0/ (accessed 26 March 2013).
2 Boris Panteleev is a researcher at the Scientific Research Institute of the General Office of Public Prosecutor of the Russian Federation.
3 *Torgovlja ludjmi: tri grani problemy dlja Rossii* [Trafficking in human beings: Three dimensions of the problem in Russia]. Online. Available at: http://news.bbc.co.uk/hi/russian/news/newsid_3435000/3435743.stm (accessed 26 March 2005).
4 As Dahl states, 'the most easily accessible candidate for intervening in a market economy in order to alter an otherwise harmful outcome, and the most effective, is ... the government of the state' (Dahl 1998:175).
5 This lack of accent on political and social activity successfully remained invisible for the Western world. Thus, British researchers visiting the Soviet Union in the early 1980s wrote that for them before the trip Soviet women seemed to be 'away from the pressures of the capitalist market, from conspicuous consumption, from advertising with its glamorous models and from street-corner pornography. Instead there are images of women successful as a result of their achievements, as factory workers, scientists, even astronauts' (Holland 1985:10).
6 The authors of the programme deny both feminist and patriarchal approaches. Thus, on the one hand, the party promises that women will get the opportunity to be in the family; on the other, the programme promises to create mechanisms that would help women to adapt to the conditions of the market economy (International Helsinki Federation for Human Rights 2001).
7 For details see Aivazova and Ketman (2001).

References

Airaksinen, T. and Siitonen, A. (2004) 'Kant on Hobbes, peace and obedience', *History of European Ideas*, 30:315–28.

Aivazova, S. and Ketman, G. (2001) *Zhenschiny na randevu s rossiiskoi demokratiei* [Women on the Rendezvous with Russian Democracy]. Online. Available at: http://bd.fom.ru/report/map/articles/aivazova/ke0110091 (accessed 26 March 2013).

Askola, H. (2001) 'Trafficking in women in the Baltic States: Legal aspects', *Trafficking in Women and Prostitution in the Baltic States: Social and Legal Aspects*, Helsinki: International Organization for Migration (IOM), pp. 13–92.

Bakacsi, G., Takács, S., Karácsonyi, A. and Imrek, V. (2002) 'Eastern European cluster: Tradition and transition', *Journal of World Business*, 37(1):69–80.

Barry, K. (1984) *Female Sexual Slavery*, New York: New York University Press.

Bridger, S., Kay, R. and Pinnick, K. (1996) *No More Heroines? Russia, Women and the Market*, London and New York: Routledge.

Brunovskis, A. and Tyldum, G. (2004) *Crossing Borders: An Empirical Study of Transnational Prostitution and Trafficking in Human Beings*, FAFO-report 426, Norway: Interface media. Online. Available at: http://www.fafo.no/pub/rapp/426/426.pdf (accessed 26 March 2013).

Carter, A. and Stokes, G. (eds) (1998) *Liberal Democracy and its Critics: Perspectives in Contemporary Political Thought*, Cambridge: Polity Press.

Dahl, R. O. (1998) *On Democracy*, New Haven and London: Yale University Press.

Doyle, M. W. (1996) 'Kant, liberal legacies, and foreign affaires', in M. Brown, S. M. Lynn-Jones and S. E. Miller (eds) *Debating the Democratic Peace*, Cambridge, MA: MIT Press, pp. 3–57.

—— (1997) *Ways of War and Peace: Realism, Liberalism and Socialism*, New York and London: W. W. Norton and Company.

Federal Law of the Russian Federation (2003) 'On introduction of changes and addenda in the Criminal Code of the Russian Federation', 8 December 2003, No. 163-F3.

Gavrilov, A. (2003) 'The improvements of the Criminal Code', *Rossyiskaya Gazeta*, 15 April.

Gibson, J. L., Duch, R. M. and Tedin, K. L. (1992) 'Democratic values and the transformation of the Soviet Union', *The Journal of Politics*, 54(2):329–71.

Holland, B. (ed.) (1985) *Soviet Sisterhood*, Bloomington, IN: Indiana University Press.

Holschevnikova, T. (2005) 'Formirovanie zakonodatel'stva Rossiiskoi Federatsii o protivodeistvii torgovle lud'mi' [The formation of the legislation of the Russian Federation on counteraction to trafficking in human beings], paper presented at the Regional Summit 'The cooperation between state and public organizations on the counteraction to trafficking in women and girls in Murmansk region', 24–25 February, Murmansk, Russia.

Hughes, D. M. (2002) *Trafficking for Sexual Exploitation: The Case of the Russian Federation*, Geneva: IOM.

Huntington, S. P. (1991) *The Third Wave: Democratization in the Last Twentieth Century*, Julian J. Rothbaum distinguished lecture series, 4, Norman, OK and London: University of Oklahoma Press.

International Helsinki Federation for Human Rights (2001) *A Form of Slavery: Trafficking in Women in OSCE Member States*. Report to the OSCE Supplementary Human Dimension Meeting on Trafficking in Human Beings, Vienna, 19 June 2000, International Helsinki Federation for Human Rights (IHF).

Keohane, R. O. (1990) 'International liberalism reconsidered', in J. Dunn (ed.) *The Economic Limits to Modern Politics*, Cambridge: Cambridge University Press, pp. 165–94.

Klimenkova, T. (1994) 'What does our New Democracy offer society?', in A. Posadskaya (ed.) *Women in Russia: A New Era in Russian Feminism*, London: Verso, pp. 14–36.

Lukin, A. (1999) 'Forcing the pace of democratization', *Journal of Democracy*, 10:35–40.

Lvova, M. (2005) *Trafficking in Women and Democracy: The Case of the Russian Federation*, Master Thesis, Tromsø: Centre for Peace Studies, Faculty of Social Science, University of Tromsø.

Mattar, M. (2004) 'Zamechania po povodu razrabotki zakonodatel'stva po protivodeistviju torgovle lud'mi predstavlennye Gosdume PF, 29 Nojabrja 2004, Moskva, Rossiiskaja Federatzija' [Observations on the creation of legislation combating trafficking in human beings represented to the State Duma of the Russian Federation, 29 November 2004, Moscow, the Russian Federation], paper presented at the Regional Summit 'The cooperation between state and public organizations on the counteraction to trafficking in women and girls in Murmansk region', 24–25 February 2005, Murmansk, Russia.

Mjasnikov, A. G. (1998) 'O glavnoi arhitipicheskoi strukture russkogo obschestvennogo samosoznanija' [On the main archetypical structure of Russian public self-consciousness], conference paper, conference 'Otechestvennaja filosofia: russkaja, rossiiskaja, vsemirnaja' [Native philosophy: Russian before the October Revolution, Russian, Universal], Nizniji Novgorod, pp. 66–7.

On Countering Trafficking in Persons (2003) Draft of Federal Law, the Russian Federation, 13 February 2003.

On the Protection of the Victims of Trafficking (2003) Draft of Federal Law, the Russian Federation, 30 October 2003.

Owen, J. M. (1996) 'How liberalism produces democratic peace', in M. Brown, S. M. Lynn-Jones and S. E. Miller (eds) *Debating the Democratic Peace*, Cambridge, MA: MIT Press, pp. 116–54.

Posadskaya, A. (ed.) (1994) 'Introduction', in *Women in Russia: A New Era in Russian Feminism*, London: Verso, pp. 1–7.

Preliminary survey report on sexual trafficking in the CIS (1999) *Preliminary survey report on sexual trafficking in the CIS, June*. MiraMed Institute. Online. Available at: http://www.miramed.org/pdf/Trafficking%20Survey.PDF (accessed 26 March 2013).

Rai, S. M. (ed.) (2000) *International Perspectives on Gender and Democratization*, Basingstoke: Macmillan.

Ross, C. (2000) 'Federalism and democratization in Russia', *Communist and Post-Communist Studies*, 33:403–20.

Rummel, R. J. (1983) 'Libertarianism and international violence', *The Journal of Conflict Resolution*, 27(1):27–71.

Schumpeter, J. A. (1994) *Capitalism, Socialism and Democracy*, with new introduction by Richard Swedberg, London and New York: Routledge.

Schweller, R. L. (1992) 'Domestic structure and preventive war: Are democracies more pacific?' *World Politics*, 44(2):235–69.

Shulman, C. (1977) 'The individual and the collective', in D. Atkinson, A. Dallin and G. Warshofsky Lapidus (eds) *Women in Russia*, Palo Alto, CA: Stanford University Press, pp. 375–84.

Shulyndin, B. P. (1998) 'Rossiiskii mentalitet i osobennosti russkoi filosofii' [Russina mentality and the specific of Russian philosophy], conference paper, conference 'Otechestvennaja filosofia: russkaja, rossiiskaja, vsemirnaja' [Native philosophy: Russian before the October Revolution, Russian, Universal], Nizniji Novgorod, pp. 106–16.

Sundstrom, L. M. (2002) 'Women's NGOs in Russia: Struggling from the margins', *Demokratizatsiya*, 10(2):207–29.

Terchek, R. J. and Conte, T. C. (eds) (2001) 'Introduction', in *Theories of Democracy: A Reader*, Lanham, MD: Rowman & Littlefield.

Touraine, A. (1997) *What is Democracy?*, Boulder, CO: Westview Press.

Tsygankov, P. A. and Tsygankov, A. P. (2004) 'Dilemmas and promises of Russian liberalism', *Communist and Post-Communist Studies*, 37:53–70.

Tverdova, Y. V. (2011) 'Human trafficking in Russia and other post-Soviet states', *Human Rights Rev.*, 12:329–44.

United Nations (UN) (2000) 'Protocol to prevent, suppress and punish trafficking in persons, especially women and children, supplementing the United Nations Convention against Transnational Organized Crime'. Online. Available at: http://www.uncjin.org/Documents/Conventions/dcatoc/final_documents_2/convention_%20traff_eng.pdf (accessed 20 March 2013).

Voronina, O. (1994) 'The mythology of women's emancipation in the USSR as the foundation of the policy of discrimination', in A. Posadskaya (ed.) *Women in Russia: A New Era in Russian Feminism*, London: Verso, pp. 37–56.

Waters, E. (1992) 'Victim or villain: Prostitution in post-revolutionary Russia', in L. Edmondson (ed.) *Women and Society in Russia and the Soviet Union*, Cambridge: Cambridge University Press, pp. 178–201.

Weede, E. (1984) 'Democracy and war involvement', *The Journal of Conflict Resolution*, 28(4):649–64.

Weiler, J. D. (2002) 'Human rights in post-Soviet Russia', *Demokratizatsiya*, 10(2):257–76.

Chapter 15

Conclusion

Revisiting Arctic security

Marina Goloviznina and Gunhild Hoogensen Gjørv

The question of the future configuration of Arctic security is still open (Huebert *et al.* 2012). This question more than ever includes a broader understanding of what security means. The interest in the subject of exploring human security in the Arctic has increased considerably since 2004 when a few of us came together in Tromsø and started a discussion about it almost a decade ago. Many more themes could have been included here, exploring more deeply the connections between human activity and society and the environment, a connection that is crucial in the Arctic setting. The work here can be thus considered a foundation, a starting point, rather than the final answer on the subject.

Almost all of the authors who have contributed to this work either live their daily lives in the region called the Arctic, or are often back to the Arctic as the lives, experiences and activities of the region are their primary focus of work. Most of the contributors were there in 2004 at the first meeting that generated the interest in this book.

The Arctic is a dynamic region, and changes are happening both quickly (for example climate change which is impacting this region faster than any other on the planet) and gradually (for example governance structures such as international law, bilateral agreements and the development of institutions such as the Arctic Council). These chapters have given us a glimpse into different ways of thinking about security, of people's lives, both political and personal – from state to individuals/communities across the Arctic, with a focus on Canada, Russia and Norway.

Human security has been considered problematic as a concept for the Arctic. Why? Largely because the region is not beset with large-scale, open conflict (to be blunt). As well, resistance to using the concept in the Arctic bases itself upon a further assumption about how much insecurity is warranted to be able to really classify it as 'human insecurity' or something that deserves the attention of human security analysis, although that threshold may be dependent upon the eye of the beholder (see Owen 2004). Thus the assumptions that are supposedly embedded within 'freedom from fear, freedom from want' are often not perceived as relevant to the region.

This book is not a call to turn our heads away from global problems and challenges and human suffering, not least that which happens in regions outside of the Arctic. Nor is it a contest – which region in the world has the greatest human security issues? By its very nature, human security and insecurity cannot be so easily encapsulated. This is why definition is so difficult, because at the level of individuals and communities, values, priorities and identities upon which human security perspectives are based can be so different. Human security and insecurity depends very much upon the context upon which we speak.

This book is instead a call to awareness and dialogue. We see that, in the few examples we have provided in this book, human security and insecurity cannot be assumed in broad, sweeping generalizations. As Gabrielle Slowey points out, even when Canada ranked in the top 10 countries in the world to live, it ranked much lower when taking into account the experiences of its indigenous peoples. We see how indigenous and non-indigenous experiences differ from one another in the region, but also that, between indigenous groups, there are significant human security differences, where the experiences of the Canadian Inuit are significantly different than that of the Norwegian Sami. Russian women fare considerably differently than Norwegian women, and are generally at much greater risk of violence, economic insecurity and political marginalization. Large-scale conflict may not be the issue, but violence in communities, and not least the home, is all too commonplace.

New challenges in the shape of greenhouse gas emissions, transboundary pollutants, climate change and environmental hazards form not only a new context for understanding the notion of 'security' in the region, but also expose us to another comprehensive security concept which is increasingly finding relevance within both global intergovernmental policy discourse, as well as in the Arctic – environmental security. The debates about how environmental and human security issues interact – an examination of social systems and ecosystems as, Oran Young argues – need even greater focus. Part of this interaction includes the ways in which the human and environmental security concepts themselves interact.

While analysts differ on drawing images of contemporary security in the Arctic, and different writers have focused on different aspects within the myriad of connections between two forces – nature and politics – the consensus seems to indicate that a new security environment in the region is emerging (Huebert *et al.* 2012:6). The complex contemporary Arctic context emphasized a complicated, and sometimes contradictory, image of a new security environment, which simultaneously includes elements of traditional, environmental and human security discourses. Traditional security concerns (military and geopolitical) in the region have not lost their relevance, and have even re-emerged and been re-shaped in some sense, whereby some Arctic states still have a variety of overlapping sovereignty claims that have yet to be resolved, helping keep the traditional security debates in the Arctic alive and well (Yalowitz *et al.* 2009:xi), but that is not to the exclusion of other emerging human and environmental securities in the Arctic.

Changes in Arctic regions are occurring extremely rapidly in both social and environmental spheres, with the two inextricably linked (AMAP 1998; ACIA 2005; Stroeve *et al.* 2008; Serreze *et al.* 2009; SWIPA 2011). According to the 2004 Arctic Climate Impact Assessment report, the Arctic experiences some of the most rapid and severe climate change on Earth. Undergoing rapid climate change is one of the important environmental challenges the Arctic communities must face and which will affect both the circumpolar region and its population as well as the rest of the world through increased global warming and rising sea levels (ACIA 2005). Melting Arctic ice will potentially improve access to the significant northern fishing areas, and thus can significantly affect the regional economies, wildlife resources and the livelihoods of many Arctic populations (both positively and negatively). Russian policy makers, among others, link the ongoing climate change in the Arctic with the hope to see the development of a Northern Sea Route that might compete with the Suez Canal route for commercial maritime traffic, and while analysts differ on how quickly it will become commercially viable, the consensus seems to indicate that the passage will be largely ice free by the summer of 2015 (Gorenburg 2011:11). Rapid climate change is additionally leading to serious risks to human health, food security and possibly to survival of some indigenous communities that heavily depend economically and culturally upon hunting, herding, fishing and gathering (Anderssen and Gabrielsen 2005). Booming ecotourism has many implications for northerly communities, currently in transition, where new economic approaches are met with both excitement and trepidation, with the old ways of pursuing a livelihood disappearing (Hovelsrud and Smit 2010), where implications for agriculture and other types of land use are uncertain, habitat and diet are changing, and vulnerability to chemical contaminants and pollutions are increasing (AMAP 2004).

Climate change is leading to rapid melting Arctic ice cover and, thereby, expanding the opportunities for Arctic natural resources extraction, making the Arctic a potentially new resource frontier. The much-cited US Geological Survey generated excitement with the realization that the Arctic contains 30 per cent of the world's undiscovered gas and 13 per cent of its undiscovered oil, mostly offshore in less than 500 metres of water (Gautier *et al.* 2009). Access to hydrocarbon resources in the Arctic is both a core economic benefit, as well as a challenge. The Arctic states are laying claim to their national interests (in part through militarization) in preparation for the extraction of hydrocarbon resources in the region. Many governments, including Norway (Finnmark/Barents Sea Offshore), Greenland (West coast by Baffin Bay), the US (Alaska/Beaufort and Chukchi Seas) and Russia (Prirazlomnoje oil field/Barents Sea Offshore) are approving drilling in deeper, riskier, more ecologically sensitive Arctic environments, justifying this through the mantra of 'energy security' (Simpson 2007). Thus, today's Arctic demonstrates how the security concept can be easily manipulated and how the rhetoric of energy security has often been used as an excuse for governing elites to pursue centralized industrialization and large energy projects, sometimes at the expense of marginalized local populations (Simpson 2007). It also demonstrates how a multiple-actor understanding of security is extremely important for the purpose of recognizing the multiple 'bottom-up' voices of individuals and local communities with their perceptions, interests, values and concerns towards (in)security.

Although extractive industries provide significant and real economic benefits to local people both directly and indirectly in forms of taxes paid for the local budget, building roads, development infrastructure and facilities, and the creation of jobs (all not just important but necessary for the existence of communities), it also increases the prevalence of contaminants, pollutions, waste and ecological threats. It is essential that increasing extraction activities as well as their aftermaths can ideally have more positive than negative consequences for local peoples and communities and contribute towards sustainable development before, during and after the extraction projects. But one of the key paradoxes of the security debate is that national security interests does not always mean secure people, as the natural resources-led wealth of the state does not lead automatically to human security and development for the citizens (Hoogensen 2005).

It is thus both a political choice and active decision to employ a concept of human security in the Arctic. It is a political choice to employ the concept of human security in the Arctic because, as mentioned, the notion of 'security', which is already in large use in the Arctic (often uncritically), invokes power, and is usually employed and legitimated for the purpose of prioritizing certain activities and practices over others. Any argument for taking a human security approach to the Arctic region must acknowledge the difficulty of confronting the dominant security discourse that associates any notion of security with the use of force, particularly by the military. To say that security is purely a military issue is to reject the many other perceptions of security that exist or to prove that security is not relevant at local levels. To say that 'human security' should not be employed in the Arctic context is to say that a tool of communication, and a powerful one, is withheld from the possibilities for local communities.

Using the language of security catapults Arctic community interests and concerns to the same level as energy security interests or state security, for example; it can capture, explain and deliver the concerns of Arctic peoples to powerful audiences. It is important to hear from the local level about what security means to communities, and how their conceptions might not be consistent with state-based notions, and that community- and individual-based notions of security need to be recognized when understanding a holistic picture of security in the Arctic.

It is also an *active decision* to use the concepts of human and environmental security in analysis of the Arctic context in order to learn more, in some cases to 'map' different non-state actors' values or concerns and, not least, to collect, measure and analyse data necessary for the broad governance and sustainable development assessments that play a significant policy role in Arctic politics. While the practice of scientific assessment of development has become relatively familiar in the natural sciences during the past 20 years (Parson 2003), there has been less of a tradition of conducting qualitative assessments of human development in the social sciences, particularly towards the Arctic (AHDR 2004:21). Taking into account criticisms emerging regarding the assessments, there is still some value in examining indirect indicators and subjective measures of social processes based on public opinion surveys and subjective evaluations by independent experts, which become informative and useful tools in puzzling the whole picture of security and governance together in the region. Moreover, systematic collection of information on embodied experience of individuals and of groups as well as their assessments of different risks and threats invites those experiencing insecurity to identify the roots of their vulnerability, to participate in development solutions and to look for ways to incorporate them into a political agenda. In this sense, assessment surveys on human and environmental security issues can contribute to empowerment of local peoples and communities by supporting their capacity to actively participate in the sustainable development of the Arctic.

Loader and Walker (2007) highlight a number of dimensions of security, including instrumental, social and constitutive dimensions. In conclusion we wish to focus upon the constitutive dimension, which is perhaps the most neglected in contemporary security debates. The constitutive dimension claims that the promotion of security is fundamental to establishment and sustenance of a sense of the social and collective. The desire for protection from threats, and the ways in which collectives realize such desires, plays a specific and significant role in establishing and maintaining trust, social identity and a sense of community. In this sense, the pursuit of security 'helps to construct and sustain our "we feelings" – our very felt sense of common publicness' (Loader and Walker 2007:191). Highlighting the beneficial public dimensions of security, the dominant state-centred discourse is only complete through other discourses *beyond* – and *below* – the state and, thus, is a step forward to the development of broad and democratic security discourse in the Arctic region.

References

ACIA – Arctic Climate Impact Assessment Report (2005) ACIA Secretariat and Cooperative Institute for Arctic Research, Fairbanks: University of Alaska.

AHDR – Arctic Human Development Report (2004) Akureyri: Stefansson Arctic Institute.

AMAP – AMAP Assessment (1998) *Arctic Pollution Issues*, Oslo: Arctic Monitoring and Assessment Programme.

—— (2004) *Persistent Toxic Substances, Food Security and Indigenous Peoples of the Russian North: Final Report*, Oslo: Arctic Monitoring and Assessment Programme.

Anderssen, J. and Gabrielsen, G. W. (2005) *Helsetrusselen som vi overser (The Health Threat We are Overlooking)*, Oslo: Aftenposten.

Gautier, D., Bird, K., Charpentier, R., Grantz, A., Houseknecht, D., Klett, T., Moore, T., Pitman, J., Schenk, C., Schuenemeyer, J., Sørensen, K., Tennyson, M., Valin, Z. and Wandrey, C. (2009) 'Assessment of undiscovered oil and gas in the Arctic', *Science*, 29 May, 324(5931):1175–9.

Gorenburg, D. (2011) 'Russia's Arctic security strategy', *Russian Analytical Digest (RAD)*, 96:7–10.

Hoogensen, G. (2005) 'Bottoms up! A toast to regional security?', *International Studies Review*, 7(2):269–74.

Hovelsrud, G. K. and Smit, B. (2010) *Community Adaptation and Vulnerability in Arctic Regions*, Dordrecht: Springer Science+Business Media.

Huebert, R., Exner-Pirot, H., Lajeunesse, A. and Gulledge, J. (2012) *Climate Change & International Security: The Arctic as a Bellwether*, Calgary: Center for Climate and Energy Solutions.

Loader, I. and Walker, N. (2007) *Civilizing Security*, Cambridge: Cambridge University Press.

Owen, T. (2004) 'Human security – conflict, critique and consensus: Colloquium remarks and a proposal for a threshold-based definition', *Security Dialogue*, 35(3):373–87.

Parson, E. (2003) *Protecting the Ozone Layer: Science and Strategy*, Oxford: Oxford University Press.

Serreze, M., Barrett, A., Stroeve, J., Kindig, D. and Holland, M. (2009) 'The emergence of surface-based Arctic amplification', *The Cryosphere*, 3:11–19.

Simpson, A. (2007) 'The environmental energy security nexus: Critical analysis of an energy "love triangle" in Southeast Asia', *Third World Quarterly*, 28(3):539–54.

Stroeve, J., Serreze, M., Drobot, S., Gearhead, S., Holland, M., Maslanik, J., Meier, W. and Scambo, T. (2008) 'Arctic sea ice extent plummets in 2007', *Eos, Trans. Amer. Geophys. Union*, 89:13.

SWIPA – Snow, Water, Ice and Permafrost in the Arctic (2011) AMAP Secretariat, Norway: Oslo.

Yalowitz, K., Collings, J. and Virginia, R. (2009) *The Arctic Climate Change and Security Policy Conference: Final Report and Findings*, 1–3 December 2008, New Hampshire: Dartmouth College, Hanover. Dickey Center for International Understanding at Dartmouth College, the Carnegie Endowment for International Peace, and the University of the Arctic Institute for Applied Circumpolar Policy.

Index